A Course in Mathematical Modeling

Visit the MAA Bookstore on MAA Online (www.maa.org) to find a description of this book, and the link to the authors' website where you can download data sets, *Mathematica* files, and other modeling resources that perform the models described in the text.

Cover photo of Sandhill Crane by Don Baccus.
Cover photo of Randall Swift, courtesy of Stuart Burrill, Western Kentucky University.
Cover design by Freedom by Design.

© *1999 by*
The Mathematical Association of America (Incorporated)
Library of Congress Catalog Card Number 98-85688
ISBN 0-88385-712-X
Printed in the United States of America
Current Printing (last digit):
10 9 8 7 6 5 4 3 2 1

A Course in Mathematical Modeling

Douglas D. Mooney
The Center for Healthcare Industry Performance Studies

and

Randall J. Swift
Western Kentucky University

MORTON COLLEGE
LEARNING RESOURCES CENTER
CICERO, ILL.

Published and distributed by
The Mathematical Association of America

CLASSROOM RESOURCE MATERIALS

Classroom Resource Materials is intended to provide supplementary classroom material for students—laboratory exercises, projects, historical information, textbooks with unusual approaches for presenting mathematical ideas, career information, etc.

Committee on Publications
James W. Daniel, *Chair*

Andrew Sterrett, Jr., *Editor*

Frank Farris   Edward M. Harris
Yvette C. Hester   Millianne Lehmann
Dana N. Mackenzie   William A. Marion
Edward P. Merkes   Daniel Otero
Barbara Pence   Dorothy D. Sherling
Michael Starbird

101 Careers in Mathematics, edited by Andrew Sterrett
Calculus Mysteries and Thrillers, R. Grant Woods
Combinatorics: A Problem Oriented Approach, Daniel A. Marcus
A Course in Mathematical Modeling, Douglas Mooney and Randall Swift
Elementary Mathematical Models, Dan Kalman
Interdisciplinary Lively Application Projects, edited by Chris Arney
Laboratory Experiences in Group Theory, Ellen Maycock Parker
Learn from the Masters, Frank Swetz, John Fauvel, Otto Bekken, Bengt Johansson, and Victor Katz
Mathematical Modeling in the Environment, Charles Hadlock
A Primer of Abstract Mathematics, Robert B. Ash
Proofs Without Words, Roger B. Nelsen
A Radical Approach to Real Analysis, David M. Bressoud
She Does Math!, edited by Marla Parker

Acknowledgments

This project began with the authors' desire to develop a course in mathematical modeling at Western Kentucky University. The mathematics department headed by James F. Porter has been very supportive of this project. We were given freedom in the development process, opportunities to team teach the course, access to computing equipment and software, and matching support for grants written for the course development. The Kentucky PRISM-UG program provided grant funding to assist the course development.

The particular flavor and direction of the book owes a lot to the 1993 NFS workshop in biomodeling organized by Robert McKelvey. Special inspiration came from workshop instructors Anthony Starfield and William Derrick.

Many individuals contributed to improving the drafts of this manuscript. We appreciate the careful reading of early drafts by colleagues John Boardman, Claus Ernst, Frank Farris and James Porter. Richard and June Kraus provided valuable input on the copy edit draft of the text. Andrew Sterrett, editor of the Mathematical Association of America's Classroom Resources Series, read many versions of this text and we are grateful for his valuable comments. The production staff at The Mathematical Association of America were extremely helpful in preparing the final version of this text. In particular, we would like to thank Elaine Pedreira and Beverly Ruedi for their assistance, advice, and constant reassurance.

Finally, the students of our modeling course suffered through rough drafts of this text and their careful comments and honest complaints have greatly enhanced the final product. Student Johnathan Jernigan deserves special mention for his help with many tedious technical details.

To Wilma Mitchell and Judy Wilson
D.D.M

To my wife Kelly,
and my children
Kaelin and Robyn
R.J.S.

Preface

This book is primarily intended as a one-semester course in mathematical modeling accessible to students who have completed one year of calculus. It is based on a course that the authors have taught at Western Kentucky University, a regional state-supported university with a small masters program and no engineering school. This course is offered as an elective and has been well received by the students taking it. There is more material in this text than can be covered in a one-semester course, giving the instructor options based on the character of a particular class. Most of the material in Chapters 0–4, for example, is accessible to precalculus students. Selections from these chapters have been used in five week summer courses for middle- and secondary-school teachers.

There are a number of issues to resolve and balance when teaching a modeling course or in writing a modeling textbook. The first is whether to teach modeling or models. The distinction here is that modeling is the process of building a model whereas a model is something someone else has already built. We chose to emphasize modeling. Models, however, are not excluded. They serve as examples of the modeling process and in many cases are building blocks that can be used when modeling.

How then does one learn modeling? An obvious answer is by actively building models. Our course and consequently this book are based on modeling projects. In the text, modeling problems will be posed. The appropriate mathematics will be introduced. Use of computer software will be discussed. Models involving this problem will be presented. Finally modeling projects will be presented. As in other mathematics classes, the reader must be an active participant and is encouraged to work through the example models at a computer and to tackle the projects at the end of each section. When teaching this course, we require students to turn in completed projects in typewritten reports. A course of this type provides noncontrived opportunities to bring writing into the mathematics curriculum.

What is the appropriate level at which to learn modeling? The answer is, of course, at any level. There are courses and texts at the high school/precalculus, calculus, junior, senior, and graduate levels, and many researchers study models and modeling. This book is aimed at students who have completed two semesters of calculus. Leafing through the text, you will see difference and differential equations, probability and statistics, matrices, and other mathematical topics. No background is assumed except that students have an intuitive idea of an average, understand what it means for something to happen 30% of the time, know what a matrix is and that matrices can be multiplied, and finally know that dx/dt is a rate of change and understand what that means. One of our hopes is that students coming out of a course taught from this book will have been exposed to advanced mathematical topics, will have seen how they can be used, and will be motivated to take more mathematics.

It is reasonable then to ask how someone can build models using ideas from advanced courses which they may not have taken. A problem with learning modeling at this level is the lack of mathematical tools possessed by the beginning modeler. Frequently, even the tools that a student does have are not understood well enough to be used competently. An additional complication is that if a particular problem which is solvable by hand is altered even slightly, the result is a problem which may be intractable by nonnumerical analysis. One example is projectile motion. The simple model can be handled quite well with an understanding of quadratic equations, and slightly more advanced questions can be answered using calculus. Add, however, to this basic model some factors involving wind speed and air resistance, and suddenly the problem has become extremely difficult to solve. Even with years of training a professional mathematician can be as stumped as a beginner when trying to analyze a model. This problem is handled in this book by using software packages. One of the benefits of technology is that it allows topics to be taught at lower levels than before. Graphing calculators, for example, enable min/max problems to be effectively taught at the precalculus level. Here we will work with nonlinear difference equations, using our math skills to set them up and letting software give us graphical and tabular solutions. Markov chains are presented and then solved using software which can multiply matrices. Thus we are able to build and study models at an early undergraduate level—models which would have been studied in a graduate course or not at all several years prior to this writing. Software changes rapidly, and to keep this book from becoming quickly outdated we are not connecting it to any particular piece of software. We will talk in general about Computer Algebra Systems (*Mathematica* and Maple being the current leaders), Systems Oriented Modeling Software (currently Stella II), spreadsheets, and statistical packages. This software is to varying degrees user-friendly, which is in keeping with the proposed level of the course. While serious simulation modeling often involves programming, numerical packages, and issues of numerical analysis, we felt that these details would obscure the ideas of modeling at the level of the intended audience. Further, easy-to-use software and quick results can be a useful first step to more serious work.

In all things, balance is important. While we have just talked extensively about the usefulness of computer simulation modeling, this book does not neglect analytic or theoretical models. Both have their uses and their place. An analytic model allows us to study it with all the mathematical tools that we know. The result is great understanding. Their limitation is that in order to be tractable, they usually have to be so simplified that their relevance to the real world becomes questionable. Computer simulation models, by contrast, can handle as much real-world complication as you wish to add (within limits of memory and processing time). It is much harder, however, to gain from them the type of understanding one can obtain from an analytic model. To test the effect of parameter variations for example, multiple runs must be performed. In general though, the philosophy of this book is that while pathologies can be found and one should be conscious of them, mathematics and software works well most of the time. (Discussion topic: Is the obsession of mathematicians with the abstract and pathological as opposed to the concrete and useful a reason that mathematicians are in such low demand?)

Modeling is a rich area. Due largely to the interests of the authors, this book deals almost exclusively with dynamical systems models and modeling. This means the problems studied involve changes over time. In this text, we will study discrete and continuous, linear and nonlinear, deterministic and stochastic, and matrix models and modeling. A goal of our study is to develop a modeling toolbox. Modeling techniques will be studied as components to be used in larger models. For example, the logistics equation will be studied, not as an end in itself, but as a building block for a more complex model.

As for the choice of our models, we will limit ourselves primarily to problems in population biology and ecology with a few problems involving finance and sociology. This choice requires little prior background. We assume that every college age student has intensively investigated the processes involved in reproduction and death if for no reason other than an innate curiosity.

This is a modeling book. The mathematics in it was chosen based on what was useful to the models. This book is not intended to be a book on difference equations, differential equations, linear algebra, probability and statistics, dynamical systems, or calculus despite the fact that all these topics are mentioned. This is one of the distinctions between this book and a book in, say, calculus that has modeling in it. In this book the modeling determines the math to be discussed; in a mathematics book, the models are chosen to match the mathematics being discussed. Leafing through the text, you will notice that there is a wide range in terms of the difficulty of the material. Some of it is accessible to precalculus students, while some requires some maturity. Again our motivation was to teach certain types of models and to let the mathematics taught be driven by the needs of the modeling.

This modeling text has a website which contains other modeling resources and information. This site will be continually updated and contains additional modeling projects, data sets, information about modeling software, and other relevant issues about modeling at the undergraduate level. The site also has the text's *Mathematica* appendix as downloadable files. The site can be reached through a link in the MAA Bookstore on MAA Online (www.maa.org).

We have observed some interesting things in the students taking this course. Motivation is the primary component to doing well. A number of "average" students have excelled in this class, while some "exceptional" students who have learned the game of "getting an A" have struggled as have those who play the game of "getting by."

Why study modeling? It integrates much of undergraduate mathematics. It provides real problems for students to cut their mathematical teeth on. It teaches mathematical reasoning. It encourages flexible thinking and multiple approaches in problem solving. It delves into interesting nonmathematical topics. Most importantly, it's fun.

We conclude with a bit of fine print. We stress in Chapter 0 that the purpose of a model is essential in determining whether it is a good model or not. This is a textbook, and all the models included have the purpose of teaching modeling skills. A model that is good for conveying a point may not be the best model for a real-world application. We made a great effort to pick, by and large, realistic examples which use realistic data and parameters. To do this, we have scavenged models from many sources. Sometimes we dramatically simplified a complex model that had a component that matched what we were teaching. Sometimes we have creatively interpreted data. Sometimes a model we present is out of favor with some faction or another. We justify this by saying the models are good for our purposes. For other purposes, we strongly urge a skeptical approach to every model and piece of data. We hope public policy is not decided based on our sandhill crane model, but that the methods of the book may be useful to someone building models for making such public policy. In other words, "your mileage may vary."

To the Teacher

As mentioned in the Preface, it is difficult to accurately represent a course in a textbook. Our course is typically very dynamic in nature, while a book is static and set in stone. In this section we hope to convey a sense of how the course is taught so, perhaps, the dynamic nature will come through somehow.

For the first half of the semester, the class meetings followed a three-step cycle: math; model; and modeling project. More specifically, one day we would lecture on the mathematics needed for a certain type of model, say solving linear recurrence relations with constant coefficients. The next we would present a model building on the math taught the previous day, say the annual plants model. Usually the model discussion mixed chalkboard derivation with numerical analysis (we always had a computer demonstrator in class). Frequently after the initial points were made, discussion from the students directed in-class explorations of a model. Finally, we would assign a modeling project based on the preceding two days' work, Great Lakes pollution for example. Our university has a unique scheduling system of two classes one week followed by three classes the next with Friday being the swing day. On Fridays (i.e., every other week) we would go to a computer lab and give the students a demonstration of the software they would be using, then give them either a short computer exercise model to reinforce ideas while we could answer questions, or let them start on the current project. Thus we taught this class with four days of lecture followed by one day of lab. While this class could be taught without occasional lab days, we highly recommend them. They get students introduced to software quickly, with immediate assistance when they get stuck. We found that generally this amount of guidance was sufficient. Further, they create an environment where the teacher and student connect in a way that is not possible in the traditional lab. It is much easier for us to write recommendation letters for students we have had in this course than for students in our more traditionally taught courses.

For the first half of the semester, we assign projects once a week. We also move through material quickly in class, getting to the useful aspects of the math without getting bogged down by the mathematical details which are frequently stressed in a non-modeling mathematics class. The book is intended to be read by the student and has more details than we present in class. The sections on the statistics of regression, for example, are relatively long in the text. In class this material can be covered in two or at most three days. The lecture is conducted with a computer demonstrator displaying a regression or ANOVA table for simple regression. The meaning of each element is discussed, and key points are present on the chalkboard. Data sets are changed, and the statistics are examined again. By the end of the lecture, the student knows all the elements and how to use the F-ratio to test for significance. The second lecture introduces multiple regression and how to use the t statistic. Again most of the lecture consists of examples. We build regression models by adding variables and by removing variables, and we hint about the ideas of forward, backward, and stepwise regression. These ideas are generally considered to be beyond our scope. In the text, there are formulas that are helpful for understanding where things come from, but generally they do not get discussed in class.

By midsemester we are done or nearly done with Chapter 3. At this point, we assign a midterm project that synthesizes a number of ideas. The impala and caribou projects have been used as midterm projects. We have let the class work on these projects in groups of two or three and generally have given the groups several weeks to complete them. We require these projects to be written in a report format, and usually we create some scenario with the students being consultants hired to address certain questions with the given information. Though mathematically not difficult, the complexity of these projects form a defining moment in the class where the students see how the ideas covered fit together to construct models.

We also require students to individually build a model to solve a problem on a topic of their choosing. These are considered final projects, and we have the students present them to the class at the end of the course and turn in a written report to us. We begin preparing the students for this project after the midterm project and cut back on the weekly projects (either making them less involved or less frequent—sometimes the in-lab project suffices as the project of the week). More material gets covered during the last portion of the course because we spend less time teaching software, have fewer projects, and meet in the lab less. We still follow a math/model format, but often the lectures may span several class periods.

Testing is a peculiar issue in this class. With the emphasis on modeling and projects, tests seem inappropriate. The one time we taught this class without testing, however, resulted in students ignoring any material that did not show up on a project, and, frequently, they were not able to discern which material that would be. This especially became a problem after the midterm project with the cutting back on the weekly projects. Since then, we have given a midterm and final exam. These exams are based on basic exercises over mathematical techniques. For example we have them find fixed points and test for stability, construct box plots and find regression lines with small data sets, construct and analyze small (two-step) Markov chains, and similar tasks. Together the midterm and final comprise 10% of the course grade. We are clear with the students what to expect on these exams to prevent panic in studying the broad nature of the material in this class. We have been pleased by the results of adding these exams.

One final comment on the static nature of the book versus the dynamic nature of the class—the most recent offering of this course was taught directly from the book, and we found this to be constraining. This is a fun and exciting course, and part of the fun is discovering models on one's own. During the different offerings of the course, we have used different models and projects each year. There only a few models that we repeat each year. (We have always assigned the Blood Alcohol Model in Chapter 1 and have done the m&m simulation in Chapter 2.) In future offerings we will continue to present the material that is here, but will probably use different model applications to illustrate this. We strongly encourage those using this text to do the same. Material that we are excited about because it is new to us always goes over better than canonical examples dutifully presented.

Contents

Acknowledgements v

Preface ix

To the Teacher xiii

0 Modeling Basics:
   Purpose, Resolution, and Resources 1
   0.1 Further Reading 6
   0.2 Exercises 7

1 Discrete Dynamical Systems 9
   1.1 Basic Recurrence Relations 10
   1.2 Spreadsheet Simulations 11
      1.2.1 Spreadsheet Basics 11
      1.2.2 Relative and Absolute Addressing 13
      1.2.3 Best, Medium, and Worst Cases 14
      1.2.4 Hacking Chicks Example 14
      1.2.5 Effects of Initial Population Example 15
   1.3 Difference Equations and Compartmental Analysis 16
      1.3.1 Specialty Software for Compartmental Analysis 19
   1.4 Closed-Form Solutions and Mathematical Analysis 19
      1.4.1 Exponential and Affine 19
      1.4.2 Fixed Points and Stability 21
      1.4.3 The Cobweb Method. 24
      1.4.4 Linear Recurrence Relations with Constant Coefficients 25
   1.5 Variable Growth Rates and the Logistic Model 31
      1.5.1 The Logistic Model 31
   1.6 Systems of recurrence relations 35
      1.6.1 Example: A Host-Parasite Model 35
   1.7 For Further Reading 37

1.8 Exercises 38
1.9 Projects 38

2 Discrete Stochasticity **47**
  2.1 Stochastic Squirrels 48
    2.1.1 Example 48
    2.1.2 A Simple Population Simulation: Death of M&M's 49
  2.2 Interpreting Stochastic Data 50
    2.2.1 Measures of Center and Spread I:
        Mean, Variance, and Standard Deviation 51
    2.2.2 Frequency Distributions and Histograms 53
    2.2.3 Example: *The X-files* 56
    2.2.4 Measures of Center and Spread II: Box Plots and Five-Point Summary ... 58
  2.3 Creating Stochastic Models: Simulations 60
    2.3.1 Distributions 60
    2.3.2 Three Useful Distributions 66
    2.3.3 Environmental versus Demographic Stochasticity 70
    2.3.4 Environmental Stochasticity—A Closer Look 78
  2.4 Model Validation: Chi-Square Goodness-of-Fit Test 82
    2.4.1 Testing the Stochastic Squirrel Model 84
    2.4.2 Testing Stochastic Models and Some Background Theory 85
    2.4.3 Validity of Sandhill Crane Model 87
  2.5 For Further Reading 89
  2.6 Exercises 90
  2.7 Projects 93

3 Stages, States, and Classes **101**
  3.1 A Human Population Model 101
  3.2 State Diagrams 104
    3.2.1 Human Population 104
    3.2.2 Money in a bank account I 104
    3.2.3 Money in a bank account II 105
  3.3 Equations From State Diagrams 106
  3.4 A Primer of Matrix Algebra 106
  3.5 Applying Matrices to State Models 110
  3.6 Eigenvector and Eigenvalue Analysis 112
  3.7 A Staged Example 120
  3.8 Fundamentals of Markov Chains 122
  3.9 Markovian Squirrels 127
  3.10 Harvesting Scot Pines:
     An Absorbing Markov Chain Model 130
  3.11 Life Tables 133
  3.12 For Further Reading 135
  3.13 Exercises 135
  3.14 Projects 139

4 Empirical Modeling 151
- 4.1 Covariance and Correlation—A Discussion of Linear Dependence 152
- 4.2 Fitting a Line to Data Using the Least-Squares Criterion 155
- 4.3 R^2: A Measure of Fit 157
- 4.4 Finding R^2 for Curves Other Than Lines 159
- 4.5 Example: Opening *The X-Files* Again 161
- 4.6 Curvilinear Models 164
  - 4.6.1 A Catalog of Functions 164
  - 4.6.2 Intrinsically Linear Models 167
  - 4.6.3 Intrinsically Nonlinear Models Which Can Be Linearized 167
- 4.7 Example: The Cost of Advertising 169
- 4.8 Example: Lynx Fur Returns 175
  - 4.8.1 Intrinsically Nonlinear Models Which Cannot Be Linearized: The Logistic Equation 180
- 4.9 Interpolation 187
  - 4.9.1 Simple Interpolation 187
  - 4.9.2 Spline Interpolation 190
  - 4.9.3 Example: The Population of Ireland 191
  - 4.9.4 Interpolating Noisy Data 195
- 4.10 The Statistics of Simple Regression 197
  - 4.10.1 Reading a Regression Table: I 199
  - 4.10.2 The Sums of Squares and R^2 199
  - 4.10.3 Regression Assumptions 200
  - 4.10.4 F-tests and the Significance of The Regression Line 201
  - 4.10.5 The t-test for Testing the Significance of Regression Coefficients 204
  - 4.10.6 Standard Errors of the Regression Coefficients and Confidence Intervals 205
  - 4.10.7 Verifying the Assumptions 207
  - 4.10.8 Reading a Regression Table: II 207
- 4.11 Example: Elk of The Grand Tetons 210
- 4.12 An Introduction to Multiple Regression 214
  - 4.12.1 Example: A Testing Model 215
- 4.13 Curvilinear Regression 219
  - 4.13.1 Example: A Corn Storage Model 220
- 4.14 For Further Reading 224
- 4.15 Exercises 226
- 4.16 Projects 228

5 Continuous Models 239
- 5.1 Setting Up The Differential Equation: Compartmental Analysis II 240
- 5.2 Solving Special Classes of Differential Equations 242
  - 5.2.1 Separable Equations 242
  - 5.2.2 Example and Method: US Population Growth and Using Curves Fitted to Data 245
  - 5.2.3 Linear Differential Equations 251

        5.2.4 Example: Drugs in the Body 252
        5.2.5 Linear Differential Equations with Constant Coefficients 253
        5.2.6 Systems of First-Order Homogeneous Linear
              Differential Equations With Constant Coefficients 258
        5.2.7 Example: Social Mobility 264
    5.3 Geometric Analysis and Nonlinear Equations 266
        5.3.1 Phase Line Analysis 266
        5.3.2 Gilpin and Ayala's θ-logistic Model 270
        5.3.3 Example: Harvesting Models 272
        5.3.4 Example: Spruce Budworm 277
        5.3.5 Phase Plane Analysis 283
        5.3.6 The Classical Predator-Prey Model 287
    5.4 Differential Equation Solvers 294
        5.4.1 The Answer Is Known, But Not By You 295
        5.4.2 Numerical Solvers When No One Can Solve the Differential Equation .. 298
    5.5 For Further Reading 305
    5.6 Exercises 307
    5.7 Projects 309

6 **Continuous Stochasticity** **323**
    6.1 Some Elements of Queueing Theory 323
    6.2 Service at a Checkout Counter 324
    6.3 Standing in Line—The Poisson Process 328
    6.4 Example: Crossing a Busy Street 331
    6.5 The Single-Server Queue 334
        6.5.1 Stationary Distributions 336
        6.5.2 Example:
              Traffic Intensity—When to Open Additional Checkout Counters 338
    6.6 A Queueing Simulation 341
    6.7 A Pure Birth Process 346
        6.7.1 A Remark on Simulating the Pure Birth Process 351
    6.8 For Further Reading 351
    6.9 Exercises 351
    6.10 Projects 354

Appendices:
    A   **Chi-Square Table** **369**
    B   **F-Table** **371**
    C   **t-Table** **375**
    D   *Mathematica* **Appendix** **377**
References **423**
Index **427**

0

Modeling Basics: Purpose, Resolution, and Resources

If we were to ask several people for examples of models, we would get a variety of responses which might include mathematical equations, toy trains, prototype cars, or fashion models. What these very different objects have in common is that they are representations of reality. The equations may represent the growth of a population, the toy train is representation of a real train, the prototype is a representation of a future car, and a fashion model is a representation of how clothes will look when worn and maybe other things that one has to live in the world of fashion to appreciate. A model is, then, a representation of reality. This, however, is not a sufficient definition. How can one evaluate different models? A fashion model and a toy train are both representations of reality, but so vastly different that they are incomparable. A street map of a city and a road map of the whole United States are representations of reality which are similar, but neither can substitute for the other. Intrinsic in a model is a sense of *purpose*. We define a *model* to be a purposeful representation of reality. A street map is a representation of reality for the purpose of navigating streets in a particular city. It is useless for driving across country or even for locating traffic jams or construction within a city, but that does not make it a bad model. Other models are needed for those purposes. A successful fashion model serves the purpose of selling clothes, perhaps by creating some fantasy about what people will look like. An artist's model serves a different purpose. For this purpose, perhaps someone who is less glamorous, but with more idiosyncrasies might be a better model. There are different types of toy trains: some are for young children, some are for older children, and others are for adults. Again different purposes result in different models.

This book deals exclusively with *mathematical models* which are models built using the tools and substance of mathematics (including computers and computer software). We do not deal with train sets or fashion models. Further, a major emphasis of this book is on *modeling* which is the process of building mathematical models. Models are presented of course, but more as examples of the results of the modeling process and less as ends in themselves.

Doing a modeling project will help illustrate several modeling concepts. You, the reader, are encouraged to get the following materials and do the tasks that are requested of you. You need a jar, preferably of an interesting shape, and a bag of M&M candies or some similarly sized objects (jelly beans, marbles, or pennies). Here is your project:

Project: Take exactly one minute and determine how many M&M candies will fit in your jar.

Warning: Not doing this project will result in inferior learning, low self esteem, and poorer jobs with lower pay.

Now that you have done this, let us think about what you have done. First the concept of a *resource constraint* should have meaning to you. If you honestly attempted to do this problem in one minute, you may have been frustrated thinking that you could do this problem if you only had more time. There may have been other frustrations involving resources; for example, you could have done this if you had a ruler or some other measuring device. While this deficiency of resources may seem to be an unfair constraint it is something modelers have to cope with all the time. For example, your boss may tell you one morning to determine the effects of some new strategy for a meeting scheduled that afternoon. The boss does not care about, "I could do it if only ..." The model you construct given a couple hours will be very different from the model you would construct given several days or years. But it is the best you can do within the constraints. Other resource constraints are money, personnel, computing power, and expertise. It may be impractical for you to buy a computer, hire a staff, or go back to school just to solve a problem.

One thing missing from the project above, which is a justifiable complaint, is that the purpose of the project was not stated. Why do you need to know how many M&M's fit in the jar? If the purpose is to determine if one bag of M&M candies would fill the jar, a quick visual inspection of the jar and of the bag would give the answer, and 60 seconds would be more than enough time. If the purpose is to figure out how many bags of M&M's would fill the jar to a point that most anyone would consider to be full, then one needs to do some figuring. Being off by a couple handfuls of candies, however, will not make any difference. Finally, if the purpose is a contest and the person who can fit the most M&M candies into the jar so that the lid can be fully tightened without squashing any candy wins valuable prizes, then it is very important to figure out exactly how many candies fit in the jar. Again the purpose is a vital part of any model that you build.

These ideas bring up the notion of *resolution*. This term may be familiar from photographs or computer images. It is hard to determine details in a photograph with low resolution, but the details are very sharp in a high-resolution picture. Models have resolution also. In the previous paragraph, we discussed three purposes for the problem of finding how many candies fit in a particular jar. In each case, the answer had a different resolution: more or less than one bag; to within a couple handfuls; the exact maximum number.

0 Modeling Basics: Purpose, Resolution, and Resources

If the purpose of a model requires only a low-resolution answer, then a low-resolution model is completely acceptable. Even if resources permitted a better answer, the better answer would not be any more useful. On the other hand, if a high-resolution answer is desired, low-resolution models still have their place. Often one starts with a low-resolution model to get a ballpark estimate and successively refines the model until either the desired resolution is achieved or resource limitations prevent one from proceeding further.

Let us go back to our modeling project and revise it a little.

Project 2: A jigsaw puzzle company wants to fill your jar with M&M candies so that a photograph can be taken with the jar looking more or less full (a handful shy will not make any difference). A bag of M&M's holds 50 candies. How many bags should they buy? You have 15 minutes to complete this project. Before you start you may acquire any standard measuring instruments you think you will need. One constraint: do not start with more than one bag of candy (several bags are okay, but you are constrained so that you cannot fill up the jar, dump it, and count the candies).

Assuming you have done this project, let us discuss what you have done. If you thought this would be an easy problem if you just had more time earlier, do you still think this is the case? There is no "right way" to approach this problem. We mention a few. Did you compute the volume of the jar? If you did this, did you approximate the jar as a cylinder, or as a small cylinder surmounting a larger one? Maybe you forgot the formula for a volume of a cylinder and used rectangular boxes instead of cylinders. Maybe you remembered the formula for the volume of a truncated cone and used that. If you had access to a well equipped kitchen you might have found the volume exactly by filling the jar with water and pouring it into a measuring cup. Perhaps your jar had the volume printed on it in some way. Now how about the M&M's? Did you treat them as ellipsoids with volume πabc? Did you treat them as little cylinders, or as little boxes?

Here is one approach to the problem. Approximate the volume of the jar using two cylinders taking care that the volume of the two cylinders is greater than or equal to the true volume. Call this volume V. Next treat an M&M candy as an ellipsoid and suppose its volume is v. Assuming that the volumetric units are the same, V divided by v gives the absolute maximum number of M&M's needed from which we can determine the number of bags needed. Of course this assumes that every millimeter of the jar is covered by candy. If you look at M&M candies poured into a jar, however, you notice that there is a lot of empty space. We can obtain a lower bound by considering the M&M's as cylinders and determining how many lie in one layer at the bottom of the jar and then figuring the number of layers the jar will hold. We may have to use multiple cylinders to represent the jar because of the neck. In any case, we can get a good estimate for how many M&M's would fit in the jar if they were stacked uniformly one on top of the other. Take a moment to convince yourself that, assuming one had the patience to stack them, one bump would knock the M&M's out of their columns and some would fit into the space between M&M's on the previous layer. This is called packing, and when it happens, which is always, more M&M's will be needed to fill the jar. Thus we have two numbers giving us a lower bound and an upper bound, and the true number is somewhere

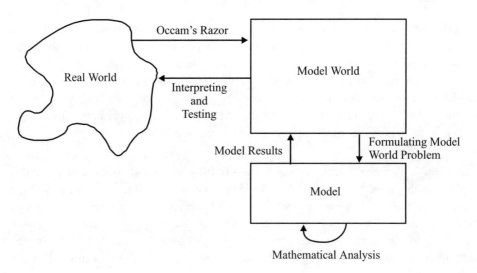

FIGURE 0.1 Model World Diagram.

in between. Despite the simple sounding nature of this problem there is a great deal of complexity. Complexity is associated with any real problem. Consider Figure 0.1.

The problems that modelers wish to solve exist in the real world. The real world is a nasty place with all sort of complications. Bottles are irregularly shaped. Candies are shaped so they can fit together in many ways, but always having some space between them. To illustrate this, the real world is given a nebulous shape. The first step in the modeling process is to simplify the real world to create a model world. As the picture depicts, the model world leaves out much of the complexity of the real world. For example, a bottle becomes two cylinders. The M&M's become cylinders. The original question gets translated into a question involving the model world. In the model world, the question is how many M&M cylinders fit inside the two bottle cylinders. Next we construct a model of the problem in the model world. All the tools and techniques of mathematics can now be applied to the problem in the model world to get an answer in the model world. A very important mistake to avoid is to assume that the solution to the problem in the model world answers the question in the real world. The final step is to interpret the answer found for the model world problem, back in the real world. Here we notice that the M&M's can pack and that the shape of the jar is not two cylinders. Depending on our purpose, resolution, and resources, we may need to work through the cycle one or more times to get a satisfactory answer.

In most mathematics classes, we start with an equation and solve it. Occasionally we start with a problem in a model world which we solve and answer back in the model world. In this book, we do both of these, but also carry the process further to begin with problems in the real world and work our way completely through this diagram to get answers in the real world.

The process of converting the real world into a model world is probably the most important step in the modeling process. This process is referred to as using "Occam's razor." William of Occam, a fourteenth-century philosopher, coined a phrase in Latin

0 Modeling Basics: Purpose, Resolution, and Resources

which literally translates to: "things should not be multiplied without good reason." In other words, don't make things harder than they need to be. In a modeling setting, exclude the details which are irrelevant given the purpose, or which cannot be handled given the constraints. We are cutting the world down to manageable size, as if with a razor, hence the term Occam's razor. Occam's razor is a sensitive and dangerous tool. Cut out too much, and the model solutions have nothing to do with reality. Cut out too little, and the problem is too difficult to solve with the available resources.

A model that every calculus student has probably seen is that projectiles follow parabolic trajectories. This is derived by starting with a constant acceleration rate g and initial conditions (s_0 and v_0) and integrating the equation $a(t) = g$ twice to get the displacement as a function of time $s(t) = \frac{gt^2}{2} + v_0 t + s_0$. Let us examine this model in light of Figure 0.1 and Occam's razor. The problem in the real world was originally predicting the motion of cannon balls and perhaps things like bullets and stones. (A great deal of technological advances have been made exploring more efficient ways of killing people.) This problem was moved into the model world. Occam's razor sliced away and left a world which was flat, with constant gravitational force, in a vacuum, with no other forces of any kind. Within this world, the model for parabolic motion was set up and solved. Then it was tested in labs and on battlefields. For these purposes, the model was very good. One danger is to assume that because the model predicts that all projectiles have parabolic motion, in reality they all will. Try throwing a feather. Try throwing a light plastic ball into the wind. Try dropping a BB into a jar of honey. Missile designers realized that they could not assume constant gravitational force or neglect air resistance. However, for throwing rocks and firing cannon on windless days, the parabolic model is a good model.

Occam's razor creates some assumptions which should be spelled out when reporting on a model that has been built. With the projectile motion, we spelled out that the model world was flat, was a vacuum, and had constant gravity and no other forces. This is a vital part of a model. When someone throws a feather and says, "Ah ha, your model is wrong," we can reply, "Well, the feather would have followed parabolic path if you had thrown it in a vacuum." The accuser would have to accept this. The next complaint might be that your model is unrealistic because we do not live in a vacuum. To which we reply that for objects of certain masses projected certain distances, the effects of not being in a vacuum are negligible. Of course we would have to back this up, and this is what the lab and field tests verify. Assumptions and model testing are two very important lines of defense against skeptics and critics.

Here is another modeling project which we will only think about and not actually solve. It serves to illustrate many of the concepts of this book. Suppose we are starting a limousine service, driving people from a particular town to the nearest airport. For the sake of illustration, suppose that this is 70 miles one way. We want to compute the cost in gasoline of these trips for budgeting purposes. Our car has an average gas mileage for this type of driving of 25 miles per gallon.

A first pass at this problem would be to divide the distance to the airport by the mileage and multiply by the current gasoline price. However, after thinking about things, we realize that the current gasoline price is changing. We look up records for the last five years and compute a function which predicts the price of gas into the future, understanding that while reality may be very different from this prediction, it will probably be good for the short term. We repeat the model with this gas price function. Now we have a *time dependent* model. The questions it answers are not just how much will it cost, but how much will it cost at different times. Both of the models we have constructed are *deterministic* models. They include no random factors and consequently predict one number or one number at each point in time. Do we believe these numbers? We might suspect that different trips to the airport might result in different gas mileages due to random factors such as traffic and weather. If we include these into our model, we can run simulations and end up, not with a single number, but with a distribution of numbers. Models which include randomness are called *stochastic*. We may think of other random factors such as varying distances based on where customers are picked up or where the driver is able to park. At some point the model may become unwieldy and we must decide (use Occam's razor) whether all this detail is important to our purpose and worth the necessary resources.

The first chapter of this book deals with deterministic time dependent models where the change in time occurs in jumps (gas prices change weekly or daily, but not from instant to instant). The second chapter adds stochasticity into models and examines ways of analyzing the output. The third chapter adds additional complications such as a limousine service with multiple cars with different gas mileages. The fourth chapter deals with ways of looking at data such as gas prices over time and fitting functions to them. The fifth chapter deals with time dependent models where the change over time is continuous (for example if the gas mileage steadily decreased between service intervals). The last chapter covers randomness which occurs continuously. All these chapters provide different ways of analyzing essentially the same problem. Occam's razor must be used with the issues of purpose, resolution, and resource constraints to decide which approach is best for a particular situation.

While we do not often think about it, we do modeling all of the time. Here is a another project. How long will it take to finish reading this chapter? This is a problem we may solve frequently when reading before bed and get tired. We could answer this question by computing our reading rate for this type of material and determining the number of words left to go. Maybe we could include a factor for reading rates slowing down as we get tired. But if we are trying to figure out whether to finish the chapter now or later, this degree of resolution is not necessary. For this purpose, a sufficient answer is "not long."

0.1 Further Reading

These first two books are good continuations of the ideas presented in this chapter. Both go considerably deeper:

John Harte, *Consider a Spherical Cow: A Course in Environmental Problem Solving*, University Science Books, Mill Valley, California, 1988.

Anthony M. Starfield, Karl A. Smith, and Andrew L. Bleloch, *How to Model It: Problem Solving for the Computer Age*, McGraw-Hill Publishing Company, New York, 1990.

The next book is an introductory modeling classic. It is recommended additional material for the whole book so we list it at the beginning and will not repeat it every chapter. The material is at a similar level to the material in this book, but it places greater emphasis on engineering-type problems:

Frank R. Giordano and Maurice D. Wier, *A First Course in Mathematical Modeling*, Brooks/Cole Pub. Co., Monterey, CA, 1985.

0.2 Exercises

1. How many 1,000 cubic-inch cylindrical tin cans can be formed from a sheet of metal one yard wide and two yards long?

2. How many restaurants are there in the United States? Clearly it is impossible, for all practical purposes, to find the right answer as restaurants open and close daily. First establish an acceptable error range for your answer and then try to answer this question. You may consult any source for information (such as population size, or number of cities) with the exception of anything which directly deals with this question. Clearly state the assumptions you are making. Next find a source which states the number of restaurants there are. Compare your answer to this answer. Realizing that any "official" number is also an estimate and not truth, which answer are you more confident in and why?

3. Estimate the number of dots in Figure 0.2? Your method should not involve counting a large number of dots; counting a sample is acceptable. Suppose this figure represents

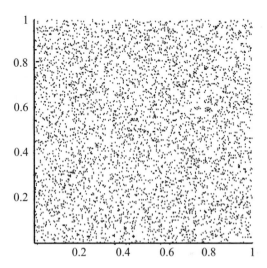

FIGURE 0.2 How Many Dots?

aerial photograph of a tree farm with each dot representing a tree. Your purpose is to tell a lumber company how many trees are in this tract of land so they can make revenue projections.

1

Discrete Dynamical Systems

Consider for a moment the difference between the growth of a population of maple trees and a population of humans. If we were to measure the population growth of some type of maple tree, we might go out and stake out a plot of land in some wilderness area and count the number of maple trees within it. If we went out day after day, we would notice very nearly the same number of trees in the plot. During a few weeks in the spring, however, the number would jump as the new trees sprouted from seeds. From then on the maple population would remain relatively constant until the next spring. A few might die and a few might sprout late, but the bulk of the activity takes place over a short period of time. After a few years of daily census taking, we might conclude that an annual census would probably be sufficient for most purposes. If we were measuring a human population, on the other hand, and were using an area that included a large number of people, there would be many birth and deaths each day. In fact, if we took a large enough area, say the entire United States, we would find changes in the population each minute and, perhaps, each second. Even this once-a-second census is an approximation to a continuously changing population. Human population growth is, then, modeled as a *continuous process*, while maple population growth is modeled as a *discrete process*. The equations we use to model each of these need to reflect this distinction.

The first three chapters of this book deal exclusively with discrete modeling. Chapters 5 and 6 deal with continuous models. (Chapter 4 concerns fitting models to data, and its results can be used in either type of model.) It is also possible to construct models with both a continuous part and a discrete part. For example, we might decide that trees sprout according to a discrete model, but die according to a continuous model. These could be combined into a single model which is continuous through the summer, fall, and winter, but discrete in the spring. Such models are called *metered* models.

This chapter is devoted to discrete time, deterministic (no randomness) models. The mathematics which describes this behavior is called *discrete dynamical systems*.

1.1 Basic Recurrence Relations

A fundamental mathematical concept for working with discrete time models is the *recurrence relation*. A function x on the nonnegative integers (or any well ordered set) is recursively defined if $x(n)$ is a function of the values of x at some or all of the numbers less than n, that is $x(n) = f(x(n-1), x(n-2), \ldots, x(2), x(1), x(0))$. In order for this definition to make sense, some of the $x(i)$'s must be specified as initial conditions. The number of time steps that a recursion relation makes reference to is called its *order*. For example $x(n) = f(x(n-1))$ is a first-order recurrence relation, while $x(n) = f(x(n-1), x(n-2), x(n-3))$ is a third-order recurrence relation. Also $x(n) = 3\,x(n-4)$ is a fourth-order recurrence relation, illustrating that it is the number of time steps involved, not the number of prior states.

Here are several examples with which you may be familiar. Let x be the factorial function. Recall that 0! is defined to be 1, so $x(0) = 1$. From then on $x(n) = nx(n-1)$. Thus $x(3) = 3\ x(2) = 3\ 2\ x(1) = 3\ 2\ 1\ x(0) = 3\ 2\ 1\ 1$. The factorial function is a first-order recurrence relation. A second example is the Fibonacci sequence $\{1, 1, 2, 3, 5, 8, \ldots\}$. Each term is the sum of the two previous terms. Since the current term relies on the two previous terms, we need to specify the first two terms to get the process going. Here $x(0) = 1$ and $x(1) = 1$, and the recurrence relation is $x(n) = x(n-1) + x(n-2)$. The Fibonacci recurrence relation is second order.

Some authors write recurrence relations using subscripts instead of functional notation. For example, the recurrence relation for the Fibonacci sequence is written $x_n = x_{n-1} + x_{n-2}$. This book avoids this notation since in later chapters $X(n)$ refers to a vector of values at time n, and the elements of the vector are denoted as $\{x_1(n), x_2(n), \ldots, x_m(n)\}$. Thus to avoid the confusion of multiple subscripts or changing notation midway through, functional notation is used throughout.

For our first modeling example we look at the population growth of the Florida sandhill crane. This endangered species has been extensively studied, and we make use of some of this research data to construct a series of increasingly sophisticated models over the next several chapters. The data and parameter values come from the book *Filling the Gaps in Florida's Wildlife System* by Cox et al [13].

We begin with a simple model using only annual growth rates. These are reported for several environmental conditions which we call best, medium, and worst. Under the best environmental conditions the growth rate r is 0.0194 or 1.94% annually. Suppose we start with a population of 100 birds. Thus $x(0) = 100$. Next year we will have the initial 100 plus the change in population, so $x(1) = x(0) + .0194x(0) = 101.9$ or 102 birds, rounding to the nearest whole bird. The year after we will have $x(2) = x(1) + .0194x(1) = 103.9$, $x(3) = 105.9$, and so on. In general, $x(n) = x(n-1) + r\ x(n-1)$, or, if you like, $x(n) = (1+r)x(n-1)$. This is a recursive definition for the crane population year by year. The reader may already know or see a way to solve this to avoid the recursion, but we wait until Section 1.4 to talk about closed-form solutions. Recursive equations are easy to analyze numerically. With a pencil, paper, and a scientific calculator, we can compute as many years of model output as we need. When computing solutions for many recursion runs, a programmable calculator or computer is of great assistance. The next section discusses the use of spreadsheets in running recursion models and analyzes

1 Discrete Dynamical Systems

the sandhill crane model as an example. The use of *Mathematica* to evaluate recurrence relations is found in the *Mathematica* appendix.

1.2 Spreadsheet Simulations

In this section, we numerically analyze the sandhill crane model and several variations of it. We use this opportunity to introduce spreadsheets as a modeling tool and spend a little time discussing how they are used. It is our opinion that spreadsheets are one of the single most important tools in everyday modeling. They are used in the work place far more than any other mathematical tool. The current generation of spreadsheets typically have the power to create quality graphics easily and have the power to do most of the statistical analyses discussed in this book (even multiple regression). The output from other programs, including C or FORTRAN programs, can be exported to a spreadsheet for further analysis or graphing. While the problems in this book can be solved without using a spreadsheet, we highly recommend their use and consider knowledge of spreadsheets part of computer literacy. So, if spreadsheets happen to be unavailable to you or if you are determined to use some other tool for solving recurrence relations, the material of this section is easily converted, but at least be aware of the spreadsheet as a valuable modeling tool.

Spreadsheets are an ideal environment for analyzing recurrence relations. It is easy to set up the equations, and the output is readily available in both numerical and graphical form.

1.2.1 Spreadsheet Basics

A spreadsheet is a grid of rectangles called cells. In each cell, we can enter numbers, text, or a formula, and the formula for one cell may depend on the contents of other cells. To facilitate this, each cell has an address that allows us to refer to it and its contents. The spreadsheet program used in our examples labels columns by letters of the alphabet and the rows by numbers. (Other spreadsheet programs may do the opposite.) Cell $B3$ is the cell in the second column and third row. In Figure 1.1, we constructed a spreadsheet for the sandhill crane model described above. Notice that one row, listed with time 0, has the initial population entered. The cells in the rows below it have formulas entered based on the model equations, and the computation of these equations is what is displayed in each cell. The initial population is in cell $B7$. Cell $B8$ is defined to be $= B7 + 0.0194 * B7$. The $=$ sign signifies that the contents are a formula to be evaluated rather than a number or text. Variations of this formula can be put in the successive cells beneath $B7$, each incrementing the cell number by one. Rather than entering these equations by hand, spreadsheets allow us to copy a cell to the right or left or up or down. Any formulas that depend on other cells have these cells' addresses automatically updated. For example if $B8$ is the formula $= B7 + 0.0194 * B7$ and we copy the formula down two cells, then $B9 = B8 + 0.0194 * B8$ and $B10 = B9 + 0.0194 * B9$.

In our spreadsheet we copied the second row down to take our simulation through 15 years (0–14). The addresses of the cells were automatically updated to keep the equations correct. Spreadsheets allow for ready graphing of output. Each does it differently, but usually it involves selecting the cells to be graphed with the cursor, specifying the type of graph, and specifying where the graph should be drawn. A graph of the simulation we just ran is shown in Figure 1.2.

	A	B
1		
2		
3		
4		
5		
6	Time	Population
7	0	100
8	1	102
9	2	104
10	3	106
11	4	108
12	5	110
13	6	112
14	7	114
15	8	117
16	9	119
17	10	121
18	11	124
19	12	126
20	13	128
21	14	131

FIGURE 1.1 Sandhill Crane Simple Spreadsheet

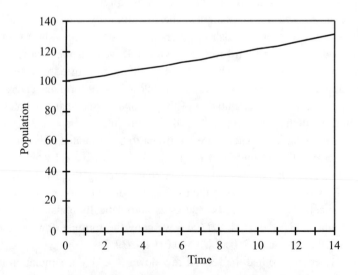

FIGURE 1.2 Sandhill Crane Growth Curve

1.2.2 Relative and Absolute Addressing

Frequently we set up a model and then decide to change part of it. Sometimes this is called creating "what if " scenarios. For example, "what if" the growth rates are decreased by one percent? If the model parameters are incorporated with the model equations, then the spreadsheet needs to be extensively rebuilt. To prevent this, all of the model parameters (birth and survival rates) should be located in one section of the spreadsheet. We can write formulas that refer to the appropriate parameter cells instead of putting the parameters into the equations. Unfortunately when we copy the formulas, the addresses for the parameters get changed also. This is called *relative addressing*, and in most cases it makes spreadsheet construction easier. Fortunately, spreadsheets allow us to specify an *absolute address* for a cell so that when we copy a formula the address remains the same. With the spreadsheet program we used, if a formula refers to cell $E9$ as $\$E\$9$ (notice the $ signs), then whenever this formula is copied, the new cell has a formula which also refers to $E9$. The new formula does not change $E9$ to some other address. Other spreadsheets do this in different ways, but any quality spreadsheet is able to do this somehow. Absolute addressing results in elegant model construction. Figure 1.3 shows a rebuilt spreadsheet with growth rate in cell $B2$. Formulas for several of the cells are also presented. Now by simply changing the growth rate in the single cell $B2$, the values throughout the entire spreadsheet change, and the graph updates as well.

With a little imagination, you should be able to come up with many useful variants of these ideas. Three examples are shown next.

	A	B
1		
2	Growth Rate	0.0194
3		
4		
5		
6	Time	Population
7	0	100
8	1	=B7+B7*B2
9	2	=B8+B8*B2
10	3	106
11	4	108
12	5	110
13	6	112
14	7	114
15	8	117
16	9	119
17	10	121
18	11	124
19	12	126
20	13	128
21	14	131

FIGURE 1.3 Sandhill Crane Model with Absolute Addressing

	A	B	C
1			
2			
3			
4	Best	0.0194	
5	Medium	-0.0324	
6	Worst	-0.0382	
7			
8			
9	Year	Best	Medium	Worst
10	0	100	100	100
11	1	102	97	96
12	2	104	94	93
13	3	106	91	89
14	4	108	88	86
15	5	110	85	82
16	6	112	82	79
17	7	114	79	76
18	8	117	77	73
19	9	119	74	70
20	10	121	72	68
21	11	124	70	65
22	12	126	67	63
23	13	128	65	60
24	14	131	63	58

FIGURE 1.4 Sandhill Crane Model with Three Growth Rates

1.2.3 Best, Medium, and Worst Cases

The spreadsheet that we developed so far is easily modified to compare behaviors among the best, medium, and worst case growth rates. To do this, we add title cells for the medium and worst cases; copy the cell $A10$ to the two cells to the right; copy $A11$ to the two cells to the right, making the appropriate changes in the absolute addresses; and copy row 11 down the desired number of years. This will give numerical runs for all three cases simultaneously. This spreadsheet is shown in Figure 1.4, and a graph based on it is shown in Figure 1.5.

1.2.4 Hacking Chicks Example

We noticed in the previous example that in the medium and worst cases the population was declining. Suppose that a manager institutes a program of *hacking* chicks, that is, hatching chicks in captivity and releasing them to the wild. The term *hacking* comes from the ancient sport of falconry where wild hawk and falcon eggs were hatched in captivity and raised for sport hunting. Since adding chicks to an already growing population is not necessary, we examine only the effects of adding five chicks each year to the medium and worst case models. We also assume that the hacked chicks have no undue problems of survival or assimilation. To do this, we simply add 5 to each of the formulas in the cells of the previous model. Or, if we want to be slick about it, we add a cell for the

1 Discrete Dynamical Systems

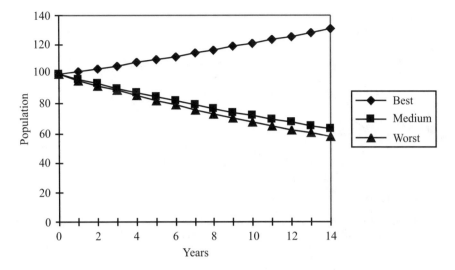

FIGURE 1.5 Graph of Sandhill Crane Model with Three Growth Rates

number of chicks to be hacked each year and use absolute addressing. If we do this, we can easily change the number of chicks to see how this affects the solution curve.

A graph of these results is shown in Figure 1.6. Notice that we ran the simulation through 130 years. We observe that the population levels appear to stabilize.

1.2.5 Effects of Initial Population Example

What effect do the initial population values have on the hacking model? We rerun the analyses from the Hacking Chicks example, using only the medium growth rate and three values of the initial population. These results are shown in Figure 1.7. Notice that in all

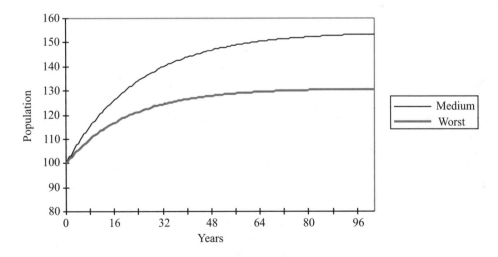

FIGURE 1.6 Effects of Hacking Five Chicks per Year to Medium and Worst Cases

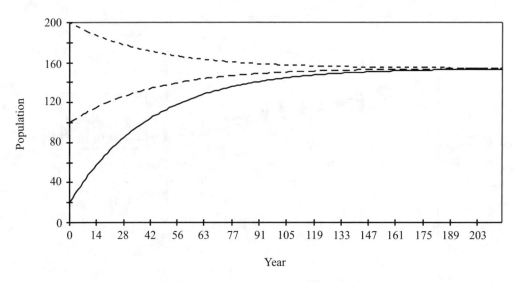

FIGURE 1.7 Effects of Differential Initial Populations on Hacking Medium Case Model

three cases the curves appear to stabilize at the same value. From this numerical evidence we hypothesize that the initial condition has no effect on where the population stabilizes. We verify this analytically in Section 1.4.

1.3 Difference Equations and Compartmental Analysis

Some people prefer to approach discrete-time problems with difference equations instead of recurrence relations. Essentially they are different ways of writing the same equation, and many books use the terms interchangeably. We make a distinction between them because each has its own use. We consider only first-order difference equations for the time being.

A *first-order difference equation* is an equation of the form $x(n) - x(n-1) = f(x(n-1))$ (or equivalently $x(n+1) - x(n) = f(x(n))$). We are computing the difference between this year's value and last year's value, hence the term "difference equation." In particular $x(n) - x(n-1)$ is the growth over one year. This is a preferable approach when one understands calculus and remembers that the derivative is defined by taking the limit of $(f(x+h) - f(x))/h$ as h goes to zero. In the case where h is not zero, $(f(x+h) - f(x))/h$ is the average growth rate over the time h, and the limit as h tends to zero is the instantaneous growth rate. In our case, $h = 1$ and $x(n) - x(n-1)$ is the average growth rate over one time increment. If the process were continuous, then $x(n) - x(n-1)$ gives the average growth rate, which is an approximation. If the process is discrete, which is this chapter's assumption, then the average growth rate is, in fact, the true growth rate. People sometimes use the derivative (instantaneous growth rate) in discrete situations because of the power that calculus gives in analyzing the problem. This is, however, just an approximation. Similarly, people use average growth

1 Discrete Dynamical Systems

rates in continuous processes when the calculus becomes intractable. Again this gives an approximation.

It is probably obvious, but we point out that a difference equation can be converted to a recurrence relation by taking the $x(n-1)$ to the other side of the equals sign. A recursion relation can be converted to a difference relation by subtracting $x(n-1)$ from both sides.

A practical reason for using difference equations is that there is a convenient way of diagramming the relationships between variables called a compartmental diagram. A *compartmental diagram* uses a rectangle or box for each variable of interest. Arrows are then drawn to indicate the flow into or out of each variable. Adjacent to each arrow is the quantity that flows into or out of a variable in one time step. The direction of the arrow indicates the direction of positive flow. For an arrow pointing towards a box, a positive flow means the variable increases by that amount, while a negative quantity means the variable decreases by that amount. Conversely if an arrow is pointing away from a box, a positive flow means the variable decreases by that amount, and a negative flow means that the variable increases by that amount. Over each time step, the change in the variable x is $x(n) - x(n-1)$ which is the amount that flows into x, less the amount that flows out. Succinctly put, $x(n) - x(n-1)$ is equal to *inflow minus outflow* It is easy to formulate the difference equation(s) by using the flow equations adjacent to the arrows.

Consider the following examples. Our sandhill crane model is shown in Figure 1.8. There is one arrow pointing into a variable representing the crane population. Over the arrow are the letters $r\,x$. This diagram indicates that at each time step $r\,x$ flows into x. If $r\,x > 0$, then x will grow; if $r\,x < 0$, then x will decline. Thus the difference equation is

$$x(n) - x(n-1) = r\,x.$$

Next consider the refinement to the crane model shown in Figure 1.9. Here there are a birth arrow going in and a death arrow going out. Above the birth arrow is $b\,x$, and

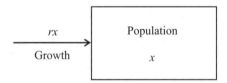

FIGURE 1.8 Compartmental Diagram for a Sandhill Crane Model with a Growth Rate

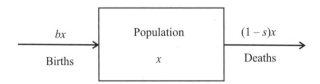

FIGURE 1.9 Compartmental Diagram for a Sandhill Crane Model with Birth and Death Rates

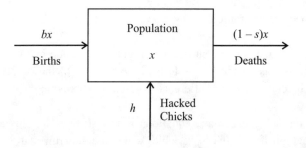

FIGURE 1.10 Compartmental Diagram for a Sandhill Crane Model with Hacking

above the death arrow is $(1-s)x$ where b is the birth rate and s is the survival rate. Again from the diagram we easily formulate the equations

$$x(n) - x(n-1) = bx - (1-s)x.$$

Hacking chicks is shown in Figure 1.10, and the equation is

$$x(n) - x(n-1) = bx - (1-s)x + h.$$

Finally, Figure 1.11 shows a model of two species interacting. From this, we set up a system of two difference equations.

While compartmental analysis seems easy, it should not be dismissed even if we are able to write down equations without its use. With a well drawn compartmental diagram, we can consult experts who are not mathematicians and inquire as to the validity of the relationships in a model. Even people who may understand a linear or logistic relationship (discussed below) may not recognize it when it is part of a bigger equation, but will recognize it when isolated as the relationship over an arrow. Finally, any model that is not being built for our own purposes needs to be communicated to others. When writing a report of a model, a compartmental diagram can replace an incomprehensibly tedious description.

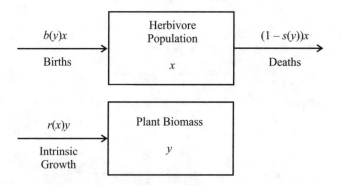

FIGURE 1.11 Compartmental Diagram for Herbivore-Plant Interaction

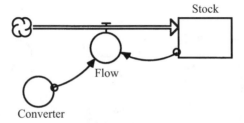

FIGURE 1.12 Elements of Stella

1.3.1 Specialty Software for Compartmental Analysis

There are some computer programs (e.g., Stella II, GPSS) which allow the user to build models from a compartmental analysis point of view. Some are self-contained computer packages; others are front ends which sit on a programming language such as FORTRAN. In general, these are very helpful to modelers by allowing them to spend more time on the model and less time on programming. The authors have used Stella II in their classes and occasionally throughout the text use compartmental diagrams created in Stella. For these reasons, we take a few moment to introduce Stella II (by High Performance Systems). We note that this is luxury software in some sense. It makes compartmental modeling easier, and students like it, but it is not a necessary part of this book. Everything it does can be done some other way. If resources are available, however, it is a wonderful tool.

Stella has four programming elements, a stock (indicated by a box), a flow (indicated by a thick arrow with a circle representing a flow regulator under it), a converter (indicated by a circle), and a connector (thin arrow) (see Figure 1.12). A stock is a compartment, and a flow is, of course, a flow in the compartmental diagram sense. Initial values are put into stocks, and formulas are put into flows to construct a simulation which can then be run. As it runs, the values in the stocks rise and fall. A converter is used to store parameters or to manipulate parameters (convert from metric to English, for example). The connectors link objects that relate to one another. If, for example, the flow into a population depends on the current size of the population, a connector must link the population to the inflow. Good programming practice is to put parameters into converters so they can be changed easily (like absolute addressing in a spreadsheet). Any flow that depends on such a parameter must also be linked to the parameter with a connector. Connectors are not part of a traditional compartmental diagram, but they help the user of Stella keep interrelationships straight. An example of a Stella drawn compartmental diagram is seen in Figure 1.13. We point out again that programs like Stella are actually simulation programs, not just programs to make compartmental diagrams.

1.4 Closed-Form Solutions and Mathematical Analysis

1.4.1 Exponential and Affine

In Section 1.1, we looked at the recurrence equation for the Florida sandhill crane, and in Section 1.2 we numerically analyzed it using a spreadsheet. Symbolically this recursion

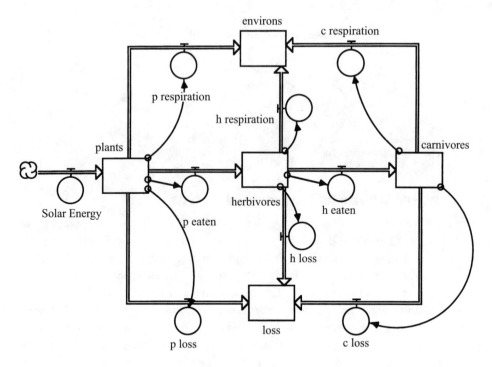

FIGURE 1.13 A Stella Model of a Cedar Bog Energy Flow

equation is $x(n) = Rx(n-1)$, where $R = r + 1$. Notice that to evaluate $x(100)$, it is necessary to compute $x(99)$, but to compute $x(99)$, it is necessary to compute $x(98)$, and so on. The problem with recurrence relations is that it is necessary to compute all or many of the preceding values just to obtain the desired value. For small values of n and with a computer assisting, this is not a problem, but for large values of n the process is prohibitively time-consuming. Sometimes it is possible to solve an equation to avoid computing previous values. This is called a *closed-form solution*.

Observe that $x(1) = Rx(0)$. It follows that $x(2) = Rx(1) = R^2 x(0)$. Continuing we have $x(3) = Rx(2)$, but from the previous step $x(3) = R^3 x(0)$. After a few steps, we might conjecture that $x(n) = R^n x(0)$, which is verified by a simple induction proof. The equation

$$x(n) = R^n x(0)$$

is the desired closed-form solution. To compute $x(100)$ it is only necessary to compute $(r+1)^{100} x(0)$. It requires the same number of steps, and hence the same amount of time to compute $x(100)$ as it does $x(1,000)$. Notice that this equation is an exponential equation with base R, hence it is called an *exponential growth* model. It is probably not an overstatement to say that these are the most important models and equations in modeling. The exponential model is the basis for the rest of this book, and the exponential equation is used more than any other single equation. The following simple fact is so important, we give it the distinction of being a proposition so we can refer to it later.

Proposition 1.1. *The function $y = R^n$ grows exponentially if $R > 1$, decays exponentially if $0 < R < 1$, is a damped oscillation if $-1 < R < 0$, is an undamped oscillation if $R < -1$, and is a constant or oscillates between two points if $R \in \{0, 1, -1\}$.*

1.4.2 Fixed Points and Stability

In Section 1.2, we ran the exponential growth model for the sandhill crane for three choices of the growth parameter r. One of these grew according to an exponential growth curve, and two declined according to an exponential decay curve. The exponentially decaying models were then modified to include a term to account for adding additional chicks. The recursive equation for this model is $x(n) = Rx(n-1) + a$. Any recursive equation of the form $x(n) = bx(n-1) + b$ with a and b non-zero constants will be called *affine*. (In general an affine equation involves adding a constant to a linear equation.) Our model equation is, then, an example of an affine recurrence relation. equation. One of the exercises is to find a closed-form solution for an affine equation. We notice from the spreadsheet runs that the successive values of $x(n)$ appear to stabilize. While the values given by the spreadsheet are useful for the number of birds at specific times, we must be careful when making inferences about the long-term behavior patterns based on relatively short runs. If we ran these simulations longer, would the apparently stable values remain stable? Intuitively we answer this question by noting that for the declining models, the point at which the population loss equals the population gain remains stable forever. For an exponentially growing population, adding to the population will result in an even larger population, and hence we do not expect it to stabilize. On the other hand if we subtracted a certain amount from a growing population, a stable point might be reached. If the reader has a computer handy, try running this affine model with several values of bird removal and see if you observe this behavior. Some mathematical analysis is needed to answer these questions conclusively.

When a recurrence relation or a dynamical system eventually settles down, the point at which it stabilizes is called a *fixed point*, and the system is said to be in a *steady state*. For the time being we concern ourselves only with first-order recurrence relations. Formally, a fixed point $\bar{x}$ of $x(n) = f(x(n-1))$ is a point such that $f(\bar{x}) = \bar{x}$. This definition is used when finding fixed points. Suppose $x(n) = -0.3x(n-1) + 2$. A fixed point occurs where $x(n)$ and $x(n-1)$ are equal, and we call their common value $\bar{x}$. Thus

$$\bar{x} = -0.3\bar{x} + 2.$$

Solving for $\bar{x}$, we conclude that $\bar{x} = 2/1.3 = 1.538$.

The power of mathematics lies in the fact that we are able to do this analysis in general once, for all choices of parameter values R and a. A general affine equation has the form

$$x(n) = Rx(n-1) + a.$$

Replacing $x(n)$ and $x(n-1)$ by $\bar{x}$, we have $\bar{x} = R\bar{x} + a$. This has solution

$$\bar{x} = \frac{a}{1-R} = \frac{a}{-r}.$$

Here is an important observation to make concerning the solution $\bar{x} = a/(1 - R)$. What role does the initial value $x(0)$ play in the value of the fixed point? Since there is no $x(0)$ in the equation, we conclude that $x(0)$ plays no role. No matter what the initial population of cranes happens to be, this model predicts the same steady state value. We have mathematically proven the hypothesis we made looking at the results of the simulation. We say that the solution is *robust* with respect to initial conditions. On the other hand, the solution does depend on a and R (or r). We say the solution is *sensitive* to a and R. An analysis of how sensitive an equation is to changes in a parameter (or other changes) is called a sensitivity analysis. Example 1.2.5 demonstrated the robustness of a particular affine equation to initial conditions. The analysis just completed verifies this robustness for all affine equations.

Now consider again the specific parameter values we have been using in the crane model. If R is greater than one, then the population is growing exponentially and $1 - R$ is negative. If we add birds to this situation $a > 0$, then $\bar{x}$ is negative so the steady state exists but is not physically realized. A similar result holds if we subtract birds from a decreasing population. The steady state solution $\bar{x}$ is positive if either R is greater than 1 and $a < 0$ or R is less than 1 and $a > 0$. The graph of the case $r = -0.0324$ and $a = +5$ was shown in Example 1.2.5. The graph of the case $r = 0.0194$ and $a = -5$ is shown in Figure 1.14.

Notice that in the first case the steady state is observed, but in the second it is not. Somehow in the case of exponential growth, even though a theoretical steady state exists (and its value is 257.73), we do not observe it occurring. This brings up the notion of stable and unstable fixed points. Consider a single car from a roller coaster on a track with lots of peaks and dips. If one starts the roller coaster car out near the bottom of

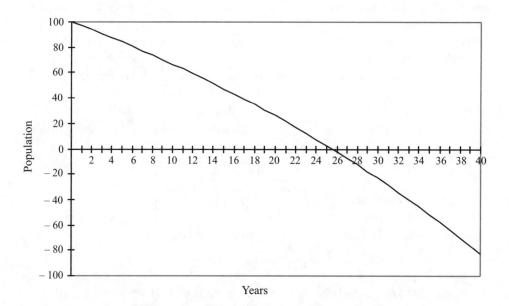

FIGURE 1.14 Removing 5 Chicks per Year from an Intrinsically Growing Population

one of the valleys, it will go back and forth slowly losing energy until it comes to rest at the bottom of the valley. Unless the amusement park retrieves it, the car will sit there forever. Now consider the car placed at the top of a peak. If perfectly balanced, the car will remain there forever, but the slightest perturbation will cause the car to roll down the track in one direction or the other. If one tries to give a car a shove to get it to the top, there is one perfect shove which will get it to the top so that it will stay. All other shoves will cause the car to go near the top and roll backwards, or go over the top and down the other side. The odds of giving it a shove of exactly the correct force are essentially zero. The bottom of a valley is a stable fixed point. The top is an unstable fixed point. In the case of the birds, if we started the population out at exactly the fixed point (which in this case is fractional), the population would stay there forever. Any deviation, even due to rounding in the 300th decimal place, will cause the population to leave the fixed point and grow or decline. Also, if by accident the growing population hit on the fixed point, it would stay there, but the odds of this happening are zero (i.e., the odds of picking a single particular number at random from any interval of the reals is zero). We can force it, by choosing parameters in clever ways, but this is mathematics not modeling.

Evidently, it is important to know whether a fixed point is stable or not. Fortunately there is a simple test. The proof is provided (one of the few in this book) because it makes use of the exponential recurrence relation.

Theorem 1.2. *(Conditions for stability) If $\bar{x}$ is a fixed point of the first-order recurrence equation $x(n) = f(x(n-1))$, then $\bar{x}$ is a stable fixed point if and only if $|f'(\bar{x})| < 1$.*

Proof. Let $e(n)$ be the difference between $x(n)$ and $\bar{x}$, i.e., $e(n) = x(n) - \bar{x}$. Then

$$e(n+1) = x(n+1) - \bar{x} = f(x(n)) - \bar{x} = f(\bar{x} + e(n)) - \bar{x}.$$

By Taylor's theorem, it follows that

$$e(n+1) \approx f(\bar{x}) + f'(\bar{x})e(n) - \bar{x}.$$

But $\bar{x}$ is a fixed point so $\bar{x} = f(\bar{x})$. Thus

$$e(n+1) \approx f'(\bar{x})e(n).$$

Notice that since $f'(\bar{x})$ is a constant, this is the recurrence relation (approximately) for exponential growth, so the error $e(n)$ decays to zero if and only if $-1 < f'(\bar{x}) < 1$ if and only if $|f'(\bar{x})| < 1$. $\square$

Sometimes if $|f'(\bar{x})| < 1$, we call the fixed point *attracting* (instead of stable) and if $|f'(\bar{x})| > 1$, we say the fixed point is *repelling* (instead of unstable). The cobweb method makes this terminology clear.

The *cobweb method* is a graphical method for finding and testing fixed points for stability. Its graphs are called *cobweb diagrams*, or simply *web diagrams*. This method is exceptional for the degree of visual insight that it gives, although to find a fixed point this way requires very precisely drawn graphs.

1.4.3 The Cobweb Method.

The problem is to find the fixed point of $x(n) = f(x(n-1))$. We think about this equation as $y = f(x)$. The fixed point occurs when $x = f(x)$, or $x = y$. But this is the same as $y = x$. So on the same coordinate system draw curves for the function $y = f(x)$ and for the line $y = x$. A fixed point occurs at the intersection of these two. To test a fixed point for stability pick a point from each side of the fixed point (if there is more than one fixed point, take care to pick the point between the fixed point we are testing and the next nearest one). Start at this point and draw a line segment vertically to the curve $y = f(x)$. From this point of intersection, move horizontally to the line $y = x$, from this point of intersection vertically to $y = f(x)$, from this intersection point horizontally to $y = x$, and so on. If this path converges to the point of intersection and the path from the point on the other side of the fixed point does also, then the fixed point is stable or attracting. Otherwise the fixed point is unstable. Unstable fixed points encompass a wide variety of behaviors including periodic, quasiperiodic, and chaotic.

Figures 1.15 and 1.16 show cobweb diagrams for the recursion functions $y = -.3x + 2$ and $y = 1.5x - 2$. We purposely chose these examples over the crane examples because the slopes of these lines make the method more evident. The crane examples would, of course, work the same, but the slopes of the lines are nearly 1, so visually they appear

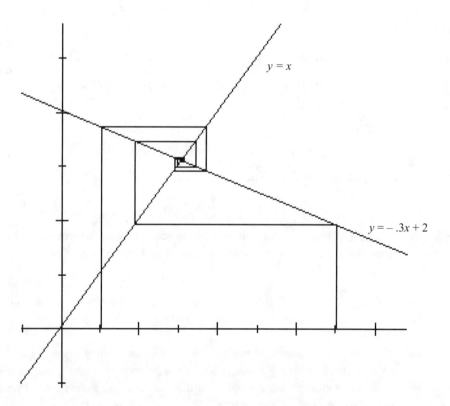

FIGURE 1.15 Cobweb Diagram for a Stable Fixed Point

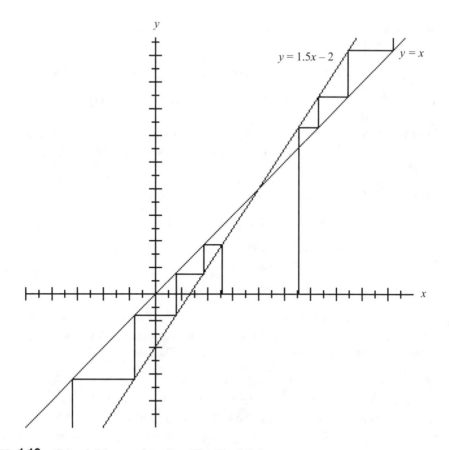

FIGURE 1.16 Cobweb Diagram for a Repelling Fixed Point

to be parallel to $y = x$. Notice that the Figure 1.15 exhibits an attracting or stable fixed point, while Figure 1.16 has a repelling or unstable fixed point.

1.4.4 Linear Recurrence Relations with Constant Coefficients

This section examines a method to solve a special class of difference or recurrence equations.

A *linear recurrence equation with constant coefficients* is a recurrence relation which is a linear combination of $x(i)$ terms, that is, it can be rearranged in the form

$$a_0 x(n) + a_1 x(n-1) + a_2 x(n-2) + \cdots + a_{m-1} x(n-(m-1)) + a_m x(n-m) = b.$$

In particular, the $x(i)$'s occur only raised to the first power and are multiplied by a constant. In this section we are interested, exclusively, in *homogeneous linear equations* which are linear equations that have constant b equal to zero. These can be written as

$$a_0 x(n) + a_1 x(n-1) + a_2 x(n-2) + \cdots + a_{m-1} x(n-(m-1)) + a_m x(n-m) = 0.$$

Several examples in earlier sections have been of this type. For example, the exponential recurrence equation and the Fibonacci recurrence equation can be written as

$$x(n) - Rx(n-1) = 0 \quad \text{and} \quad x(n) - x(n-1) - x(n-2) = 0$$

respectively. Notice that the affine equation is a linear non-homogeneous equation, and the method of this section does not apply.

The idea behind the method is that since the exponential equation has a solution of the form $x(n) = R^n x(0)$, perhaps a linear homogeneous equation of higher order has a similar solution.

The Method: Consider the equation

$$a_0 x(n) + a_1 x(n-1) + a_2 x(n-2) + \cdots + a_{m-1} x(n - (m-1)) + a_m x(n-m) = 0. \quad (1)$$

1. Assume a solution of the form $x(n) = C\lambda^n$.
2. Substitute $x(n) = C\lambda^n$ into equation (1) to get

$$a_0 C\lambda^n + a_1 C\lambda^{n-1} + a_2 C\lambda^{n-2} + \cdots + a_{m-1} C\lambda^{n-(m-1)} + a_m C\lambda^{n-m} = 0.$$

3. Factor out the common factors of C and λ^{n-m}. Since we are not interested in the cases $C = 0$ and $\lambda = 0$ (these give a zero constant function which is fine but uninteresting), we set the rest equal to zero. The result is called the characteristic equation

$$a_0 \lambda^m + a_1 \lambda^{m-1} + a_2 \lambda^{m-2} + \cdots + a_{m-1} \lambda + a_m = 0.$$

4. Solve the characteristic equation for λ. The characteristic equation has m roots, some may be repeated, and some may be complex. These roots are called *characteristic roots* or *eigenvalues*, and they are the growth rates of the system. In fact the R used in the exponential equation was an eigenvalue, and we could have called it an eigenvalue at that time. Eigenvalues are denoted by $\lambda_1, \ldots, \lambda_m$. In practice, we solve second-degree polynomials using the quadratic formula. Third- and fourth-degree polynomials can be solved using the cubic and quartic formulas, but probably a computer algebra system like *Mathematica* or *Maple* is the best bet for solving them. For quintic polynomials and above, there is no general formula (thanks to work by Galois). So unless the equation has a special form, or we know the factorization, or it has enough rational roots to get us down into the quadratic, cubic, quartic range (recall the rational root test), it can not be solved exactly either by hand or with a computer algebra system. If this is the case, numerical root-finding routines or graphical zooming on zeros are used to find approximations of the eigenvalues to any desired precision. Depending on resources, these methods are sometimes used for lower-degree polynomials as well.
5. Each equation of the form $x(n) = C\lambda^n$ is a solution of the original equation. The *general solution* is all linear combinations of these solutions or

$$x(n) = C_1 \lambda_1^n + C_2 \lambda_2^n + \ldots + C_m \lambda_m^n.$$

6. The C_i's can be found, provided enough initial information is given. Usually specific values for $x(0), \ldots, x(m-1)$ are given. They are used to find m equations in m

1 Discrete Dynamical Systems

unknowns, which are solved by any standard method such as Gaussian elimination. Again computers or calculators are used if the m is large or if it is just easier to use them than solving the system by hand.

This method yields a closed-form solution:

$$x(n) = C_1\lambda_1^n + C_2\lambda_2^n + \ldots + C_m\lambda_m^n.$$

Since each term is an exponential expression, we can analyze its pieces using Proposition 1.1. Over the short term the various eigenvalues or growth constants can contribute significantly to $x(n)$, but eventually the contribution due to the eigenvalue which has the largest absolute value dominates the contribution of the others. For obvious reasons the eigenvalue λ_i with largest absolute value, that is $|\lambda_i| > |\lambda_j|$ for all $i \neq j$, is called the *dominant eigenvalue*. Thus just by looking at the list of eigenvalues for an equation, we are able to determine the nature of its long-term behavior.

Example: Fibonacci Sequence As an example of solving linear recurrence equations, consider the familiar Fibonacci sequence. Interestingly, it was once considered as a model for rabbit growth. Recall that the sequence is $1, 1, 2, 3, 5, 8, \ldots$ which is be generated by the recurrence relation

$$x(n) = x(n-1) + x(n-2).$$

This can be rewritten in standard linear recurrence relation form as

$$x(n) - x(n-1) - x(n-2) = 0.$$

Substituting $x(n) = C\lambda^n$, we obtain

$$C\lambda^n - C\lambda^{n-1} - C\lambda^{n-2} = 0.$$

Factoring out $C\lambda^{n-2}$ gives the characteristic equation

$$\lambda^2 - \lambda - 1 = 0.$$

Using the quadratic formula, we get growth rates, or eigenvalues, or characteristic roots of

$$\lambda_1 = \frac{(1+\sqrt{5})}{2} \quad \text{and} \quad \lambda_2 = \frac{(1-\sqrt{5})}{2}.$$

Notice that $\lambda_1 \approx 1.618$ and $\lambda_2 \approx -0.618$, so λ_1 is the dominant eigenvalue, and the long-term behavior resembles exponential growth with a base of $\frac{(1+\sqrt{5})}{2}$.

The general solution is

$$x(n) = C_1 \left(\frac{(1+\sqrt{5})}{2}\right)^n + C_2 \left(\frac{(1-\sqrt{5})}{2}\right)^n.$$

We know the initial values are $x(0) = 1$ and $x(1) = 1$. Using these we get the simultaneous equations,

$$C_1 + C_2 = 1, \quad \text{and} \quad C_1\frac{(1+\sqrt{5})}{2} + C_2\frac{(1-\sqrt{5})}{2} = 1.$$

Solving gives $C_1 = \frac{5+\sqrt{5}}{10}$ and $C_2 = \frac{5-\sqrt{5}}{10}$. Thus the closed-form solution to the Fibonacci recurrence relation is

$$x(n) = \frac{5+\sqrt{5}}{10}\left(\frac{1+\sqrt{5}}{2}\right)^n + \frac{5-\sqrt{5}}{10}\left(\frac{1-\sqrt{5}}{2}\right)^n.$$

One amazing feature of this solution is that despite the fact that all of its factors are irrational, it takes on integer values for every n.

1.4.4.1 A Model for Annual Plants.
Loosely speaking, linear models are appropriate when the model parameters are constants. More precisely, a model is linear if the amount of some item is a constant times the amount last year, or the sum of constants times the amount of the item for several past years. One very nice example of this type is a model for the propagation of annual plants. This example is taken from *Mathematical Models in Biology* by Edelstein-Keshet [15].

Consider an annual plant which germinates in the spring, blooms in early summer, and produces seeds in early fall. A fraction of these seeds survive the winter (i.e., do not rot, are not eaten, etc.) and a fraction of these seeds germinate the next spring, flower, and produce more seeds. Of the seeds that do not germinate, a fraction survive the winter, and a fraction of these germinate. This process could continue, but we assume that the number of seeds which survive more than two winters and germinate are negligible.

We make the following parameter and variable assignments:
1. $p(n)$ is the number of plants in year n.
2. γ is the average number of seeds produced per plant during a year.
3. σ is the fraction of the seeds which survive a winter
4. α is the fraction of one-year-old seeds which germinate in the spring
5. β is the fraction of two-year-old seeds which germinate in the spring

Next we set up the equation for the annual plant model. How does a new plant get formed? Either a one-year-old seed germinates or a two-year-old seed germinates. Thus $p(n)$ = plants from one-year-old seeds plus the plants from two-year-old seeds. A plant germinates from a one-year-old seed if last year a plant produced seeds, which survived the winter, and then germinated. Last year there were $p(n-1)$ plants, that produced $\gamma p(n-1)$ seeds, of these $\sigma \gamma p(n-1)$ survived the winter, thus $\alpha \sigma \gamma p(n-1)$ new plants germinated from one-year-old seed. Similarly to be a new plant from a two-year-old seed, a plant two years ago ($p(n-2)$) produces seeds (γ), which survive a winter (σ), do not germinate ($1-\alpha$), survive another winter (σ), and then germinate (β). Thus the number of new plants germinating from two-year-old seed is $\beta\sigma(1-\alpha)\sigma\gamma p(n-2)$. Putting these two parts together yields

$$p(n) = \alpha\sigma\gamma p(n-1) + \beta\sigma(1-\alpha)\sigma\gamma p(n-2).$$

This is, of course, a second order linear homogeneous recurrence equation with constant coefficients. To simplify notation, let $a = \alpha\sigma\gamma$ and $b = \beta\sigma(1-\alpha)\sigma\gamma$. The equation is now $p(n) - ap(n-1) - bp(n-2) = 0$. Its characteristic equation is $\lambda^2 - a\lambda - b = 0$

1 Discrete Dynamical Systems

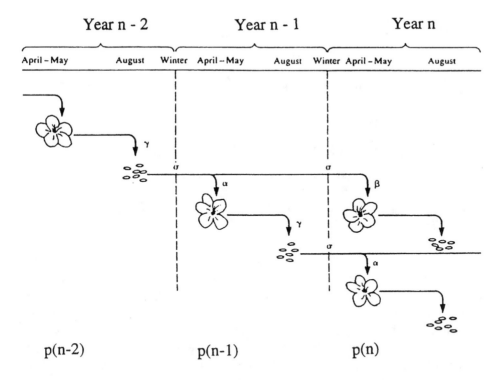

FIGURE 1.17 Annual Plants. Figure reproduced from *Mathematical Models in Biology* by Edelstein-Keshet, 1988, McGraw-Hill, Inc. Used with Permission of The McGraw-Hill Companies.

and its growth rates or eigenvalues are

$$\lambda = \frac{a \pm \sqrt{a^2 + 4b}}{2}.$$

Replacing a and b yields

$$\lambda = \frac{\alpha\sigma\gamma \pm \sqrt{(\alpha\sigma\gamma)^2 + 4\beta\sigma^2(1-\alpha)\gamma}}{2}.$$

With specific values of the parameters, we now find numerical values for the growth rates (eigenvalues) and determine from them whether the plant population is growing or declining, find the constants C_1 and C_2 to complete the closed-form solution, and use either the recursive- or closed-form equation to project the population of the plants.

Suppose, for example, $\alpha = 0.5$, $\beta = 0.25$, $\gamma = 2$, and $\sigma = 0.8$. We compute $\lambda_1 = -0.166$ and $\lambda_2 = 0.966$. The dominant eigenvalue is λ_2, which is less than one, so the population is declining. If all we are interested in is whether the plants will survive, we are finished. They will not. If we are interested in predicting the population size over the next several years, maybe to determine when half of the plants remain, then we need to find either the closed-form solution, or run a simulation. Suppose that currently ($t = 0$) there are 95 plants and that last year there were 100 plants. We want to track the population 10 years into the future. We know that

$$p(n) = C_1(-0.166)^n + C_2(0.966)^n.$$

Using our initial conditions gives the equations

$$95 = C_1(-0.166)^0 + C_2(0.966)^0 = C_1 + C_2,$$

and

$$100 = C_1(-0.166)^{-1} + C_2(0.966)^{-1}.$$

Solving for C_1 and C_2 gives $C_1 = -0.234629$ and $C_2 = 95.2346$. Thus

$$p(n) = -0.234629\,(-.166)^n + 95.2346\,(.966)^n.$$

This equation can be plotted by using a computer program or calculator. The ten-year plot is shown in Figure 1.18; a hundred-year plot is shown in Figure 1.19. The ten-year plot is what we originally asked for. From a modeling point of view, we see that even though this plant is dying out, it will be around for a while. If we are trying to save it, there is plenty of time to change conditions (fertilizing the soil or relocation of the plants). If we are trying to get rid of it (a weed, for example) there will still be plenty around for years. Notice the strange corner in the graph. This is due to the competition between the two eigenvalues. The hundred-year graph is more of mathematical interest than modeling interest since it is unlikely that the various parameters would remain constant that long. Of mathematical interest is the fact that this curve looks like a pure exponential decay curve. Over this time scale, the behavior of dominant eigenvalue has dominated the behavior due to the other eigenvalue.

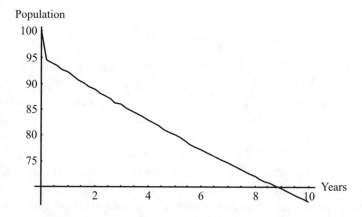

FIGURE 1.18 Annual Plants over Ten Years

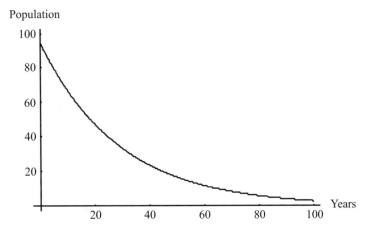

FIGURE 1.19 Annual Plants over One Hundred Years

1.5 Variable Growth Rates and the Logistic Model

"In 1977 there were 37 Elvis impersonators in the world. In 1993 there were 48,000. At this rate, by the year 2010 one out of every three people will be an Elvis impersonator."
—communicated by the National Institutes of Health to Columbia University to Harvard University to Chicago Law School picked up on e-mail then published in Playboy and anonymously put in the mailbox of one of the authors.

One of the problems with the exponential model (constant growth rate) for population growth is that even when critters (or Elvis impersonators) are piled four feet deep all over the surface of the earth, the model is still predicting growth at the same rate as when there were relatively few. There are several ways to deal with this problem. One of the best is to realize that even though the equation holds in the model world, in the real world there is a limited range over which the model solution is valid. Another, which we consider in this section, is to replace the constant growth rate with a variable one. This might extend the range over which a model is valid, but requires more information to construct.

Until now we were interested in the equation $x(n) - x(n-1) = r\,x(n-1)$. In this section, we consider some special cases of making r a function. Hence we are looking at equations of the form $x(n) - x(n-1) = r(x(n-1))\,x(n-1)$.

1.5.1 The Logistic Model

As a population increases, fixed resources must be shared between more and more individuals. A reasonable assumption is that as the population increases, the growth rate declines because of some combination of increased deaths or decreased births. Further there is probably some population level which cannot be exceeded called the *carrying capacity*. If a population were near the carrying capacity, its growth rate would be zero. If the population were small, the growth rate would be at its largest. Thus the growth rate function should pass through $(0, R)$ and $(K, 0)$ where R is the intrinsic growth rate and K is the carrying capacity. The simplest function that can be put through two points is a line. The model that makes all these assumptions is called the logistic model. Various assumptions can be questioned and refined, but the result is a different model.

The growth rate function for the logistic model is found using the point-slope equation for a line through $(0, R)$ and $(K, 0)$, which is

$$r(x) - R = -\frac{R}{K}(x - 0).$$

Solving for $r(x)$ yields

$$r(x) = R - Rx/K = R(1 - x/K).$$

Recall that the basic equation form of this chapter is

$$x(n) - x(n-1) = r(x(n-1))\,x(n-1).$$

Substituting our expression for $r(x)$ gives us

$$x(n) - x(n-1) = Rx(n-1)\left(1 - \frac{x(n-1)}{K}\right). \tag{2}$$

This equation is the *logistic difference equation*. It (together with its continuous cousin presented in Chapter 5) is one of the most important equations in ecology, second only to the exponential equation.

Next we apply some of the analysis techniques from earlier sections to this equation. We begin with the mathematical analysis of fixed points and their stability. Rewriting equation (2) as a recurrence relation gives

$$x(n) = x(n-1)\left(R\left(1 - \frac{x(n-1)}{K}\right) + 1\right). \tag{3}$$

Thus

$$f(x) = x\left(R\left(1 - \frac{x}{K}\right) + 1\right). \tag{4}$$

To find the fixed points, we look at

$$\bar{x} = \bar{x}\left(R\left(1 - \frac{\bar{x}}{K}\right) + 1\right). \tag{5}$$

This has two solutions, namely $\bar{x} = 0$ and $\bar{x} = K$. These make sense. One is extinction; the other is the carrying capacity. Next let us look at the stability of these fixed points using Theorem 1.2. Computing the derivative of $f(x)$ we obtain

$$f'(x) = R - \frac{2Rx}{K} + 1. \tag{6}$$

Substituting $\bar{x} = 0$ gives $f'(0) = R + 1$. This is stable if

$$|R + 1| < 1$$

which is the same as

$$-2 < R < 0.$$

1 Discrete Dynamical Systems

This makes sense because the model is set up so that if there is a non-zero population, it will grow (move away from the fixed point) if the growth rate is positive. If the growth rate is negative, the population dies out, which means it moves towards the fixed point. If $R \leq -2$, there are some strange mathematical artifacts, and the reader is encouraged to explore these with simulations. Next we look at the other fixed point, $\bar{x} = K$. In this case, $f'(K) = 1 - R$. Applying Theorem 1.2 we know this fixed point is stable for

$$|1 - R| < 1 \quad \text{or} \quad 0 < R < 2.$$

For negative growth rates, the population declines, thus moving away from the carrying capacity. For positive growth rates under 200% the carrying capacity is stable, which indicates that the population will grow up to the carrying capacity and then stabilize. For large growth rates this analysis predicts instability. To examine this further we turn to spreadsheet simulations.

Figures 1.20, 1.21, and 1.22 show the results of simulations for a logistic model with carrying capacity $K = 100$ and growth rates $r = 0.5$, $r = 2.2$, and $r = 2.7$. From our fixed point analysis we expect the carrying capacity to be a stable fixed point in the first case and unstable in the other two. Notice that the unstable cases exhibit very different behavior. When $r = 2.2$, the behavior is periodic which is not technically stable, but is well behaved. On the other hand, when $r = 2.7$, the behavior appears random. In Chapter 2, we examine models with random output, but notice that there is nothing random about the logistic equation. If we were to run this simulation 1,000 times, we would get exactly the same graph every single time. This is an example of *chaos* which can informally be thought of as deterministic or non-random behavior that looks random.

We conclude this section with some philosophical thoughts concerning the logistic equation. As we discussed, a population obeying a logistic model starts out growing exponentially, but grows more and more slowly, until it reaches the carrying capacity

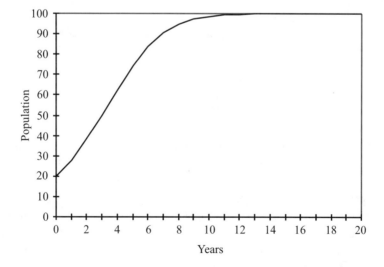

FIGURE 1.20 Logistic Solution Curve with $R = 0.5$

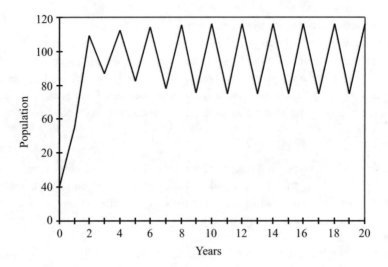

FIGURE 1.21 Logistic Solution Curve with $R = 2.2$

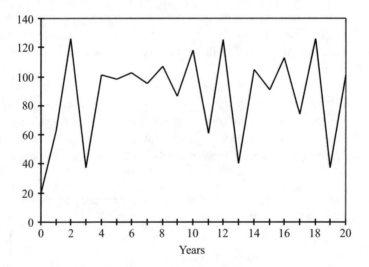

FIGURE 1.22 Logistic Solution Curve with $R = 2.7$

where growth stops and the population remains stable. Naively we might think that a stable population at the carrying capacity is good. But let us think a little further. Think about things from the animal's point of view. (For instance, pretend you are a human and the human population will ultimately reach a carrying capacity). Far below the carrying capacity, food is plentiful, and lives are long and healthy. Things are so good that the population grows at its maximum rate. As the population nears the carrying capacity, food is more scare, adults starve, infant death rates soar. There are just enough resources so that deaths are replaced by births. All in all it is a pretty dismal existence. The carrying capacity is the level of most miserable existence without a decline in population.

1 Discrete Dynamical Systems

The moral here is that if a species could control its reproduction rate so that its population stabilizes well below the carrying capacity, it would be much better off than if it let itself grow to the carrying capacity. Most species cannot control their growth rates, but humans can. Humans are also able to control the population sizes of other species. In Chapter 5 we look at several harvesting models and see that certain harvesting methods result in the population's stabilizing below the carrying capacity.

1.6 Systems of recurrence relations

The ideas of this chapter extend to two or more time-dependent variables, and the result is a system of recurrence relations. Systems arise in a variety of settings, some examples of which are competition, cooperation, or other interaction between species, companies, or countries. For the scope of this book, we content ourselves with finding fixed points and constructing simulations, and do not address stability issues.

Finding the fixed points of a system is nearly identical to the one equation case, except that we now have a system of potentially non-linear equations to solve. Sometimes the substitution method works with such a system, but frequently analytic methods are unsuccessful and a numerical equation solver must be employed.

If we have a system of three recurrence relations

$$x(n) = f(x(n-1), y(n-1), z(n-1))$$
$$y(n) = g(x(n-1), y(n-1), z(n-1))$$
$$z(n) = h(x(n-1), y(n-1), z(n-1)),$$

then we replace all occurrences of $x(n)$ or $x(n-1)$ by $\bar{x}$, and, similarly, replace all y's and z's by $\bar{y}$ and $\bar{z}$ respectively. Thus

$$\bar{x} = f(\bar{x}, \bar{y}, \bar{z})$$
$$\bar{y} = g(\bar{x}, \bar{y}, \bar{z})$$
$$\bar{z} = h(\bar{x}, \bar{y}, \bar{z}),$$

which we solve for $\bar{x}$, $\bar{y}$, and $\bar{z}$.

1.6.1 Example: A Host-Parasite Model

Many insects have one growth period per year, and their population dynamics are modeled well by a discrete-time model. The following model is from Krebs [34] and explores the interaction of an insect parasite (predator) and an insect host (prey). One example of such an interaction is the azuki bean weevil (*Callosobruchus chinensis*) host with a larval wasp (*Heterospilus prosopidis*) parasite. This model is an example of what we term a *theoretical model* which means its purpose is to explain observed behavior, rather than to predict accurate numbers. In particular the data (originally from Utida [70]), shows that the populations oscillate. This model's purpose is to see if oscillating populations are predicted from reasonable and standard assumptions or whether further mechanisms are required. This model is the result of Occam's razor streamlining and simplifying until a bare minimum exists for this purpose. Hopefully this model will also make the reader aware that many models are not ends in themselves and can be used as building blocks for more complicated models. Here we observe the logistic used as a component in a more complicated model.

The first assumption is that in the absence of predation, the prey population follows a logistic equation

$$N(n+1) - N(n) = R\left(1 - \frac{N(n)}{K}\right) N(n), \tag{7}$$

where $N(n)$ is the population of prey in year or generation n, R is the growth rate when $N(n) << K$, and K is the carrying capacity.

In the presence of predators, equation (7) must be modified to take predation into account. The predation term is a significant part of this model, and it is important that this term reflect the reality of the predation process. It is common to assume that the prey lost due to predation is proportional to the product of the number of predators and the number of prey. This term is discussed further in the predator-prey model discussed in Chapter 5. We use it here, but again emphasize that this is an assumption, not a law. With the predation term, the prey equation becomes

$$N(n+1) - N(n) = R\left(1 - \frac{N(n)}{K}\right) N(n) - C\,N(n)P(n), \tag{8}$$

where $P(n)$ is the size of the predator population at time n, and C is a proportionality constant which can be thought of as a measure of predator efficiency.

Another common assumption in predator-prey models is to assume that the predator population is proportional to the product of the predator population and the prey population. In other words, the growth rate of the predator population is proportional to the number of prey killed. Thus

$$P(n+1) = Q\,N(n)P(n), \tag{9}$$

where Q is a proportionality constant which can be thought of as measuring the efficiency of utilization of prey for reproduction by predators.

Thus we have the system of equations:

$$N(n+1) = R\left(1 - \frac{N(n)}{K}\right) N(n) + N(n) - C\,N(n)P(n) \tag{10}$$

$$P(n+1) = Q\,N(n)P(n). \tag{11}$$

Replacing all N's and P's by $\bar{N}$'s and $\bar{P}$'s yields

$$\bar{N} = R\left(1 - \frac{\bar{N}}{K}\right) \bar{N} + \bar{N} - C\,\bar{N}\bar{P} \tag{12}$$

$$\bar{P} = Q\,\bar{N}\bar{P}. \tag{13}$$

This is a non-linear system of equations which, in general, can be difficult or impossible to solve. This one, however, solves easily. Equation (13) can be rearranged to have the

1 Discrete Dynamical Systems

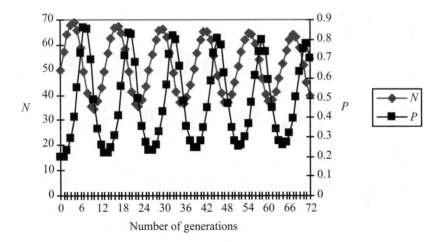

FIGURE 1.23 Simulation of a Discrete Predator-prey Model with $K = 100$, $R = 1.5$, $Q = 0.02$, $N(0) = 50$, and $P(0) = 0.2$.

form $\bar{P}(1 - Q\bar{N}) = 0$ which has solutions $\bar{P} = 0$ and $\bar{N} = \frac{1}{Q}$. Substituting $\bar{P} = 0$ into equation (12) gives the solution $\bar{N} = K$. This solution makes sense for it asserts that in the absence of predators, the prey stabilizes at the carrying capacity. While valid, this solution is not of interest. Substituting $\bar{N} = \frac{1}{Q}$ into equation (12) gives $\bar{P} = \frac{R}{C}(1 - \frac{1}{QK})$. This is a second fixed point $\left(\frac{1}{Q}, \frac{R}{C}(1 - \frac{1}{QK})\right)$, one that is of interest this time. Equation (12) also has solution $\bar{N} = 0$, which, upon substitution into the second equation, gives $\bar{P} = 0$. Thus a final fixed point is $(0, 0)$, which again is uninteresting.

Thus far our analysis has used general parameters. Now we choose some specific parameter values and perform some numerical analysis.

Following Krebs, we assume that $R = 1.5$ and $K = 100$. Further we assume that under best conditions the predators double their numbers each generation; thus when $N \approx K$, $P(n+1) \approx 2P(n)$. Thus $QK = 2$ and $Q = \frac{2}{K}$. Since $K = 100$, we have $Q = \frac{1}{50} = 0.02$. We further assume that initially there are 50 predators and a prey density of 0.02. Running a simulation with these parameters produces the graph shown in Figure 1.23. Indeed, as predicted, we obtain oscillatory behavior from these modeling assumptions.

1.7 For Further Reading

For those interested in biological modeling, we highly recommend:

Leah Edelstein-Keshet, *Mathematical Models in Biology*, McGraw Hill, Inc., Birkhäuser Mathematics Series, New York, 1988. This book includes not only difference equation models, but differential and partial differential equations as well. It contains lots of biological examples ranging from microbiology to ecology.

An accessible book devoted to the mathematics of discrete dynamical systems:

James T. Sandefur, *Discrete Dynamical Systems: Theory and Applications*, Clarendon Press, Oxford, 1990. Sandefur has a second book on discrete dynamical modeling, but we prefer this one.

A comprehensive though terse book for further study in both discrete and continuous models:

J.D. Murray, *Mathematical Biology*, Springer-Verlag, Berlin, 1989.

1.8 Exercises

1. Use an induction proof to verify that $x(n) = R^n x(0)$ is a solution of $x(n) = R x(n-1)$.

2. Find a closed-form solution for the affine recurrence relation
$$x(n) = R x(n-1) + a.$$

3. Find fixed points for the following recursion relations, and test for stability. Draw a cobweb diagram for parts c and d.

  a. $x(n) = \dfrac{x(n-1)}{1 + x(n-1)}$.

  b. $x(n) = x(n-1) e^{r\, x(n-1)}$ where r is a constant.

  c. $x(n) = x(n-1)^2 - 6$.

  d. $x(n) = x(n-1)^2 + .7x(n-1) + .02$.

4. Construct a recursion relation that has a stable fixed point at $x = 1$ and an unstable fixed point at $x = 3$.

1.9 Projects

Project 1.1. The Credit Card Problem

The motivation for this project comes from the following problem which is paraphrased from a calculus book.

> A man has a credit card with a $1,000 balance, 18.9% annual interest, and no grace period. He makes no payments for a year and is surprised when his statement shows a balance of more than $1,189. Use the idea of monthly compounding to help him out.

This project starts with this simple problem and successively refines it to incorporate payment schemes (a credit card with no payments for a year?) and variable interest rates.

1. Start with the compartmental diagram shown in Figure 1.24. Assuming an initial $1,000 balance, 18.9% annual interest rate, monthly compounding, no grace period and no payments, determine the balance after one year and five years.

2. Introduce payments to the above model by adding a flow out of the balance box. Assume a minimum monthly payment of 3% of the balance, determine what the balance will be after five years. How long will it take to get the balance down to under 10 dollars?

1 Discrete Dynamical Systems

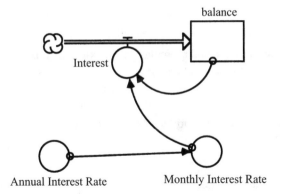

FIGURE 1.24 Credit Card Compartmental Diagram

3. One of the authors, has a credit card which has a minimum payment of 3% of the balance or $10, whichever is larger. With this assumption, how long will it take to pay off the credit card? Note: When defining payments you may use an IF THEN ELSE construction.

4. Keeping with the assumptions in 3, suppose we are interested in how much borrowing $1,000 costs. Add a mechanism for keeping track of total payments. A compartmental diagram is shown in Figure 1.25.

5. Suppose we pay $10 more than the minimum payment each month. What effect does this have on the total payments and the time to pay off the card?

Project 1.2. Bobcats

Most species of wild cats are endangered including the bobcat. This project explores the behavior of a bobcat population using growth rate data from the state of Florida (from Cox, et. al., *Closing the Gaps in Florida's Wildlife Habitat Conservation System* [13]). We know annual growth rates for bobcats under best ($r = 0.01676$), medium ($r = 0.00549$), and worst ($r = -0.04500$) environmental conditions. For this project,

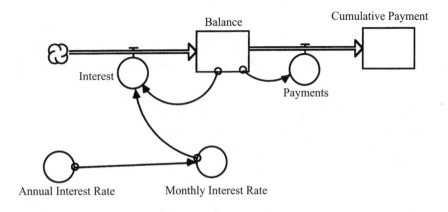

FIGURE 1.25 A More Refined Model

assume that growth rates are constant from year to year (complications will come later). For now consider each growth rate as representing a different region in Florida.

1. Construct a spreadsheet (or other simulation) tracking three bobcat populations, each initially consisting of 100 individuals, over a period of 10 years under the three types of environmental conditions. Plot all three simulations on a single graph. The graph should have dates on the horizontal axis (starting with the current year), it should have a legend identifying which curve is which, and the style should have each curve identifiable when printed in black ink.

2. Repeat part **1** over a period of 25 years.

3. You should have noticed that under the best conditions the population is growing. Several management plans have been discussed. The first is to allow one bobcat per year to be hunted. The second is to allow five bobcats per year to be hunted. The third is to allow one percent of the animals to be hunted. The last is to let five percent of the animals be hunted. Construct a simulation which compares these strategies over 10 and 25 years. Which of these strategies result in a stable population?

4. Continuing with the theme of part **3**, experiment to find strategies which cause the population to stabilize. Find a strategy (which can be more involved than just subtracting a fixed number or percentage) which causes the population to rise to 200 animals (give or take an animal or two) and then stabilize.

5. You should also have noticed that under the worst conditions the population is declining. Proposed management plans include adding three animals per year, adding 10 animals per year, adding one percent of the population each year, and adding five percent of the population each year. Compare these strategies for 10 and 25 years. Which strategies cause the population to stabilize?

6. Experiment to find strategies which will cause the population under the worst conditions to stabilize at 50 and at 200 animals.

7. Which of the strategies in parts **3** and **5** are affine? By hand, analyze these equations (which are affine) to find fixed points and test for stability. Use the theory of affine equations to verify the numerical results.

Project 1.3. Blood Alcohol Model

Data and Assumptions: (Data are from Davis, Porta, Uhl, *Calculus&Mathematica: Derivatives*[14] ©1994 Addison Wesley Longman Inc. Reprinted by Permission of Addison Wesley Longman.)

The average human body eliminates 12 grams of alcohol per hour.

An average college age male in good shape weighing K kilograms has about $.68K$ liters of fluid in his body. A college-age female in good shape weighing K kilograms has about $.65K$ liters of fluid in her body. People in poor shape have less.

One kilogram = 2.2046 pounds.

Threshold for legal driving: If your body fluids contain more than one gram of alcohol per liter of body fluids (or 0.1 gm/100 ml which is the usual way of reporting it),

1 Discrete Dynamical Systems

Type of Drink	Grams of Alcohol
12 ounce regular beer	13.6
12 ounce light beer	11.3
4 ounce port wine	16.4
4 ounce burgundy wine	10.9
4 ounce rose wine	10.0
1.5 ounce 100-proof vodka	16.7
1.5 ounce 100-proof bourbon	16.7
1.5 ounce 80-proof vodka	13.4
1.5 ounce 80-proof bourbon	13.4

TABLE 1.1 Grams of Alcohol for Different Types of Drinks.

then you are too drunk to drive legally in most states. Find out the level for your state and use it in this project.

A blood alcohol concentration of 4.0 gm/l is likely to result in coma. A blood alcohol concentration of 4.5-5.0 gm/l is likely to result in death.

Alcohol content of various beverages: see Table 1.1

Project: Construct the basic model from the compartmental diagram shown in Figure 1.26. This diagram was drawn using Stella; recall that circles either hold parameters or perform operations on parameters. Pick an appropriate time step. We suggest one minute. (Technically this is a continuous model, but we are treating it as discrete with a short time step.) Pick a hypothetical weight and sex.

1. Assume your hypothetical person arrives at a party and instantaneously downs a six-pack of beer (i.e., high initially alcohol level, no input flow). Graph alcohol concentration

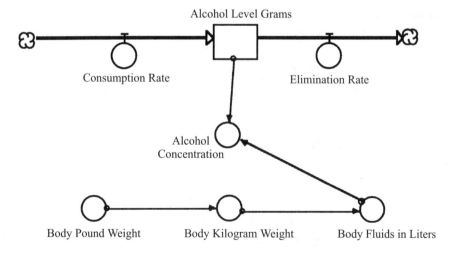

FIGURE 1.26 Compartmental Diagram for the Blood Alcohol Model.

as a function of time. Plot the legal driving level on the same graph. Be very careful of the scales. How long will it be before Hypothetical can drive home legally?

2. Construct a more realistic manner of consuming six beers (i.e., zero initial alcohol level, positive flow). You may want to use if-then statements to turn off the drinking after a time, or to change rates of drinking throughout the evening. Again plot alcohol concentration against time and include the legal driving level on the graph.

3. Try some other alcohol input functions. Here are some ideas. A piecewise defined function can be used to model periods of drinking and non-drinking. A function that steps up for a short time, then back down (some programs even have a special pulse function) can be used for simulating shots. One could construct a function which relates the rate of drinking to alcohol concentration or to the number of people at a party.

Graphs should include the legal driving, coma, and death levels.

Prepare an information sheet or sheets for distribution to students (your peers) showing the effects of drinking styles on blood alcohol concentration.

We acknowledge a high school teacher in the TORCH program, whose name we no longer recall, for originally introducing us to the ideas of this project. The book *Calculus&Mathematica* by Davis, Porta, and Uhl has a problem on blood alcohol levels which significantly influenced this project.

Project 1.4. Compound Interest

When you save money in a bank account, interest is compounded periodically, which means that periodically the interest you earn is added to your account so that you start earning interest on your interest.

1. Suppose you deposit \$1,000 into a bank account that has a 5% annual interest rate. Further suppose that interest is compounded monthly. Thus initially you have \$1,000; at the end of one month your account has $\$1,000 + 0.05/12 \times 1,000$ or \$1,004.17; at the end of two months your account has $\$1,004.17 + 0.05/12 \times 1,004.17$ or \$1,008.35; and so on. Notice that even though the model's scale is in years, the time step is one month. Notice further that since the interest rate is given in years (the time scale), we need to divide by 12 to compute the interest rate per time step. Set up a recurrence relation for the amount of money $x(t+1)$, given that we know the amount of money in the account after t time steps $x(t)$. Track the amount of money in the account over two years.

2. Find a closed form solution for $x(i)$, the amount in the account after i time steps, using data and assumptions from part **1**.

3. Find the recurrence relation for the amount of money in the account x if P dollars are invested at an annual interest rate r which is compounded n times per year for i time steps.

4. Solve to find a closed-form solution for the recurrence relation in part **3**. Express the function in terms of t years, instead of i time steps.

5. Suppose now that you deposit \$1,000 in an account that is compounded monthly but which has a variable interest rate. Track your account over two years using the monthly interest rates given in Table 1.2. The rate at $t=0$ corresponds to the rate from the moment of deposit until the first compounding period.

1 Discrete Dynamical Systems

Month	Rate	Month	Rate
0	0.05	13	0.049
1	0.051	14	0.049
2	0.052	15	0.048
3	0.053	16	0.048
4	0.052	17	0.048
5	0.052	18	0.047
6	0.053	19	0.048
7	0.053	20	0.047
8	0.053	21	0.046
9	0.052	22	0.045
10	0.052	23	0.045
11	0.051	24	0.044
12	0.051	25	0.044

TABLE 1.2 Monthly Annual Interest Rates.

Project 1.5. Inflation

You should work through Project 1.4 before working this problem. Both terminology and results of Project 1.4 are needed. As you are probably aware, inflation is the annual increase in the prices of goods and services. Suppose inflation raises prices this year by 3% over last year, and next year inflation raises prices 3% over this year's prices. Notice that each year the inflation is raising not the original price, but the inflated price. This is compound interest compounded annually with a slight perspective shift.

1. Assuming an annual inflation rate of 3%, how much will an item that currently costs $100 cost in 1, 5, 10, and 30 years? (Assume a compound interest compounded annually type model. Some items obey this, others do not. We are assuming this one does.)

2. Still assuming that an annual inflation rate of 3% held for the past 10 years, how much did an item that costs $100 today cost 10 years ago? (Again assume a compound-interest-compounded-annually-type model.)

3. Look up the annual inflation rate for the past 30 years. (One source is the Consumer Price Index in the Statistical Abstract of the United States published by the Census Bureau; online versions exist.) If an item cost $100 thirty years ago, how much did it cost after 1, 5, 10, and 30 (today) years? What is the average inflation rate over this time period. Assuming a constant annual inflation rate equal to the 30-year average, how much did a $100 item cost after 30 years? Compare your answers from the two models. You should address the effect of using an average value instead of the actual value. Once again figure the cost of a $100 item after 30 years assuming the inflation rate of 30 years ago remained constant. Compare this result with the results of the first two models. How good were the predictions? This is essentially what we do when we plan for our retirement using current inflation rates. Thirty years from now, we will see how good our choice of inflation rate was.

Project 1.6. Plant-Herbivore System

The purpose of this project is to model the dynamics involved in introducing a population of 100 deer into an area consisting of a patchwork of grassland and forest. By "dynamics" we mean we want to predict the type of behavior seen (growing, declining, oscillating, chaotic), and we are not interested in predicting actual numbers of animals in a given year. Rather than count individual plants, we measure the plant density in appropriate density units. In the absence of deer, the plants observe a logistic type growth with a growth rate of 0.8 when the plant density is low and a growth rate near 0 when the plant density is near 3,000 density units. Initially the area is ungrazed, so we assume that the initial plant density is 3,000 units. The deer herd eats the plants, of course, and when the vegetation is burned out or grazed flat, the herd will decline at a rate of 1.1. (That is, the number of deer lost in a year is 1.1 times the number of deer. This seems impossible, but it only means that the herd would be gone in less than a year with no food.) When food is present, this decline is offset by a growth term which has a rate of 0 when plant density is 0, 1.5 when plant density is 3,000, and is roughly linear (assume it is linear) in between. (Note this is the growth rate, not the actual number of deer.) Similarly when deer are present, the plant density is diminished. Each deer can eat as much as 1.2 density units per year when food is plentiful (plant density of 3,000), but as food becomes scarcer, the deer will eat less with no food eaten when there is no food available. Again assume a linear relationship between eating no food when the plant density is zero and 1.2 density units when the density is 3,000.

Part I
1. Construct a simulation model and run it for 30 years under the following three conditions
   a. Discrete time steps of one year
   b. Discrete time steps of one day
   c. Optional (If the software that is being used for the simulation easily does this; otherwise it will be discussed in Chapter 5): Continuous time (fractional time steps)

2. For each of the cases in **1**, make a graph with the deer and plant density plotted as functions of time on the same axis (the scales may be different). Also make a phase plane plot for each case (deer on one axis and plants on the other). Include comments about existence and stability of fixed points and the differences between the various cases. Which case is most appropriate for this scenario?

Part II
The situation above is a modification of a model by Caughley which is identical except that the interaction terms for the deer and plants do not vary linearly. His model had the decrease in plant density due to deer equal to 1.2 deer population $(1 - \exp(-.001 *$ plant density$))$, and the increase in deer due to the plants as 1.5 deer population$*(1 - \exp(-.001 *$ plant density$))$.

Replace the plant-deer interaction terms with Caughley's terms and repeat the work of Part I, including the discussion.

Part III
Compare the models in Part I and Part II. Part of this comparison should involve plotting (1-exp(-.001 x)) from 0 to 5,000 and explaining what this term contributes to the second model over the line used in the first model. Which model does the best job of explaining the observed behavior?

1 Discrete Dynamical Systems

Project 1.7. Great Lakes Pollution

Problem statement: Most of the water flowing into Lake Ontario is from Lake Erie. Suppose that the pollution of these lakes ceased suddenly. How long would it take for the pollution level in each lake to be reduced to 10 percent of its present level?

Assumptions, simplifications, and data: Assume that all of the water in Lake Ontario comes from Lake Erie and all of the water that leaves Lake Erie goes into Lake Ontario. (i.e., ignore feeder and outlet streams, evaporation, and other little irregularities). Let $a(n)$ and $b(n)$ be the total amount of pollution in Lake Ontario and Lake Erie, respectively, in year n. Since the pollution has stopped, the concentration of pollution in the water coming into Lake Erie is zero. It has also been determined that, each year, the percentage of the water replaced in Lakes Erie and Ontario is approximately 38 and 13 per cent respectively. This means that each year, 38 per cent of the water in Lake Erie flows into Lake Ontario and is replaced by rain and pure water flowing in from other sources. Also each year 13 per cent of Lake Ontario's water flows out and is replaced by the water flowing in from Lake Erie. Assume that the concentration of the pollution in each lake is constant throughout that lake. We are not interested in the true levels of pollution, only in the decrease, so assume that $a(0) = 1$ unit of pollution and $b(0) = 1$ unit of pollution.

1. Construct a compartmental diagram for these two lakes. Explain in words comprehensible to a college-educated non-mathematician why the following equations hold:

$$a(n+1) = 0.87a(n) + 0.38b(n)$$
$$b(n+1) = 0.62b(n).$$

2. Show, in enough detail that a bright precalculus student can follow, how to rearrange and substitute and do other algebraic kinds of things to the two equations in part 1 to get the following single equation: $a(n+2) = (0.87 + 0.62)a(n+1) - 0.87(0.62)a(n)$.

3. Remembering that you know a(0) and can find a(1), find the closed-form solution for this equation.

4. Set up two columns on a spreadsheet (or other program), one which computes $a(n)$ recursively and one which computes $a(n)$ using the closed form solution and verify that you get the same results. How long will it take for Lake Ontario's pollution level to drop 10%?

5. Find a closed form solution for $b(n)$. Hint: This is much easier than $a(n)$ so think before you work too hard. Use this to determine analytically how long it will take for Lake Erie's pollution to decline by 10%.

The idea and data for this project come from Sandefur's book *Discrete Dynamical Systems* [57].

Project 1.8. California I: A Simple, More Complicated Model

This model extends the simple birth/death exponential model to include migration. Table 1.3 gives the populations of California and the rest of the United States in 1955 and 1960 and the changes between 1955 and 1960 due to births, deaths, and net migrations (immigration less emigration) between these two regions.

Region	Population 1955	Births 1955-60	Deaths 1955-60	Net Migration 1955-60	Population 1960
California	12,988	1,708	614	1,124	15,206
Rest of U.S.	152,082	19,499	7,417	-1,124	163,040
Total	165,070	21,207	8,031	0	178,246

TABLE 1.3 Population Change in California and the Rest of the United States Between 1955 and 1960. All Population Counts Are in Units of 1,000 People.

1. The Rates. Find the birth, death, and migration rates for California and the rest of the United States between 1955 and 1960. This information is organized in tabular form in Table 1.4 with two rates already computed.

Region	Birth Rate	Death Rate	Migration Rate
California	0.1315060		
Rest of U.S.		0.0487697	

TABLE 1.4 1955-1960 Rates.

2. The Model. Use compartmental analysis to set up a difference equation model for the population of California. Set up another model for the population of the rest of the United States. Use the 1955 data as the initial information. Verify that after one time step (five years) the models accurately give the 1960 data (these values should match exactly except possibly due to rounding errors).

3. Predictions. Run a simulation using the model from 1960 up to the present (in five-year jumps, of course). Do research to find official estimates for the population of these regions for each prediction. Plot the model results with the observed results and discuss how well the model predicts these population sizes. Discuss improvements that could be made to the model.

4. Long-Term Behavior. Again run the simulation, only this time divide the population in each region by the population of the entire United States. At each stage the model results should be numbers between 0 and 1. Run the simulation until these numbers stabilize. Discuss the meaning of the model output. Complete the statement: "Assuming that the birth/death/migration rates remain constant at the 1955-1960 values, then eventually 1 out of every _____ Americans will live in California."

5. Extra. Repeat this project for your own state or, if you are from California, using more recent data.

This project is based on Chapter 2 of the book by Rogers [55]. California data is from the California State Department of Finance, Financial and Population Research Section [8] and is reprinted by permission of the University of California Press.

2

Discrete Stochasticity

In the previous chapter, we studied recurrence and difference equations. In particular, we considered models with equations of the form

$$x(n) = r\,x(n-1),$$

where r was the growth rate of the population. This growth rate could be either a constant or function of x, n, or other variables in the model. One defining aspect of these models is that at any moment, depending on the variable values, r can take only one value. If $r(x) = 0.2x$, then when $x = 1$, $r = 0.2$—not 0.21, not 0.18, but 0.2. If we ran a simulation model 100 times, it would give exactly the same output 100 times. Such a model is called *deterministic*.

Consider for a moment the world as you know it. Sometimes there are a lot of people you know having children, sometimes there are none. Sometimes your garden produces well, sometimes it does not. Sometimes a basketball player can hit free throws, sometimes that player can not. Sometimes many houses are up for sale on your street, sometimes there are none. Sometimes you win a lottery, usually you do not. These events happen in a somewhat unpredictable manner. On average a basketball player might hit 75% of his free throws, but if he has hit three in a row, that does not mean he will miss the fourth (regardless of what the sportscasters try to make us believe). Deterministic models do not take into account the everyday uncertainties of life. On the average they may be pretty good for large samples, but on the individual level they can be absurd. For example, a deterministic model of the basketball player would have him making 3/4 of every shot, whatever that means. Models that take random events into account are called *stochastic*. Different runs of the model will produce different results. A stochastic model of the basketball player would have a 75% chance of each free throw going in, but might have a string of 10 missed free throws. This is much more like we expect from real life than the meaningless deterministic model.

This chapter introduces the ideas of stochastic models. There are two main themes: 1) How to interpret noisy data and output; and 2) How to build models with stochastic components. In this chapter, time is still moving in discrete time steps. Chapter 6 addresses randomness in continuous models.

2.1 Stochastic Squirrels

We begin with a hypothetical example that we study both deterministically and stochastically.

2.1.1 Example

In a city park, the grounds keeper notices that the park's squirrel population is on the decline. The keeper has recorded that the annual growth rate for this population of squirrels is $r = -0.50$. If the initial population of squirrels is $x(0) = 100$, determine the population size over the next five years.

This is the exponential decay model of Section 1.1. We have the following relation

$$x(n) = 0.5x(n-1),$$

which has closed-form solution

$$x(n) = (0.5)^n x(0).$$

A table of the population size shows that the population is declining and ultimately "crashes." A graph of this data, Figure 2.1.1, shows the demise of the squirrel population.

The exponential model of Example 2.1.1 is useful in predicting the squirrel population size, but requiring that the population always declines by 50% each year is not realistic. If we measured the squirrel population size, we might observe 52 squirrels after one year, 27 squirrels after two years, and so on. We need to incorporate these types of variations into our modeling toolkit. One way to model this effect is to consider the values of a model as *stochastic* in nature, that is, they vary randomly.

There are many ways to incorporate random effects in our modeling toolkit. We could assume that some intrinsic parameter of the model (such as the growth rate r in the

Year	Population Size
0	100
1	50
2	25
3	12.5
4	6.25
5	3.125

TABLE 2.1 Squirrel Population.

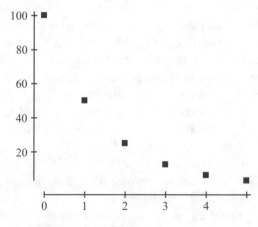

FIGURE 2.1 Squirrel Population — Deterministic Model

squirrel population) is random in nature, or we could assume that the model output is random. We will consider both of these types of assumptions as well as others in our models that follow.

An important distinction between stochastic models and the deterministic models previously considered is that there is no single value for a stochastic model. Rather, a stochastic model takes on different values every time it is observed. To understand a stochastic model we need to know things such as what value it is most likely to be and how far is it likely to be from its most likely value. There are several ways of conveying this information. One method is to present the model as a distribution. Another is to define numbers measuring the center of the distribution (mean, median) and the spread of the distribution (standard deviation, range, inter-quartile range). All these topics will be addressed shortly.

When we consider a distribution of values, we are talking about all possible values a parameter can take and the frequency of occurrence of each of these values. The precise term for this is the *population distribution*. (The term *population* in this context refers to the set of all possible values for a model or parameter; not necessarily a collection of animals, humans or other living organisms.) In all but the simplest cases, finding the population distribution requires theoretical assumptions and mathematical analysis. If we can not make the theoretical assumptions or can not perform the analysis, we experimentally gather information (called *sampling*) about the values of the model or parameter. This experimental information is known as the *sample distribution*, and we will often assume that it approximates the population distribution. Sometimes it is not even possible to sample a parameter. For example, what is the time until extinction of the Siberian tiger? If we could relive the next 100 years or so many times, we could sample this. But, we live this time line only once, as far as we are aware. An alternate to sampling is simulation. We ran deterministic simulations in Chapter 1. Now we incorporate random elements into our simulations. Repeating the simulation many times gives us a distribution (which is only as good as the assumptions that went into the model). We continue by physically simulating the squirrel demise.

2.1.2 A Simple Population Simulation: Death of M&M's

The project requires 100 M&M candies (or something similar with two distinct sides, such as 100 pennies) and a small container, such as a cup or small bag, to hold the items.

Simulation of a Population. We are going to simulate the squirrel population of Example 2.1.1. Each M&M (or penny) represents a squirrel. Initially, there are 100 squirrels in the population so we start with 100 M&M's. Place the M&M's in the container and then pour them out on a clean, flat surface. Remove the M&M candies that have the letter side facing up. Count and record the number of candies that remain. This count represents the squirrel population after one year. Put these remaining M&M's back into the container and repeat the process, each time removing those with the letter facing up and recording the number which remain. Continue this process until no M&M's remain. At this point, the data collected should be similar to Table 2.2. This data represents one simulation of the squirrel population. Each trial represents the population

Trial	M&M count
0	100
1	53
2	27
3	15
4	9
5	3
6	1
7	0

TABLE 2.2 One Simulation of the Squirrel Population.

size at the end of one year. Since each simulation will produce a different collection of numbers, several simulations need to be conducted to determine the likely behavior of the population. Repeat the above process nine more times (for a total of 10 simulations), each time starting with 100 M&M's. The data collected should be summarized in a table similar to Table 2.3.

If each run is graphed on the same axes (see Figure 2.2), we observe that each run is exhibiting the same type of behavior seen in the squirrel population in Example 2.1.1, but the numerical values obtained in the simulations are different from those in Example 2.1.1. These differences are due to the randomness involved in shaking and dumping the M&M's.

2.2 Interpreting Stochastic Data

To understand stochastic behavior, we develop some methods of data summary and interpretation.

Trial	\multicolumn{10}{c}{Simulation Number}
	1	2	3	4	5	6	7	8	9	10
0	100	100	100	100	100	100	100	100	100	100
1	53	52	47	55	49	50	51	48	50	51
2	27	28	24	25	26	23	30	24	22	27
3	15	12	15	11	13	10	11	13	12	14
4	9	7	8	5	7	6	3	5	6	5
5	3	3	4	2	3	3	2	1	2	2
6	1	0	1	0	1	1	0	1	0	1
7	0		0		0	1		0		1
8						0				0

TABLE 2.3 Squirrel Population Counts for Ten Simulations.

2 Discrete Stochasticity

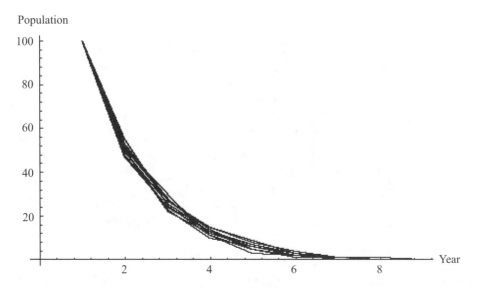

FIGURE 2.2 Ten Simulations of the Squirrel Population.

2.2.1 Measures of Center and Spread I: Mean, Variance, and Standard Deviation

The simulation runs graphed in Figure 2.2 are "varying" about the graph of the deterministic model (see Figure 2.1). Initially, each simulation starts with 100. After one time step, each simulated population size has a different value as seen in simulation data reproduced in Table 2.3.

Examining these numbers, we see that they are "centered" around 50. The simplest and most familiar measure of a data center is the arithmetic average or *sample mean*, denoted $\bar{x}$. It is defined to be

$$\bar{x} = \frac{\sum_{i=1}^{n} x_i}{n}.$$

The sample mean $\bar{x}$ of the 10 data points from trial 1 is

$$\bar{x} = \frac{53 + 52 + 47 + 55 + 49 + 50 + 51 + 48 + 50 + 51}{10}$$
$$= 50.6.$$

This is shown in the "mean" column of Table 2.4. The sample means for the other trials are computed similarly and are shown in Table 2.4. If we compare these average values with the values obtained for the squirrel population in Example 2.1.1, we see that these numbers are fairly close. Why? This observation is a key to understanding deterministic versus stochastic behavior. The sample mean "averages out" the stochastic behavior and represents the deterministic behavior fairly well.

Stochastic behavior is often summarized by the sample mean, but the mean alone does not give the sense of a distribution. A measure of the *spread* of a data set should also be reported. There are several measures of spread for a data set. The *sample variance* and the related *sample standard deviation* are the most commonly used measures. The

Trial	Mean	Standard Deviation
0	100	0
1	50.6	2.36
2	25.6	2.46
3	12.6	1.71
4	6.1	1.73
5	2.5	0.85
6	0.6	0.52
7	0.2	0.42
8	0	0

TABLE 2.4 Mean and Standard Deviations of Ten Squirrel Simulations.

sample variance, denoted s^2, is the average square distance from $\overline{x}$ and is given by the formula

$$s^2 = \sum_{i=1}^{n} \frac{(x - \overline{x})^2}{n - 1}. \tag{14}$$

The $n - 1$ term in the denominator of this expression (as opposed to n in the sense of an average) provides a better estimate to a population variance. It can be shown that s^2 is an *unbiased* estimate; that is, the average of the s^2 values over many repetitions is the population variance.

The sample variance for trial 1 is calculated by subtracting $\overline{x}$ from each of the outcomes then squaring each difference and summing the results. That is,

$$s^2 = \frac{(53 - 50.6)^2 + (52 - 50.6)^2 + (47 - 50.6)^2 + \ldots + (51 - 50.6)^2}{9}$$
$$= 5.6.$$

The *sample standard deviation* s is the positive square root of the sample variance; that is,

$$s = \sqrt{s^2}.$$

Measures of spread are often reported in terms of standard deviations. In general, the variance is used in theoretical work since it does not involve a square root which complicates calculations, while the standard deviation is used when trying to understand the data.

One practical way to think about the standard deviation is, if instead of reporting a single number, you report a number give or take an amount. For example if you weigh yourself using a bathroom scale 100 times in a row, you will get a variety of weights due to slight variations in how you stand, how the spring stretches, and other factors. Instead of reporting your weight as a single number, say 150 pounds, you can report your weight as 150 pounds give or take two pounds. Here the amount of two pounds was guessed at while looking at the data. Usually the values were between 148 and 152 pounds. Rarely,

2 Discrete Stochasticity

though occasionally, they were less than 148 or larger than 152. The two pounds gives a range in which you expect the scale to differ from 150, either over or under. In a conceptual sense the two pounds can be thought of as the standard deviation, though we just guessed at the number two. If you actually compute the standard deviation and it is, say 2.1 pounds, then it is appropriate to report your weight as the mean give or take the standard deviation; here, 150 pounds give or take 2.1 pounds. Part of the justification for this is that if your measurements are normally distributed (which we will discuss shortly), then 68% of the time your measurements will be within one standard deviation of the mean. Thus your value is most likely to be within the interval defined by the mean give or take one standard deviation. The 68% number is not necessarily true for other distributions, but similar results apply except in some degenerate cases.

The standard deviations for the squirrel simulation are shown in the last column of Table 2.4. We note that data for trials 0 and 8 are constant so that variances and standard deviations of 0 are expected.

The numbers obtained for trial 1 of each of the 10 simulations represent a possible squirrel population size. From Example 2.1.1, we expect that there will be 50 squirrels after one year. Our 10 simulations gave values in a "distribution" around 50. This distribution of values represents the chance variation of the experiment. There is a 50% chance that each M&M will land with its letter side face up, so that we should roughly see the population decreasing by *about* 50% each trial, but not *exactly* 50% as in Example 2.1.1. To facilitate our study of chance or random behavior, we need to formalize the concept of a distribution.

2.2.2 Frequency Distributions and Histograms

In Section 2.1.2, we considered a random experiment. With each trial we found a population size which was random. In that experiment, we considered the population size as a function of the trial number. This is a "random function" in the sense that it assigns different (random) values each time the experiment is performed (i.e., each time the simulation is run). Such a function is called a *random variable*. To gain insight into the behavior of a random variable, a method of data summary needs to be developed. This section considers one of the most commonly used methods of data summary known as a *frequency distribution* and a related graphical summary method called a *histogram*.

Frequency Distributions. We develop the notion of a frequency distribution by considering the following example.

Example: The Arithmetic Skills Test (A.S.T.) is a standardized basic mathematics skills test administered to general mathematics students at the authors' university. The test has 30 multiple choice questions with a minimum passing score of 21. The scores from a recent A.S.T. exam are given in Table 2.5.

If we are presented with just this set of data, it is difficult to determine how this particular class performed on the A.S.T. We begin by organizing the data into a frequency distribution. A frequency distribution provides a frequency (or count) for each occurrence of a data point. Looking through this data we notice that there is one 8, no 9, one 10, no 11, no 12, no 13, one 14, one 15, one 16, one 17, and so on. A frequency distribution

19	25	22	23	24	19	22	19	24	24
27	18	20	23	25	28	10	18	23	19
21	17	23	26	24	26	21	14	27	18
19	24	23	21	23	20	22	25	25	16
24	19	22	15	22	26	19	8	22	24

TABLE 2.5 Student Scores on the A.S.T.

which has many data points with few or no occurrences is said to be *flat* and is only slightly more informative than the original data would be if written in ascending order.

To increase the usefulness of our frequency distribution, we group the data—a process sometimes called *summarizing by grouping*. This involves breaking the domain of the data into equal-sized pieces and grouping all the data lying within a single piece. In our example, a group containing, say, 14, 15, and 16 would now have 3 counts. A group containing 14-19 would have 14 counts. Proper grouping removes the flatness from our data summary.

In when grouping data there is no right number of classes to form, however, and the size and range of the data set play a role in the decision. If our A.S.T. data set included all scores in the last five years, it would contain thousands of entries. There would be hundreds of students with a score of 7, and hundreds more with a score of 8. There would be no need to group. With the data from this one class, however, there are no 7's nor no 8's. Many other scores have only a single occurrence. On the other hand, if we group all the data into two groups, the distribution is not flat, but it is not useful either. What we want is a clear picture of the distribution. Too few classes produce a group of over-inflated frequencies, whereas too many classes produce a flattened distribution. The location where the first class begins can also influence the shape of the histogram. The purpose of a histogram is to help us understand the data through visualization, and it is advisable to try a number of different class sizes and starting points until we get the most informative histogram. Computer programs typically have some formula for a default number of classes that usually is fairly good.

In this example, we arbitrarily (subject to knowledge from trying this several times) choose seven groups or classes, with the first group beginning at the smallest data point. We next find the *range* of the data by locating the largest and smallest values (in our case 28 and 8) and subtracting the smallest from the largest. The range of our data is $28 - 8 = 20$. To find the width of each group, divide the range by the number of groups;

$$\text{width} = \frac{\text{range}}{\text{number of groups}} = \frac{20}{7} = 2\ 6/7.$$

We group all the data in the interval $[8, 10\frac{6}{7})$, all the data in $[10\frac{6}{7}, 13\frac{2}{7})$, and so on. Table 2.6 shows the frequency distribution of the grouped data (we list actual data groupings, not the interval). The table also lists the *relative frequency* of each score. The relative frequency of a data point is the frequency of the data point divided by the total number of data points (sometimes expressed as a percentage). We point out that different authors define the grouping process differently. Some consider discrete data,

2 Discrete Stochasticity

Score	Frequency	Relative Frequency
8-10	2	2/50 = 4%
11-13	0	0/50 = 0%
14-16	3	3/50 = 6%
17-19	11	11/50 = 22%
20-22	11	11/50 = 22%
23-25	17	17/50 = 34%
26-28	6	12/50 = 24%
	$n = 50$	total = 100%

TABLE 2.6 Grouped Student Scores on the A.S.T.

like our current example, differently from continuous data. We adopted our conventions to match the output from the statistical package we were using.

Now that the data has been organized, we observe several trends. The most frequent scores obtained were between 23 and 25, with a relative frequency of 34%. A majority (64%) of the students passed the exam. Ten percent scored below 17.

This frequency distribution summarizes the data in a more compact form, but we have lost some of the "resolution" of the individual data points. For example, there are 11 scores between 17 and 19, but we do not know what they are. This information is not obtainable from the frequency distribution.

Histograms. When data are grouped in intervals, they can be depicted by a *histogram*, a bar chart that shows either the frequency or the relative frequency of each interval. In order to separate one interval from the next, we create boundaries. A *boundary* is the endpoint used in separating groups. In the previous example, the first two intervals are $[8, 10\frac{6}{7})$ and $[10\frac{6}{7}, 13\frac{2}{7})$, so 8, $10\frac{6}{7}$, and $13\frac{2}{7}$ are boundary points.

Once the boundaries have been computed, they are marked off on the horizontal axis. The vertical axis represents the frequency or relative frequency of each interval. Above each interval, a rectangle is drawn whose height equals either the frequency or the relative frequency of the interval. Comparison of the heights of the rectangles shows which interval has the most (or least) data points. A histogram for the A.S.T. scores is shown in Figure 2.3.

In the previous example, the score on the A.S.T. was integer valued. The nature of this particular exam has the possible scores as an integer between 0 and 30 inclusively. When a random variable has only a finite (or countably infinite) number of possible values, the random variable is called *discrete*. If the values that a random variable is able to take, fall anywhere in the reals or some interval of the reals, we call it a *continuous* random variable. In general, continuous random variables are used to represent measurements on physical quantities, such as the amount of annual rain fall or the height of adult females. Since model parameters are often physical quantities which can be measured, we will frequently make use of continuous random variables. Continuous random variables sometimes are

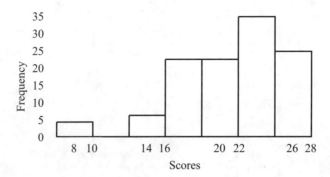

FIGURE 2.3 Histogram of A.S.T. Scores.

used to represent discrete random variables when the discrete random variable has a wide range of possible values.

2.2.3 Example: *The X-files*

The television show *The X-files* has become very popular since its first season broadcasts. Table 2.7 shows the number of viewers in millions for each of the first two seasons of this unique show. These data are from B. Lowry's book *The Truth is Out There, The Official Guide to The X-files* [40] and are used by permission of Nielsen Media Research.

Upon inspection of the data, one observes there is a noticeable difference between the number of weekly viewers for each of the two seasons. The minimum number of viewers for the first season was 5.1 million, and the maximum was 8.3 million. The second season data shows that *The X-files* gained in popularity. The minimum number of viewers for season two was 7.9 million, and the maximum was 10.8 million.

Season 1	Season 2	Season 1	Season 2
7.4	9.8	6.2	8.8
6.9	9.3	6.8	10.2
6.8	8.7	7.2	10.8
5.9	8.2	6.8	9.8
6.2	8.5	5.8	10.7
5.6	9.2	7.1	9.6
5.6	9	7.2	10.2
6.2	9.1	7.5	9.8
6.1	8.6	8.1	7.9
5.1	9.9	7.7	8.5
6.4	8.5	7.4	8.1
6.4	9.7	8.3	9
			9.6

TABLE 2.7 The Number of Viewers (in millions) of the Television Show *The X-Files*.

2 Discrete Stochasticity

Number of Viewers in millions	Frequency	Relative Frequency
$[5.1, 5.6)$	1	$1/24 = 4\%$
$[5.6, 6.1)$	4	$4/24 = 17\%$
$[6.1, 6.6)$	6	$6/24 = 25\%$
$[6.6, 7.1)$	4	$4/24 = 17\%$
$[7.1, 7.6)$	6	$6/24 = 25\%$
$[7.6, 8.1)$	1	$1/24 = 4\%$
$[8.1, 8.6]$	2	$2/24 = 8\%$
	$n = 24$	total $= 100\%$

TABLE 2.8 Frequency Distribution for the Number of Viewers (in millions) of the First Season of the Television Show *The X-files*.

The range of the first season data is $8.3 - 5.1 = 3.2$ million. We group the data into seven classes; the width of each class is $3.2/7 \approx 0.5$ million. Table 2.8 shows the frequency distribution for the number of weekly viewers for the first season. Notice that the four numbers in the second class represent numbers of viewers between 5.6 million and 6.1 million, an amount equal to the width (0.5) of the class. (We are considering discrete data to be continuous.)

A histogram for the number of viewers (in millions) of the television show *The X-files*, during the first season is shown in Figure 2.4. The histogram shows the relative frequency of the number of viewers.

An inspection of the data in Table 2.7 reveals that the popularity of the show did not increase until late in the first season. If we compute the arithmetic average for the number of viewers for the first season, we find

$$\overline{x} = \frac{\sum_{i=1}^{n} x_i}{n} = \frac{7.4 + 6.9 + 6.8 + \cdots + 8.3}{24} = 6.695.$$

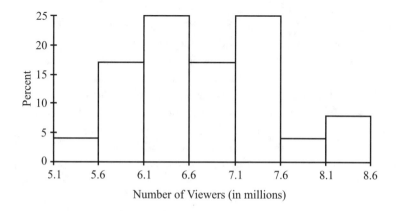

FIGURE 2.4 Histogram of *The X-files* First Season Viewers.

Does this number reflect the increased popularity of the show at the end of its first season?

We should point out that any summary statistic provides a limited amount of information. To determine the behavior of a data set, several different types of summary measures should be provided.

2.2.4 Measures of Center and Spread II: Box Plots and Five-Point Summary

The mean is often skewed by data points much lower or much higher than the rest of the data. Such data points are called *outliers*. We now introduce a measure of central tendency that is not affected by outliers. The *median* is, as the name suggests, the middle observation of an ordered data set. Outliers have much less impact on the median than they do on the mean. One common example is the reporting of housing prices. Suppose that most of the homes in a community are selling between $60,000 and $150,000, but a few homes are sold for around $500,000. The few half-million dollar houses will put the mean higher than what you might consider to be the center of the data. In fact it might put the mean up over $150,000, depending on the data. The median, however, is in the middle of the bulk of the sales and is much closer to the point you would intuitively consider to be the center of the data. See Exercise 1.

To compute the median of a data set, first arrange the data set in increasing order. If the number of data points is odd, the median is the data point in the middle. If the number of data points is even, the median is the arithmetic average of the two middle data points.

Referring to the first season of *The X-files*, we observe that it has an even number of data points. The middle two observations of the ordered data set are 6.8 and 6.8 so that the median is 6.8 million viewers. We note that the sample mean of 6.695 million is lower than the median of 6.8 million.

We have already introduced several summary statistics, namely, the sample mean and median, the sample minimum and maximum, and the range. There are several other useful measures. Percentiles are points that divide the data. The *25th percentile* or *first quartile*, denoted $\pi_{.25}$, marks the point below which 25% of the data lie. The *75th percentile* or *third quartile* denoted $\pi_{.75}$ is defined similarly. A working definition of percentile will be given shortly. The median is the 50th percentile and is sometimes called the second quartile. These statistics give rise to an additional measure of spread. We already encountered the *range* which is the difference between the maximum and minimum data points. All of the data lies in an interval the size of the range. The *inter-quartile range* is the difference between the third quartile and the first quartile. The middle half of the data lies in an interval the size of the inter-quartile range. The range and inter-quartile range convey important distribution information. If the data were evenly spread, we would expect the inter-quartile range to be roughly half of the range. If the data were concentrated near the median, the inter-quartile range would be much smaller than half the range. If the inter-quartile range is centered on the center of the range, then the distribution should be symmetric. If the inter-quartile range is not centered on the range, then distribution is skewed.

Before finding a percentile, the data must be ordered in increasing order. To find the pth percentile, $\pi_p, 0 < p < 1$, of an ordered data set with n data points, we simply find

2 Discrete Stochasticity

the point that has approximately $p \times n$ data points less than it and also $(1-p) \times n$ greater than it. One way to do this is to define the pth percentile as the $p \times (n+1)$ data point in the ordered data set. This method works if $p \times (n+1)$ is an integer. If it is not an integer, but is equal to k plus some proper fraction a / b, we use a weighted average of the kth and $(k+1)$st data points. That is, to find the pth percentile, we use

$$\pi_p = x_k + \left(\frac{a}{b}\right)(x_{k+1} - x_k)$$

where x_k is the kth data point in the ordered data set, k is the greatest integer less than or equal to $p \times (n+1)$, and $\frac{a}{b}$ is the number such that $k + \frac{a}{b} = p \times (n+1)$.

For example, suppose we want to find the first and third quartiles of the first season *The X-files* data. Since there are 24 data points in the data set, we find that $0.25 \times (24+1) = 6.25$, so that $k = 6$ and $a/b = 0.25 = 1/4$. The 25th percentile for the first season viewers of *The X-files* is

$$\pi_{.25} = x_6 + \left(\frac{1}{4}\right)(x_{6+1} - x_6)$$

$$= 6.1 + \left(\frac{1}{4}\right)(6.2 - 6.1)$$

$$= 6.125.$$

Similarly, to compute 75th percentile, we first find $0.75 \times (24+1) = 18.75$. Thus

$$\pi_{.75} = x_{18} + \left(\frac{3}{4}\right)(x_{18+1} - x_{18})$$

$$= 7.2 + \left(\frac{3}{4}\right)(7.4 - 7.2)$$

$$= 7.35.$$

The *five-point summary* of a data set consists of the minimum, the 25th percentile, the median, the 75th percentile, and the maximum, written in that order. These five numbers give information on the center and spread of a data set. The five-point summary of the first season viewers of *The X-files* is 5.1, 6.125, 6.8, 7.35, and 8.3.

A graphical method for displaying the five-point summary of a data set is known as a *box plot*. To construct a box plot, draw a vertical axis that is scaled to the data. Next to the axis, draw a box, with the bottom of the box at the 25th percentile and the top of the box at the 75th percentile. Draw a horizontal line in the box at the height of the median. Draw a line segment, called a "whisker," from the top of the box to the maximum. Similarly, draw a whisker from the bottom of the box to the minimum. The box plot graphically conveys the median, the inter-quartile range, and the range. Figure 2.5 shows the box plot for the five-point summary of the first season viewers of *The X-files*. Notice that the inter-quartile range is about half of the range, indicating that the data is not concentrated at the median.

Box plots and five-point summaries are easily obtained using most statistical software packages. Exercise 4 asks the reader to calculate the five-point summary and box plot for the second season data set.

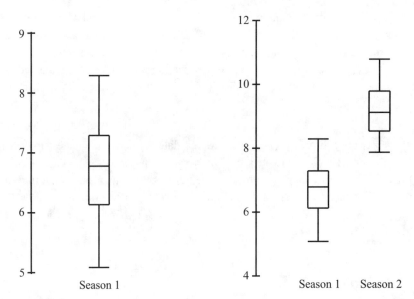

FIGURE 2.5 Box Plot of *The X-files* First Season Viewers.

FIGURE 2.6 Side-by-side Box Plots of *The X-files* Data.

We use box plots and five-point summaries to understand trends that appear in our model output. We compare the two seasons viewing of *The X-files* by examining their box plots, which are shown in Figure 2.6. Two or more box plots on the same coordinate axes are called *side-by-side* box plots. This is a very efficient way of conveying distribution information. Exercise 2 asks the reader to construct side-by-side box plots for the results of the M&M project. Whether you do this exercise or not, you should be able to see how side-by-side box plots allow us to communicate the output of a stochastic model where there is not a number at each time step, but rather, a distribution.

The trend that we suspected earlier in the data is now evident. There is clearly a larger number of viewers during the second season. Notice that the minimum of 7.9 million viewers during the second season exceeds the 75th percentile, 7.35 million, of the first season. The show became more popular during the second season. We have convinced ourselves of the trend we noticed upon visual inspection of the data sets. To borrow a phrase from the show, "THE TRUTH IS OUT THERE," and we made use of mathematics to discover it. We will make further use of *The X-files* data in Chapter 4.

2.3 Creating Stochastic Models: Simulations

2.3.1 Distributions

An important observation for frequency distributions and their histograms is that the sum of all the relative frequencies is one. This can be interpreted as the total area bounded by the graph of the histogram (of the frequency distribution), and the x-axis is one. We may further interpret these relative frequencies as approximate probabilities (approximate, since they are obtained empirically from a data set.) How good these approximations are

2 Discrete Stochasticity

depends upon many factors and is a topic considered in a statistics course. We discuss some of these ideas in Section 2.4 and again in Chapter 4. For the time being, we use these frequencies as approximate probabilities. For example, we say that there is an approximate probability of 0.17 that the number of viewers of first season episodes of *The X-files* was between 5.60 million and 6.10 million. The area interpretation may be taken a step further. For instance, the area under the histogram over the interval $[6.6, 7.6]$ is the sum of the two areas 0.17 and 0.25; hence, we may conclude that the approximate probability of there being between 6.6 and 7.6 million viewers is $0.17 + 0.25 = 0.42$.

In the previous example, if we had obtained a larger set of measurements and had grouped this large data set into a larger number of classes, the histogram would look smoother. Continuing in this fashion, assuming we had a sufficient sample size, if we reduce the width of each class and, hence, increase the number of rectangles, the histogram converges to a smooth curve. This idea is illustrated in Figure 2.7. This smooth curve is called a *probability density function* which is often abbreviated p.d.f., but we usually just refer to it as a density function or distribution.

Part of what is being observed in Figure 2.7 is the difference between a population and a sample. A *population* is everything under consideration; a *sample* is a subset of the population. For example if we wanted to find the average height of all males of age 21 or older, then the population is all males of age 21 or older. Frequently it is impossible to measure everything in a population. Even if we could measure all several billion such

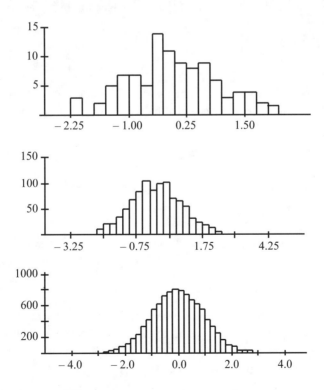

FIGURE 2.7 Convergence of Histograms.

males, by the time we were finished more would have had their birthdays. Instead we estimate the population with a sample. Generally the more random a sample is the better, and the larger the sample is the better. In Figure 2.7, we see histograms of samples converging to the distribution of the population. Generally speaking, theory deals with populations, and practice deals with samples.

Theoretical Background and Interlude. This section deals entirely with populations, not samples. It gives a quick overview of the central definitions and notations of probability theory.

In Section 2.2.2, we introduced the concept of a random variable. Roughly speaking, a random variable is a quantity which takes on random values. It does not take on just any random value, but different possible values have different probabilities of being assigned. These probabilities form the probability law or distribution underlying the random variable. This distribution is just the population probability density function discussed above. Usually it is denoted by $f(x)$. For a discrete random variable (values come from the integers) this underlying distribution can be visualized using a histogram. For a continuous random variable (takes values from the reals), the distribution can be visualized as a curve. It is customary to use capital letters to represent random variables. We follow that convention in this section, but relax it in other sections where capital symbols may be used for other reasons.

The probability that $a \leq X \leq b$ is denoted by $P[a \leq X \leq b]$. For a discrete random variable, this probability is the sum of the area under the histogram between a and b. That is,

$$P[a \leq X \leq b] = \sum_{k=a}^{b} f(k). \tag{15}$$

For a continuous random variable X, the probability that $a \leq X \leq b$ is the area under its probability density function $f(x)$ bounded by the x-axis and the lines $x = a$ and $x = b$, that is

$$P[a \leq X \leq b] = \int_{a}^{b} f(x)\,dx. \tag{16}$$

Notice that since the area under a curve over a point is zero, the following is true:

$$P[a \leq X \leq b] = P[a < X < b]. \tag{17}$$

Using this idea as motivation, we define an additional function. For any random variable X, the *cumulative distribution function*, (often abbreviated c.d.f.) is defined as

$$F(x) = P[X \leq x]. \tag{18}$$

The cumulative distribution function gives the total probability of the random variable assuming all values less than or equal to x. Cumulative distribution functions are defined for all random variables, discrete or continuous. It is not difficult to obtain some elementary properties of c.d.f.'s. For instance,

$$\lim_{x \to +\infty} F(x) = 1,$$

2 Discrete Stochasticity

since the total probability must be one. Similarly,

$$\lim_{x \to -\infty} F(x) = 0.$$

Using the interpretation of probability as area, equation (18) for continuous random variables is evaluated using

$$F(x) = \int_{-\infty}^{x} f(t)dt. \tag{19}$$

Similarly, for discrete variable, equation (18) is

$$F(x) = \sum_{k \leq x} f(k). \tag{20}$$

In equation (20), the sum is taken over all values in the range of the random variable X which are less than or equal to x.

For continuous random variables, expression (19) and the alternate version of the Fundamental Theorem of Calculus indicate that the probability density function $f(x)$ is the derivative of $F(x)$, thus

$$f(x) = \frac{d}{dx}F(x). \tag{21}$$

Example: Suppose X is a continuous random variable with probability density function given by

$$f(x) = 3x^2 \quad \text{for } 0 \leq x \leq 1.$$

Since $f(x)$ is only defined on the interval $[0, 1]$, we assume that $f(x) = 0$ for all other values of x.

Using equation (16), we calculate the probability that X has a value between 0.2 and 0.65 by

$$P[0.2 < X < 0.65] = \int_{0.2}^{0.65} 3x^2 \, dx = 0.2666.$$

The cumulative distribution function $F(x)$ for the random variable X is given by

$$F(x) = P(X \leq x) = \int_{0}^{x} 3t^2 dt = x^3 \quad \text{for } 0 \leq x \leq 1.$$

One important use of the probability density function of a random variable is to compute the mean or average of a random variable. In Section 2.2.1 we defined the sample mean $\bar{x}$, which is the arithmetic average of sample data. The sample mean approximates the actual average of a random variable X, but the sample mean changes with each sampling of the random variable that is obtained. To find the average of a random variable, we consider the following motivational example. Suppose that a class has five 100-point exams and a student scores on these exams: 70, 85, 100, 85, and 85.

The arithmetic average (and hence the sample mean) for this student is

$$\bar{x} = \frac{70 + 85 + 100 + 85 + 85}{5} = 85.$$

Regrouping the terms in this expression gives

$$\bar{x} = \frac{1}{5} \times 70 + \frac{3}{5} \times 85 + \frac{1}{5} \times 100 = 85,$$

which can be given a frequency interpretation. The student scored 70 one fifth of the time, scored 85 three-fifths of the time, and scored 100 one fifth of the time. If we consider the student's score on the test as a random variable X, then these frequencies are given as the following probabilities:

$$P[X = 70] = \frac{1}{5}$$

$$P[X = 85] = \frac{3}{5}$$

$$P[X = 100] = \frac{1}{5}.$$

Using this idea, the *expected value* or mean of a discrete random variable X with probability function $p(x)$ is defined as

$$E(X) = \sum_x x p(x) \tag{22}$$

where the sum in (22) is taken over all values of the random variable X. Using the analogy between the sum and the integral, the expected value of a continuous random variable X with density function $f(x)$ is similarly defined as

$$E(X) = \int_{-\infty}^{\infty} x f(x)\, dx. \tag{23}$$

The symbol μ is often used to denote the mean or expected value of a random variable X, that is

$$\mu = E(X).$$

Example: To compute the expected value of the continuous random variable X in the previous example, we simply use the density function

$$f(x) = 3x^2 \quad \text{for } 0 \leq x \leq 1.$$

to compute the expected value as

$$\mu = E(X) = \int_0^1 x f(x)\, dx = \int_0^1 3x^3\, dx = \frac{3}{4}.$$

An important property (see Exercise 6) of the expected value of a random variable is

$$E(aX + bY) = aE(X) + bE(Y), \tag{24}$$

where X and Y are random variables; a and b are any real numbers. Expression (24), which is interpreted as the average of a linear combination of random variables, is the linear combination of the average of each random variable. Any operation which satisfies (24) is termed "linear," hence the expected value operator E is often called a linear operator.

2 Discrete Stochasticity

One last bit of information about the expected value is needed before we go on. We can define the expected value of any function of X. If g is a function of X, its expected value is

$$E(g(X)) = \int_{-\infty}^{\infty} g(x) f(x) \, dx$$

for a continuous random variable (discrete defined with a summation). For example, suppose we want the expected value of X^2. This is found as

$$E(X^2) = \int_{-\infty}^{\infty} x^2 f(x) \, dx.$$

Following the analogy between the sample mean $\bar{x}$ and $E(X)$, we define the variance of a random variable as

$$V(X) = E((X - \mu)^2). \tag{25}$$

Recall that the sample variance of a data set was defined by

$$s^2 = \sum_{i=1}^{n} \frac{(x - \bar{x})^2}{n - 1},$$

which is interpreted as the average square distance from the sample mean. The expression (25) is the expected square distance of the random variable X from its mean μ.

The symbol σ^2 is often used to denote the variance of the random variable X, that is

$$\sigma^2 = V(X).$$

The standard deviation of a random variable X is defined as the positive square root of the variance,

$$\sigma = \sqrt{\sigma^2}.$$

The equation (25) is often tedious to compute. An alternate form for the variance of a random variable X is obtained by expanding the square in (25) and using the linearity condition (24) to obtain

$$V(X) = E(X^2) - (E(X))^2. \tag{26}$$

The expression $E(X^2)$ is computed in the same manner as $E(X)$. For instance, to find $E(X^2)$ for our example random variable X with density function

$$f(x) = 3x^2 \quad \text{for } 0 \leq x \leq 1,$$

we compute

$$E(X^2) = \int_0^1 x^2 f(x) \, dx = \int_0^1 3x^4 \, dx = \frac{3}{5}.$$

Hence, using (26), we obtain

$$\sigma^2 = V(X) = E(X^2) - (E(X))^2 = \frac{3}{5} - \left(\frac{3}{4}\right)^2 = \frac{3}{80}.$$

2.3.2 Three Useful Distributions

This section examines three distributions which are useful in simulations: the uniform, normal, and binomial.

The Uniform Distribution. There are many situations in which the outcomes of an experiment are equally likely. Let us consider the following two scenarios:

1. Consider the interval $[0, 1]$. If you drop a pin on this interval, the point x at which the tip of the pin lands could be any value between 0 and 1. Further, since we can not predict where the tip of the pin lands, each of these possible values of x would have the same chance of being selected, namely zero, since there are an uncountable number of numbers in that interval. This scenario is the same as describing the "at random" choice of a number between 0 and 1.

2. A bank opens at 8:00 am. The first customer arrives at random within the first 10 minutes that the bank is open. How can we describe the possible arrival times of the first customer? Since the customer could arrive at any time between 8:00 and 8:10 am, our description must include all these values (i.e., the 10-minute interval $[0, 10]$).

The frequency distribution for a continuous random variable where intervals of equal length are equally likely is known as the *uniform distribution*. A graph of a uniform distribution is shown in Figure 2.8. The general form of the uniform density function is

$$f(x) = \begin{cases} \frac{1}{\beta - \alpha} & \text{for } \alpha < x < \beta, \\ 0 & \text{elsewhere} \end{cases} \tag{27}$$

Notice that the non-zero part of this function is a horizontal line segment, and the area under this segment is one. Since intervals of equal length are equally probable, the mean of the uniform distribution on (α, β) is the midpoint of the interval, which is $(\alpha + \beta)/2$. In Exercise 7 it is shown that the variance is $(\beta - \alpha)^2/12$.

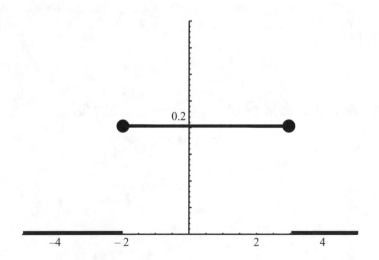

FIGURE 2.8 The Uniform Distribution on $[-2, 3]$.

2 Discrete Stochasticity

Example: Suppose a random variable is uniformly distributed on the interval $[-2, 3]$. The graph of the density function is shown in Figure 2.8.

To compute a probability for a uniform random variable, we need only to find the area of the appropriate rectangle. For example, to compute the probability that the value of x is between -1 and 1.5, written $P(-1 < X < 1.5)$, we simply compute the area of the enclosed rectangle; that is,

$$P(-1 < X < 1.5) = (1.5 + 1)\frac{1}{(3+2)} = 0.5.$$

Similarly, to find the probability of values between 0 and 2.5,

$$P(0 < X < 2.5) = (2.5 - 0)\frac{1}{(3+2)} = 0.5.$$

Notice that these intervals $(-1, 1.5)$ and $(0, 2.5)$ are the same lengths and, hence, have the same probabilities.

The Normal Distribution. We have previously seen bell-shaped frequency distributions in connection with empirically obtained data. Frequency distributions and histograms that are bell-shaped are often represented by a *normal distribution*. It is a fact that measurements from many random variables appear to have been generated from frequency distributions that are closely approximated by a normal probability distribution. For this reason the normal distribution is considered the most important probability distribution. The normal distribution has the form

$$f(x) = \frac{1}{\sigma\sqrt{2\pi}} e^{\frac{-(x-\mu)^2}{2\sigma^2}} \quad \text{for } \sigma > 0,\ -\infty < \mu < \infty,\ -\infty < x < \infty. \tag{28}$$

The normal distribution function depends upon the location and shape parameters μ and σ. The choice of symbols for these parameters is not an accident. It is an exercise to show that μ and σ^2 are the mean and variance of a normal random variable. The graph of a normal distribution in Figure 2.9 shows the characteristic bell-shape.

This curve also shows other important features. Note that $f(x)$ obtains its maximum at the point $x = \mu$, and that the curve is symmetric about $x = \mu$. The function is decreasing as x moves away from μ. It is a worthwhile exercise (see Exercise 8) to show that the

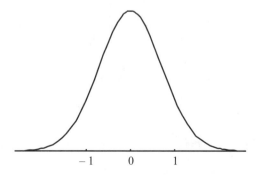

FIGURE 2.9 The Normal Distribution with $\mu = 0$ and $\sigma^2 = 1$.

normal distribution (28) obtains its maximum at $x = \mu$ and that it has inflection points at $x = \mu \pm \sigma$.

An often cited rule of thumb for determining if a data set is from a normal distribution is the "68 - 95 - 99" rule. It states that for a normal distribution, approximately 68% of the data (area) is within one standard deviation of the mean; that is,

$$P(\mu - \sigma < X < \mu + \sigma) = 0.6827. \tag{29}$$

Similarly, two standard deviations from the mean, $\mu \pm 2\sigma$, contains approximately 95% of the data. Three standard deviations from the mean, $\mu \pm 3\sigma$, contains approximately 99% of the data. A large data set is approximately normally distributed if one, two, and three standard deviations about the sample mean contain approximately these percentages of data.

The Binomial Distribution. Some experiments consist of the observation of a trial which results in one of two possible outcomes. For instance, a coin flip lands either heads up or tails up, a manufactured item is either defective or non-defective, or a free throw either hits or misses the basket. These are just some examples of a Bernoulli random variable which results when an experiment has only two outcomes, often labeled "success" or "failure."

More specifically, a random variable X that assumes only the values 0 or 1 is known as a *Bernoulli variable*, and the performance of an experiment with only two possible outcomes is called a *Bernoulli trial*. A Bernoulli random variable X is a discrete random variable since it has only two possible values. The Bernoulli distribution is given by

$$f(x) = p^x q^{1-x} \quad \text{for } x = 0, 1 \tag{30}$$

where $p = P[X = 1]$ is the probability of a "success" and $q = 1 - p = P[X = 0]$ is the probability of a "failure."

Notice that $E(X) = 0 \cdot q + 1 \cdot p = p$ and $E(X^2) = 0^2 \cdot q + 1^2 \cdot p = p$, so that the variance of X is $V(X) = p - p^2 = p(1 - p) = pq$.

An important assumption often made when performing Bernoulli trials is that successive trials are *independent*; that is, the probability of the occurrence of a success or failure on a trial does not depend upon the outcome of the previous trial (or trials).

When a sequence of n independent Bernoulli trials are conducted, an important distribution arises from counting the number of successes that occur. Let Y be the number of successes that occur on n independent Bernoulli trials. Since successes or failures occur randomly, Y is a random variable with possible values $0, 1, \ldots, n$. The distribution of Y is known as the *binomial* distribution and has the form

$$f(k) = \binom{n}{k} p^k q^{n-k} \quad \text{for } k = 0, 1, \ldots, n. \tag{31}$$

The expression $\binom{n}{k}$ is known as the *binomial coefficient* or the *combination number*, and is defined by

$$\binom{n}{k} = \frac{n!}{(n-k)!k!}. \tag{32}$$

2 Discrete Stochasticity

The symbol "!" represents the *factorial*, and is defined for each positive integer n by

$$n! = n \times (n-1) \times (n-2) \times \cdots \times 2 \times 1. \tag{33}$$

As a convention, we define

$$0! = 1. \tag{34}$$

Note that the binomial distribution is completely determined by the parameters n and p and that a sum of n independent Bernoulli random variables is binomially distributed. See Exercise 9.

To understand the binomial distribution, we first consider the binomial coefficient. The binomial coefficient is a device for counting the number of ways k items can be chosen from n items where order is unimportant. For example, $\binom{52}{5}$ is the number of 5 card hands that can be dealt from a 52-card deck.

As another example, the number of ways that two heads can occur on four coin tosses is

$$\binom{4}{2} = \frac{4!}{(4-2)!2!} = 6.$$

We easily verify this value by listing all of the possible arrangements of two heads on four tosses. If we let "H" denote heads and "T" denote tails, then two heads occur on four coin tosses in the following six ways: HHTT, HTHT, HTTH, THTH, TTHH, and THHT.

Continuing this example, we compute the probability of two heads's occurring on four coin tosses. Since a (fair) coin has probability $p = 1/2$ of landing heads and $q = 1/2$ of landing tails, each of the six outcomes listed above has probability

$$\frac{1}{2} \times \frac{1}{2} \times \frac{1}{2} \times \frac{1}{2} = \frac{1}{16}$$

of occurrence. Since these six outcomes are the only ways to have two heads occur on four coin tosses, we have

$$P[\text{ Two heads on four coin tosses }] = 6 \times \frac{1}{16}.$$

Now this is the same value we obtain from the binomial distribution, for if Y represents the number of heads which occur on $n = 4$ coin tosses, then

$$P[Y = 2] = \binom{4}{2} \left(\frac{1}{2}\right)^2 \left(\frac{1}{2}\right)^{4-2} = \frac{6}{16}.$$

The binomial distribution does not require that $p = q$ as is the case with fair coin flips. As another example, suppose that a manufacturing process produces 5% defective light bulbs. A box of $n = 20$ light bulbs will contain two defective bulbs with probability

$$\binom{20}{2} (0.05)^2 (0.95)^{18} = 0.1887.$$

In light of the above discussion, it is useful to note the form of this expression. The coefficient $\binom{20}{2}$, which is the number of ways to find two defective bulbs in a box of 20, is multiplied by the probability $(0.05)^2$ of two defectives and then multiplied by $(0.95)^{18}$, the probability of 18 nondefective bulbs.

In this last example, if a there is a 5% chance that a light bulb will be defective, then it is reasonable to expect that a box of 20 light bulbs would contain $20 \times 0.05 = 1$ defective bulb. This is indeed the case. Exercise 9 asks the reader to show that the mean of a binomial random variable Y is

$$E(Y) = np, \tag{35}$$

and the variance of Y is

$$V(Y) = npq. \tag{36}$$

2.3.3 Environmental versus Demographic Stochasticity

We began this chapter with an example of a hypothetical declining squirrel population. We assumed that the growth rate of -0.5 is actually an average growth rate for that population. These considerations led us to a brief study of some of the ideas of probability and statistics. In this section, we model a population using the tools we developed.

There are many reasons for variations in birth or death rates. Variations resulting from slight changes in the population's behavior or structure are called variations due to *demographic* stochasticity. Variations from environmental conditions (flood, drought, fire, etc.) are called variations due to *environmental* stochasticity.

Since each year the numerical values of the birth and death rates vary about some central value and values far from this central value are rare, we frequently assume that demographic stochasticity follows a normal distribution. Environmental stochasticity can take many forms. We will consider it to be a catastrophe that randomly affects the population. For instance, a flood may occur on average once every 25 years or 4% of the time. We incorporate environmental stochasticity as a Bernoulli random variable; either a catastrophe occurs, or it does not. Most software has uniform and normal random numbers built in, but a Bernoulli random number generator is rare. We can, however, construct a Bernoulli random number from a uniform random number chosen between $[0, 1]$. Suppose we want a Bernoulli random number which is a success 4% of the time. We pick a uniform random number between 0 and 1. If the number is less than 0.04, then we consider it a success; if it is greater than 0.04, we consider it a failure.

Stochasticity and the Sandhill Crane. We have previously modeled the population of the Florida sandhill crane. We considered the population under the three different growth rates determined by the best, medium, and worst environmental conditions. Each of these conditions gave a fixed average value of the growth rate r. In this example, we use both birth and death rates for the cranes. The average reproduction rate of the sandhill crane is 0.5, while the average death rate is 0.1. Demographic stochasticity affects the birth and death rates. Here we assume that the birth and death rates are normally distributed with means of 0.5 and 0.1 respectively. The standard deviations of these birth and death rates are 0.03 and 0.08 respectively. We further assume that a catastrophe will occur on

2 Discrete Stochasticity

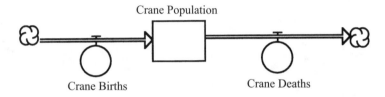

FIGURE 2.10 Compartmental Diagram for the Crane Population.

average once every 25 years. These catastrophes affect the cranes by lowering the birth rate by 40% and increasing the death rate by 25%. Given all this information, we would like to make some predictions of the sandhill crane population over a five-year period. Finally, we assume that there are initially 100 cranes.

As a first attempt at this problem, we model the population without any random elements. This will give us a deterministic result to which we can compare our subsequent modeling refinements. Further, this is an opportunity to review what we learned in Chapter 1. Figure 2.10 shows a compartmental diagram for the crane population.

As the diagram shows, cranes enter the population from crane births and leave the crane population due to crane deaths. We write this equation using the difference equation

$$x(n) - x(n-1) = \text{crane births} - \text{crane deaths}.$$

This gives

$$x(n) - x(n-1) = 0.5x(n-1) - 0.1x(n-1) = 0.4x(n-1)$$

or the recurrence relation

$$x(n) = x(n-1) + 0.5x(n-1) - 0.1x(n-1) = 1.4x(n-1),$$

which has a closed-form solution

$$x(n) = (1.4)^n x(0).$$

A table and graph of these population predictions show that the crane population is growing exponentially in absence of stochasticity.

Year	Crane Population
0	100
1	140
2	196
3	274
4	384
5	538

TABLE 2.9 Crane Population without Stochasticity.

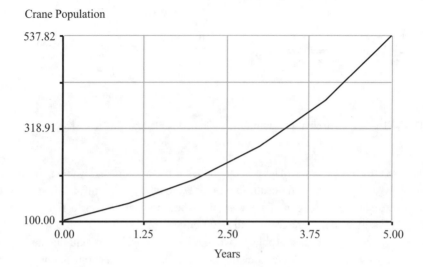

FIGURE 2.11 Graph of the Crane Population without Stochasticity.

Environmental Stochasticity. We first consider the effects of environmental stochasticity alone on the crane population. Environmental stochasticity for the crane population will take the form of a catastrophe such as a flood.

The environmental stochasticity affects the cranes by lowering the birth rate by 40% and increasing the death rate by 25%. These catastrophes occur randomly on average once every 25 years. We again assume that there are 100 cranes initially, and we consider only the effects of the environmental stochasticity. We will not incorporate the demographic stochasticity into this model.

Since the environmental stochasticity occurs on average once every 25 years, we generate a uniformly distributed random variable for each time step, from the interval $(0, 1)$. We say that a catastrophe occurred if the value of this random variable is less than 0.04. If this occurs, the birth and death rate for that time step are modified. Most software packages and graphing calculators have the ability to generate a uniform random variable. A typical method for implementing this idea at each time step is to generate a uniform random variable usually with the command Uniform$(0, 1)$. Let Cat (for CATastrophe) be this random variable. That is, let

$$Cat = Uniform(0, 1);$$

then the adjusted birth rate is computed with the following "if" statement:

If $Cat < 0.04$**, then** adjusted birthrate $= 0.6 \times$ birthrate**; else** adjusted birthrate $=$ birthrate.

The "if" statement for the death rate is similar and is

If $Cat < 0.04$**, then** adjusted deathrate $= 1.25 \times$ deathrate**; else** adjusted deathrate $=$ deathrate**.**

Since the birth and death rates are reduced during the year a catastrophe occurred, it is necessary that the uniform random variable Cat generated during each time step be used

2 Discrete Stochasticity

Run	Initial	1	2	3	4	5
1	100.00	140.00	196.00	274.40	384.16	537.82
2	100.00	140.00	196.00	274.40	384.16	451.39
3	100.00	140.00	196.00	230.30	270.60	317.96
4	100.00	140.00	196.00	274.40	384.16	451.39
5	100.00	140.00	196.00	274.40	384.16	537.82
6	100.00	140.00	196.00	274.40	384.16	537.82
7	100.00	140.00	196.00	274.40	384.16	537.82
8	100.00	140.00	196.00	274.40	384.16	537.82
9	100.00	140.00	196.00	274.40	384.16	537.82
10	100.00	140.00	196.00	274.40	384.16	537.82
11	100.00	140.00	196.00	274.40	384.16	537.82
12	100.00	140.00	196.00	230.30	322.42	378.84
13	100.00	140.00	196.00	274.40	322.42	451.39
14	100.00	140.00	164.50	230.30	322.42	451.39
15	100.00	140.00	196.00	274.40	384.16	537.82
16	100.00	140.00	164.50	230.30	322.42	451.39
17	100.00	140.00	196.00	274.40	384.16	537.82
18	100.00	140.00	196.00	274.40	384.16	537.82
19	100.00	140.00	196.00	274.40	384.16	537.82
20	100.00	140.00	196.00	274.40	384.16	537.82

TABLE 2.10 Twenty Simulation Runs of the Crane Population with Environmental Stochasticity.

for both the birth and death rates. This is our modeling assumption. We could equally well have a modeling assumption that the different catastrophes affect the birth and death rates. This would be a fine model, but with different assumptions from our model.

We ran the model for five years. Table 2.10 and the accompanying graph, Figure 2.12, show 20 simulation runs of the crane model with environmental stochasticity.

The web-like pattern of the graph in Figure 2.12 is very informative. The web branches correspond to the number of catastrophes that occurred during the simulations. The highest branch of this graph corresponds to no catastrophes' occurring. With no catastrophes, the crane population after five years will be at 538 (rounded up). Compare this to the deterministic model shown in Table 2.9. The next lowest branch corresponds to one catastrophe's occurring during the simulation. The two lower branches correspond to two and three catastrophes occurring.

If we create a histogram (see Figure 2.13) for the ending populations of the simulation runs (the last column of Table 2.10), we observe that the most frequently occurring population size was 537 which corresponds to no catastrophes. The ending population size with the fewest occurrences was 317, corresponding to three catastrophes. This makes sense if we think about it. If we expect a catastrophe once every 25 years, then most likely there will be none in a period of five years, but it is not unreasonable that there will be one. Two catastrophes could happen, but should be rare. Three catastrophes

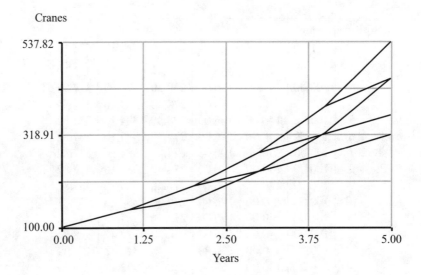

FIGURE 2.12 Graph of Twenty Simulation Runs of the Crane Population with Environmental Stochasticity.

should be rarer still, and more than three is extremely rare. This is what the histogram shows. We verify this behavior mathematically in Section 2.3.4.

Also observe that the web-like pattern of the graph, Figure 2.12, suggests that the important feature of this type of environmental stochasticity and its effect on the population is the *number* of catastrophes, not *when* they occurred. In Table 2.10, we note that simulation runs numbers 14 and 16 both had only one environmental catastrophe which occurred in the second year. Simulation run number 2 also had one catastrophe which

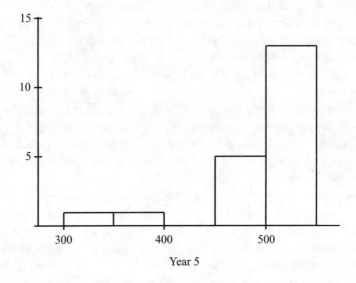

FIGURE 2.13 Histogram of Ending Populations.

2 Discrete Stochasticity

occurred in year five. These three runs ended with the same final population size. This web-like pattern is a consequence of assuming constant birth and death rates that are affected only by a catastrophe and each catastrophe alters the rates by exactly the same amount. Exercise 5 asks you to rerun this model with a catastrophe altering the birth and death rates by a random amount.

Demographic Stochasticity. Next we consider demographic stochasticity for the crane model. In the absence of catastrophes, population parameters vary from year to year due to many reasons, ranging from small environmental variations to changes in the number of fertile females. Usually, though, the rates are close to the average, and large deviations are rare. For this reason, backed by empirical evidence, the birth and death rates are generally considered to be normally distributed. Each year, the numerical values of these parameters differ, but follow normal distributions. To incorporate this into our model, at each time step we generate normal random variables for the birth and death rate parameters. A typical command for generating a normal random variable with mean μ and standard deviation σ is Normal(μ, σ).

For the sandhill crane model, we use Normal$(0.5, 0.03)$ to generate the birth rate for each time step, and we use Normal$(0.1, 0.08)$ for the death rate. The modeling diagram is the same as the deterministic model, Figure 2.10, with only the birth and death rate parameters altered.

Running this model for five time steps, produces a table of values similar to Table 2.11. This table and the accompanying graph, Figure 2.14, show that the population of cranes is growing approximately exponentially, but comparison of the population sizes with the population sizes of the exponential model (Table 2.9) shows that these two models are different. The difference is the stochastic birth and death rates. Each run of the crane model with demographic stochasticity would produce a different collection of population sizes.

Table 2.12 and the accompanying graph, Figure 2.15, show 20 simulation runs of the crane model with demographic stochasticity.

The graph and the data show some interesting aspects of this crane model. The population is still exhibiting approximately exponential growth, but the range of the population size after five years varies from a minimum of 475 to a maximum of 669.

Year	Crane Population
0	100
1	150.70
2	206.71
3	250.59
4	375.90
5	569.29

TABLE 2.11 Crane Population with Demographic Stochasticity.

The box plots introduced earlier provide a visual summary of the crane population simulations. A side-by-side box plot for the simulation data for each of the five time steps in Table 2.12, is shown in Figure 2.16.

The box plots provide a better graphical summary than do the graph of the simulations shown in Figure 2.15. First note that the median of each time step, shown as the midpoint of each box plot, closely matches the deterministic model shown in Figure 2.11. This again verifies our intuition that deterministic crane models show us the central tendency of stochastic models. Finally note that these box plots give distribution information about the simulation output. We have a sense of the center, the spread, and even asymmetries.

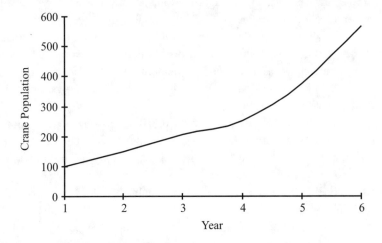

FIGURE 2.14 Graph of the Crane Population with Demographic Stochasticity.

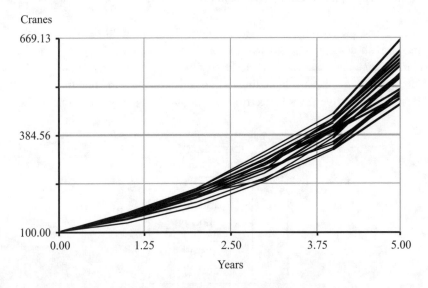

FIGURE 2.15 Graph of Twenty Simulation Runs of the Crane Population with Demographic Stochasticity.

2 Discrete Stochasticity

Run	Initial	1	2	3	4	5
1	100.00	150.70	206.71	250.59	375.90	569.29
2	100.00	135.48	185.42	254.97	344.65	508.78
3	100.00	142.11	196.89	280.39	361.76	492.60
4	100.00	142.97	217.63	312.87	408.81	590.45
5	100.00	142.50	204.76	278.68	433.34	610.81
6	100.00	124.51	172.28	245.55	340.17	475.57
7	100.00	146.95	208.32	299.30	405.43	558.78
8	100.00	135.23	196.73	268.39	401.41	566.20
9	100.00	150.80	197.64	270.12	381.91	555.35
10	100.00	141.99	198.48	297.57	393.55	500.37
11	100.00	150.56	225.20	333.14	445.33	669.13
12	100.00	148.04	216.56	272.05	398.14	510.18
13	100.00	134.13	184.87	274.16	380.83	519.28
14	100.00	142.18	210.94	283.52	354.89	513.51
15	100.00	150.14	217.59	311.21	418.84	542.21
16	100.00	154.57	225.58	321.43	430.42	636.98
17	100.00	147.78	219.66	307.02	423.35	602.18
18	100.00	149.47	217.96	297.59	399.51	566.49
19	100.00	149.08	222.34	292.37	397.82	521.36
20	100.00	137.39	198.40	297.95	430.17	624.98

TABLE 2.12 Twenty Simulation Runs of the Crane Population with Demographic Stochasticity.

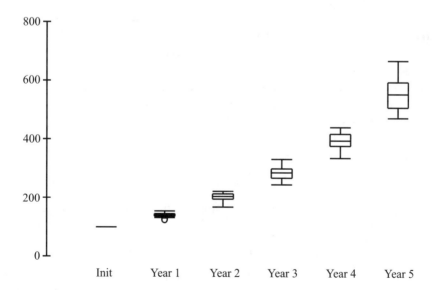

FIGURE 2.16 Side-By-Side Box Plot of the Twenty Simulation Runs of Crane Population with Demographic Stochasticity.

Some Comments. In the last two subsections, we considered the individual effects that environmental and demographic stochasticity have upon the sandhill crane population. It is not difficult to combine the two types of stochasticity, but before proceeding with that task, we must ask ourselves which is the dominant effect? Occam's razor needs to be applied to see how much information is gained by complicating the model further.

If we had begun by simultaneously incorporating both types of stochasticity into the model (which one could argue is "more realistic"), we would not have been able to easily observe the interesting web-like pattern of Figure 2.12. The "noise" from the demographic stochasticity would have obscured it. It is always best to build models with the least amount of complication. Each layer of complexity makes subsequent analysis and discoveries difficult. On the other hand, simple models may not be sufficient. Behaviors, such as the web-like pattern, may be artifacts of the assumptions and not something observable in the real world. The web-like pattern for the environmental stochasticity is examined in more detail in the next section.

2.3.4 Environmental Stochasticity — A Closer Look

In this section, we take a closer look at environmental stochasticity and the long-term effects catastrophes have on a population.

As we did in the sandhill crane model, we consider catastrophes which occur with probability p each time step. Recall that in the sandhill crane model, catastrophes occurred randomly on average once every 25 years, so that there was a 4% chance of their occurrence each year.

A conclusion we drew about the sandhill crane model with environmental stochasticity is that the *number* of catastrophes that have occurred up to time n affects the population size at time n, while the *time of occurrence* of the catastrophe does not. Using this observation, we construct a model for the population size at time n.

To begin, we consider the effect a catastrophe has on a population. In the absence of a catastrophe, we assume that the population has a growth rate of r per time step. Catastrophes negatively affect a population by lowering its growth rate. We assume that the growth rate of the population is r_c if a catastrophe has occurred, so that the negative effect of a catastrophe is reflected by requiring that $r_c \leq r$.

For example, consider the sandhill crane model. In the absence of a catastrophe, the sandhill cranes have a growth rate of

$$r = \text{birth rate} - \text{death rate} = 0.5 - 0.1 = 0.4, \tag{37}$$

while the effects of a catastrophe lower the birth rate by 40% and increase the death rate by 25%, so that

$$r_c = \text{modified birth rate} - \text{modified death rate}$$
$$= 0.6 \times 0.5 - 1.25 \times 0.1 = 0.1750. \tag{38}$$

If no catastrophes occur in n time steps, then, using the ideas from Chapter 1, a closed-form solution for the population after n time steps is

$$x(n) = (1+r)^n x(0). \tag{39}$$

2 Discrete Stochasticity

To obtain a closed-form expression for the population size after n time steps if one catastrophe occurs, we note that expression (39) gives the population size for the time steps *prior* to the catastrophe. That is, if the catastrophe occurs at time step m, where $1 \leq m \leq n$, then the population at time $m - 1$, (prior to the catastrophe) is

$$x(m-1) = (1+r)^{m-1}x(0).$$

The catastrophe changes the population's growth rate to r_c, so the population at time m is

$$\begin{aligned} x(m) &= x(m-1) + r_c x(m-1) \\ &= (1+r_c)x(m-1) \\ &= (1+r_c)(1+r)^{m-1}x(0). \end{aligned} \quad (40)$$

The expression (40) is the resulting population size after the catastrophe occurred in time step m. Since we assumed only one catastrophe occurs in the n time steps, the remaining $n - m$ time steps are free of such disasters and the population returns to its growth rate r. Thus, the population after n time steps when one catastrophe has occurred is

$$\begin{aligned} x(n) &= (1+r)^{n-m}x(m) \\ &= (1+r)^{n-m}(1+r_c)(1+r)^{m-1}x(0) \\ &= (1+r_c)(1+r)^{n-1}x(0). \end{aligned} \quad (41)$$

To obtain the closed-form expression for the population size after n time steps when two catastrophes have occurred, assume that the second catastrophe occurred at the mth time step, $2 \leq m \leq n$. The population prior to this catastrophe is given by (41) as

$$x(m-1) = (1+r_c)(1+r)^{m-2}x(0),$$

since one catastrophe has occurred prior. The population size at the mth time step is then

$$x(m) = (1+r_c)^2(1+r)^{m-2}x(0),$$

since the second catastrophe occurs at time step m. Both catastrophes have now been accounted for in the expression for the population size, so for the remaining $n - m$ time steps, the population has a growth rate of r and the resulting expression for the closed-form solution is

$$\begin{aligned} x(n) &= (1+r)^{n-m}(1+r_c)^2(1+r)^{m-2}x(0) \\ &= (1+r_c)^2(1+r)^{n-2}x(0). \end{aligned} \quad (42)$$

Continuing in this manner, an induction argument (see Exercise 10) shows that the population size after n time steps when $0 \leq k \leq n$ catastrophes have occurred is

$$x(n) = (1+r_c)^k(1+r)^{n-k}x(0). \quad (43)$$

The population size $x(n)$ is a random variable since catastrophes occur randomly with probability p on each time step. Using some terminology given earlier, the occurrence of a catastrophe is a Bernoulli random variable with probability p of occurrence on each time step. Further, as seen from equation (43), the random variable $x(n)$ depends upon the number of catastrophes which occur.

If we assume that the occurrence of a catastrophe does not depend upon whether a catastrophe occurred on the previous time step (or time steps), then the total number of catastrophes C in n time steps is a random variable with possible values $0, 1, 2, \ldots, n$, and has a binomial distribution.

More specifically, the probability of $C = k$ catastrophes', ($k = 0, 1, 2, \ldots, n$), occurring in n time steps is given by

$$P[k \text{ catastrophes in } n \text{ time steps}] = P[C = k] = \binom{n}{k} p^k q^{n-k}. \tag{44}$$

Combining expression (44) with (43) gives

$$x(n) = (1 + r_c)^k (1 + r)^{n-k} x(0) \text{ with probability } \binom{n}{k} p^k q^{n-k}, \tag{45}$$

which is the probability distribution for the random variable $x(n)$, the population size after n time steps when k catastrophes have occurred.

In the sandhill crane model, it was assumed that the initial population was 100, so $x(0) = 100$. In addition, the probability of a catastrophe occurring each year was $p = 0.04$, so $q = 1 - p = 0.96$. Using $n = 5$ years and the growth rates $r = 0.4$ and $r_c = 0.1750$ (see equations (37) and (38),) we can compute the population size and its probability of occurrence. For instance, if $k = 1$ catastrophe occurs, then the population size is

$$x(5) = (1 + 0.1750) \times (1 + 0.4)^{5-1} \times 100 = 451.39.$$

The probability of the occurrence of one catastrophe in five years is

$$\binom{5}{1} (0.04)(0.96)^4 = 0.1699,$$

or about 17%. Similar calculations give the other possible population sizes. Table 2.13 gives these possible population sizes and their respective probability of occurrence for the sandhill crane population after five years.

Number of Catastrophes	Resulting Population Size	Probability
0	537.82	0.8153
1	451.39	0.1699
2	378.84	0.0142
3	317.96	0.0006
4	266.86	1.229×10^{-5}
5	223.97	1.024×10^{-7}

TABLE 2.13 The Probability Distribution for the Sandhill Crane Population after 5 Years.

2 Discrete Stochasticity

Examining the entries in Table 2.13, we note that the probability of the number of catastrophes rapidly decreases as the number of catastrophes increase. This coincides with our intuition since the probability of a catastrophe's occurrence is small. What is the most likely number of catastrophes that will occur? There is an 81.53% chance that no catastrophes occur during a five-year period. The value of a random variable with the largest probability is called the *mode* of the random variable. Thus, the mode of the sandhill crane population after five years is 537.82.

To find the mode of the number of catastrophes random variable in general, we find, using (45), the most probable number of catastrophes. This is equivalent to finding the mode of a binomial random variable with parameters n and p. Since the binomial distribution is a discrete distribution, the powerful maximization methods of calculus are not available. However, we can still use the idea of a maximum. The maximum probability for the binomial distribution occurs at the integer $k_{\max}$, provided that

$$\binom{n}{k} p^k (1-p)^{n-k} \leq \binom{n}{k_{\max}} p^{k_{\max}} (1-p)^{n-k_{\max}} \tag{46}$$

$$\text{for } k = 0, 1, \ldots, n, k \neq k_{\max}.$$

We solve inequality

$$\binom{n}{k_{\max}-1} p^{k_{\max}-1} (1-p)^{n-(k_{\max}-1)} \leq \binom{n}{k_{\max}} p^{k_{\max}} (1-p)^{n-k_{\max}}$$

for $k_{\max}$. Exercise 9 yields the solution

$$k_{\max} = \lceil (n+1)p - 1 \rceil \tag{47}$$

where $\lceil x \rceil$ is the least integer not smaller than x. Expression (47) is the most likely number of catastrophes that occur. Using this value in (43), the most likely sandhill crane population size is

$$x(n) = (1+r_c)^{\lceil (n+1)p-1 \rceil} (1+r)^{n-\lceil (n+1)p-1 \rceil} x(0). \tag{48}$$

Table 2.13 is also of interest when compared with the values obtained in the sandhill crane simulation model. Summarizing the ending population sizes from the 20 simulation runs shown in Table 2.10 gives Table 2.14. How close are the relative frequencies in this table to the predicted probabilities in Table 2.13? At first glance, the numbers do not compare favorably. However, upon closer inspection, we note that some of the relative frequencies are higher than predicted and some are lower. In particular, the relative frequency of no catastrophes' occurring is 65%, which is lower then the predicted value of 81.53%, whereas the relative frequencies of one, two and three catastrophes are each higher than predicted. Since these relative frequencies were obtained empirically from a small set of simulation data, these differences could be due to chance. In the next section, we develop a statistical test for comparing these relative frequencies with their predicted values.

Number of Catastrophes	Number of Occurrences	Relative Frequency
0	13	13 / 20 = 65%
1	5	5 / 20 = 25%
2	1	1 / 20 = 5%
3	1	1 / 20 = 5%
4	0	0 / 20 = 0%
5	0	0 / 20 = 0%

TABLE 2.14 The Frequency Distribution for the Number of Catastrophes during a 5-Year Period for 20 Runs of the Sandhill Crane with Environmental Stochasticity Simulation.

2.4 Model Validation: Chi-Square Goodness-of-Fit Test

Rarely do the results predicted by deterministic models match the observed results. Indeed, the main point of this chapter is to deal with deviations from either random factors or unknown factors which we take to be random. An important question is: If observations deviate from model predictions, is the model a good model or not? This is called *model validation*. A model's purpose needs to be considered when determining whether or not a model is satisfactory. If the only purpose is to determine if the interaction of a plant community and an herbivore community can result in cyclical population patterns, then the numbers produced by a model are irrelevant; only the pattern of the output is important. If the purpose of a model is to obtain a population value in a given year, then a close relation between observed and predicted values is very important. In this book, we present two methods of comparing observed and predicted data: the chi-square goodness-of-fit test (presented in this section) and the R^2 statistic (presented in Chapter 4) which measures the degree to which a model accounts for the variance seen in the data.

Since models and data rarely match exactly, the important issue is how close is close enough. In this section, we use the idea of a *confidence* or *significance level*. We attempt to determine if the variation between the observed and predicted values would occur at least 90% of the time if we could re-collect the data many times. If this is the case, we say the model is valid at a better than 90% *confidence level*. This is often reported as an $\alpha = 0.1$ *significance level*, where $\alpha = 1 - 0.9$. Statisticians usually use the word significance when testing a model and confidence when claiming a model's parameters are between certain bounds called a confidence interval. For our purposes, we blur this distinction in word usage. To avoid confusion, in our usage: 1) one is always 1 minus the other (after removing the percentage of course); 2) α is always significance; and 3) confidence is reported as a percentage. Different fields and applications consider different confidence levels to be satisfactory. In the current discussion, we are taking 90% to be acceptable, but frequently 95% or 99% is used, especially in highly controlled or experimental settings. Values lower than 90% are also accepted in some situations. If computer output reports a cutoff or confidence level and does not specify the level, assume that it is 95%, or $\alpha = 0.05$.

Let us make some of the statements in the preceding paragraph more precise. To measure how close the observed values are to the values expected from the model, we make use of a what is called the *chi-square test statistic*. This statistic is defined by

$$X^2 = \sum_{i=1}^{n} \frac{(\text{observed value}_i - \text{expected value}_i)^2}{\text{expected value}_i} \tag{49}$$

where n is the number of pieces of data and observed value$_i$ and expected value$_i$ are the observed and expected values of the ith piece of data respectively.

The test statistic X^2 is a random variable which means that if we could repeat the observing process, it could (probably would) take a different value. Looking at equation (49) carefully, we note that it is a sum of the differences between predicted and observed values (divided by something). The closer the observed values are to the predicted values, the closer X^2 is to zero. Thus low X^2 values are good, high X^2 values are bad. How good is good enough depends on the desired confidence level and the number of pieces of data. There is a *cutoff value* (also known as a *critical value*), denoted by χ_0^2, (which changes for different values of α and different numbers of data items) such that if X^2 is above this level, we reject the model at this confidence level. Alternatively we can say we accept (fail to reject) the model at a better than α level.

The idea of rejecting a model at a confidence level rather than accepting it at a confidence level may seem a little stilted. The reasoning comes from a method of gaining understanding called *hypothesis testing*. In hypothesis testing we propose a theory or hypothesis which we call the null hypothesis. We then test it in various ways and if the test fails we reject it (usually at some significance level). If the test does not fail, we continue to accept the null hypothesis, but devise another test to reject it. Some scientists make an effort to emphasize that we never accept the null hypothesis, only fail to reject it. While this seems like some fancy lawyer talk, it just means the null hypothesis is never known to be true—it is just the best theory to date. For centuries physics proposed a null hypothesis which is what we now know as Newtonian mechanics. Many experiments failed to reject this theory, but that did not mean it should be accepted as truth. It only meant that scientists had failed to reject it yet. As we now know, in the early 1900s, Albert Einstein proposed an alternative theory or hypothesis called special relativity. Experimental data showed that it was a better theory than Newtonian mechanics.

In our case, we have a model that we are proposing as the null hypothesis. We are testing it against the alternative hypothesis that the model is not true. To correctly do a test, we need to determine the significance level before we do our test. If our X^2 value is above the cutoff, we reject the null hypothesis at this level, which means we conclude that the model is poor. If X^2 is below the cutoff, we will say that the model is a good model which really only means that we have no excuse to reject it. If we are careless, as many people are, we might abuse formal language and say we accept the model. We must always understand, however, that we are accepting it in absence of anything better, not that we are accepting it as truth. In fact for particular cutoff level there are many models that can be considered to be good. If we "accept the model" we will say we accept it at a better than α level, since, if X^2 is less than a $\alpha = 0.1$ cutoff, it will be less than the cutoffs for every α less than 0.1.

Critical or cutoff values can be found in a chi-square table such as the one in Appendix A. For reasons which do not concern us, chi-square tables do not list the number of pieces of data n, they use the number of *degrees of freedom*, which is $n-1$. (If we need to use the data to estimate model parameters, then the degrees of freedom is $n-1$ less the number of parameters estimated.) Typically the symbols r or ν are used to denote the degrees of freedom. For our purposes, degrees of freedom is just one less than the number of pieces of data.

An approach that has become more popular recently is to not test a hypothesis, but rather report a *p-value*. A p-value is the area under the chi-square distribution curve (with appropriate degrees of freedom) to the right of the X^2 value found. Large X^2 values have very small p-values, and small X^2 values have large p-values. P-values can be found using software and an example is given in the *Mathematica* appendix. P-values can be thought of as a refinement of a chi-square table—filling in the missing values using a computer. They tell us more information. To say X^2 has a p-value of 0.026 tells us more than saying that we reject the model at the $\alpha = 0.05$ level, but not at the $\alpha = 0.025$ level. It tells we are much closer to 0.025 than 0.05.

The statistic X^2 can be used for any type of observed values, but if we want to use the chi-square distribution to test for goodness-of-fit the observed values must be the integer number of observations that occur in each category. For noninteger data the statistic can be used to compare which of two models is better, but methods such as sums of squares presented in Chapter 4 are more appropriate.

2.4.1 Testing the Stochastic Squirrel Model

We demonstrate these ideas with our squirrel/M&M model from the beginning of this chapter. The data in Table 2.15 is from our first squirrel population simulation run, originally displayed in Table 2.2. Pretending that we do not know the mechanism for

Year	Actual Count	Expected Count Model 1	Expected Count Model 2
0	100	100	100
1	53	50	85.71
2	27	25	71.43
3	15	12.5	57.14
4	9	6.25	42.86
5	3	3.125	28.57
6	1	1.5625	14.28
7	0	0.78125	0
8	0	0.390625	0

TABLE 2.15 Observed versus Expected Squirrel Population.

2 Discrete Stochasticity

generating the data, we devise two models to explain the data. One model is the true trend without the noise, the other is a linear model which is a good fallback when we do not have reasons to suspect something else. Model 1 starts with 100 squirrels and halves the population each year and has equation $N(t) = 100\left(\frac{1}{2}\right)^t$. Model 2 starts with 100 squirrels, ends with 0 squirrels after seven years, and is linear between these two points. Thus $N(t) = 100 - \frac{100}{7}t$. The expected counts from these models are also in Table 2.15.

We first consider Model 1. We stopped the table after eight years, but the expected counts for the first model continue forever. Later we will observe that in order for the X^2 statistic to have an approximate chi-square distribution, we should have at least five expected counts in each cell. For this reason we group the data from year five on. Convince yourself that the sum of the expected counts from year 5 to infinity is exactly 6.25. We also leave out year 0 data since both observed and expected values are 100 by design, not by model prediction. We have then five cells corresponding to years 1, 2, 3, 4, and 5 or more. Our X^2 statistic is $(53-50)^2/50 + (27-25)^2/25 + (15-12.5)^2/12.5 + (9-6.25)^2/6.25 + (4-6.25)^2/6.25 = 2.86$. The degrees of freedom in this case is five minus one, or four. Looking up this value in the chi-square table in Appendix A, we see that $X^2 = 2.76$ is below the cutoff for even the $\alpha = 0.25$ level. We can also say that the p-value is greater than 0.25. What do we conclude? Recall that large X^2 values are a sign of poor fit. Here we have an X^2 smaller than all the standard cutoffs. Thus we conclude that model 1 is a good fit to the observed data (or we fail to reject this model).

Next we consider Model 2. Here we do not need to group data, but we exclude the data at times 0 and 7, since we forced the model to match the data at these points. This time $X^2 = 133.2$, and we have 6 cells hence 5 degrees of freedom. Looking at our chi-square table, we find the data gives strong evidence to reject the model at a better than $\alpha = 0.001$ level (p-value less than 0.001). Thus model 2 is a very poor fit to the data.

2.4.2 Testing Stochastic Models and Some Background Theory

We take a moment and go over the ideas of the chi-square goodness-of-fit test for the more theoretically inclined and present the material in a manner that is useful for testing stochastic models. The example after this discussion makes use of these ideas.

Suppose that when a random experiment is performed, observations fall into one of n distinct categories. If a distribution is assumed for these observations, let p_i be the probability (under this assumed distribution) that any observation falls in category i. If there are N observations altogether, the expected number that fall into category i is Np_i. This number is to be compared with the observed number N_i in category i. Note here that N_i is a random variable, since it is the outcome of an experiment.

To compute the X^2 test statistic, first compute the square of each deviation of the N_i from the Np_i. Divide each of these squares by the expected number Np_i. The goodness-of-fit statistic is the sum of these terms. Thus

$$X^2 = \sum_{i=1}^{n} \frac{(N_i - Np_i)^2}{Np_i}. \tag{50}$$

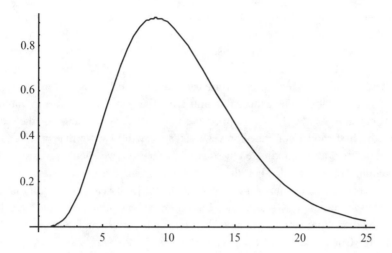

FIGURE 2.17 A Chi-square Distribution.

In our original discussion/example, the n categories are the n x-axis values of the data (the 18 decades used in the example). The number of observation N_i in each category are the observed data values (y-axis values of the data). Similarly the Np_i values are the predicted model values.

As asserted before, X^2 is a random variable which means that different experimental runs result in different X^2 values. Under certain conditions that we address below, these values approximately follow a chi-square distribution with $r = n - 1$ degrees of freedom. A graph of a chi-square distribution is shown in Figure 2.17.

To reject a model at a $\alpha = 0.01$ significance level, we require that the test statistic X^2 lie within the rightmost 10% of the area under the distribution. The point dividing the leftmost 90% from the rightmost 10% is the cutoff value χ_0^2. That is,

$$P(X^2 \leq \chi_0^2) = 0.9.$$

In general, to reject a model at an α significance level, we require that χ_0^2 divide the leftmost $1 - \alpha$ area from the rightmost α area. Thus

$$P(X^2 \leq \chi_0^2) = 1 - \alpha,$$

or equivalently

$$P(X^2 > \chi_0^2) = \alpha.$$

A few comments are in order. First, the derivation of the above result would take us far from the scope of this text. The interested reader is referred to the texts on mathematical statistics given in the references. Second, the theory described above strictly applies only in the limit as $N \to \infty$; it is not possible to say how large N must be for the chi-square goodness-of-fit test to be applicable. It is generally agreed in practice that the test is applicable when $N > 200$. Further, N should not be smaller than 50, though this condition is often relaxed. Third, and most importantly, applied statisticians usually

2 Discrete Stochasticity

require that

$$Np_i \geq 5 \text{ for } 1 \leq i \leq n. \tag{51}$$

In practice, if condition (51) is violated when the categories are first chosen, then categories should be combined to increase Np_i to satisfy (51).

A final comment is that the chi-square goodness-of-fit test is a quirky test at best. It is similar to a relative error, but squaring the numerator and not the denominator makes its behavior very different. The consequence of this is the bizarre effect that we can reject almost any model if we collect enough data. For example, suppose we have $N = 100$, five categories, and $X^2 = 4$. Further assume that this is a good measure of deviation between observed and expected, and if we repeated the measurements we would again get about 4. We would generally consider this to be a good fit. If we collect $N = 1,000$ pieces of data, X^2 will probably be about 10 times greater (of course things are random, but this is roughly what we expect) or X^2 is about 40. Now we consider the model to be a poor fit. The only thing to change, however, is the amount of data we collected. It is a very hard test for a model of the U.S. or world population to pass. It is an odd predicament. Too few counts and the chi-square approximation is not valid, too many and we will surely reject our model. It is a good rule-of-thumb test for an intermediate number of counts, and a good measure with which to compare two models. We caution against using this test as an absolute standard of model rejection.

This issue is one reason why we use only counts data with the chi-square test. If we tried to use a chi-square test to test the distance driven on U.S. highways in a day, for example, the X^2 value would be different depending on whether we measured distance in miles, kilometers, cm, or light years. Almost certainly any halfway reasonable model would be rejected if we measure in cm, and accepted if we measure in light years.

2.4.3 Validity of Sandhill Crane Model

In Section 2.3.4, we asked how well the sandhill crane environmental stochasticity simulation compared with equation (45), which is the theoretical model we obtained for the sandhill crane population size. The chi-square goodness-of-fit test can be used to answer this question. In light of the requirement that the number of data points N should be greater than 200, the sandhill crane with environmental stochasticity simulation will be performed again. This simulation will still be for five years, but now we run the simulation 500 times. Naturally, this is too much data to display here. Table 2.16 shows the frequency distribution for the number of catastrophes which occurred during the five years for these 500 simulations.

Before we apply the chi-square goodness-of-fit test to the data in Table 2.16, note how the relative frequencies of the number of catastrophes compare with the predicted probabilities of Table 2.13. There is a much closer agreement in these values than we previously had with the 20 simulation runs in Table 2.14. This improvement should not come as a surprise. After all, a sample of 20 data points is substantially smaller than 500. If the model is correct, the relative frequency of occurrence approaches the predicted probabilities as the number of data points increases. We now apply the goodness-of-fit test to test whether the model is valid.

Number of Catastrophes	Number of Occurrences	Relative Frequency
0	406	406 / 500 = 81.2%
1	88	88 / 500 = 17.6%
2	5	5 / 500 = 1%
3	1	1 / 500 = 0.2%
4	0	0 / 500 = 0%
5	0	0 / 500 = 0%

TABLE 2.16 The Frequency Distribution for the Number of Catastrophes during a 5-year period for 500 runs of the Sandhill Crane with Environmental Stochasticity Simulation.

Number of Catastrophes i	Probability p_i	Expected Number of Occurrences Np_i
0	0.8153	407.65
1	0.1699	84.95
2	0.0142	7.10
3	0.0006	0.3
4	1.229×10^{-5}	0.0061
5	1.024×10^{-7}	5.12×10^{-7}

TABLE 2.17 The Probability Distribution and the Expected Number of Catastrophe Occurrences for $N = 500$ Runs of the Sandhill Crane Model with Environmental Stochasticity Simulation.

To apply the chi-square goodness-of-fit test to compare the simulated number of catastrophes in Table 2.16 to the predicted number of catastrophes, we assume that the underlying probability distribution obtained in equation (45) is correct. That is, the probability distribution for the number of a catastrophes' occurring is binomial with $p = 0.04$ and $n = 5$. The values of this distribution are summarized in Table 2.17

Since the condition (51) that each category has at least five data points is not satisfied, the last four categories are combined. This is shown in Table 2.18.

To compute the X^2 test statistic, the simulation data in Table 2.16 must also have the last four categories combined. This gives Table 2.19.

Number of Catastrophes i	Expected Number of Occurrences Np_i
0	407.65
1	84.95
≥ 2	7.4062

TABLE 2.18 The Expected Number of Catastrophe Occurrences for $N = 500$ Runs of the Sandhill Crane Model with Environmental Stochasticity.

Number of Catastrophes i	Observed Number of Occurrences N_i
0	406
1	88
≥ 2	6

TABLE 2.19 The Observed Number of Catastrophe Occurrences for $N = 500$ Runs of the Sandhill Crane model with Environmental Stochasticity.

Using (50), for the three categories of Tables 2.18 and 2.19, the X^2 test statistic is computed as

$$X^2 = \sum_{i=1}^{3} \frac{(N_i - Np_i)^2}{Np_i}$$
$$= \frac{(406 - 407.65)^2}{407.65} + \frac{(88 - 84.95)^2}{84.95} + \frac{(6 - 7.4062)^2}{7.4062}$$
$$= 0.3832.$$

This value for the X^2 test statistic is small which indicates close agreement between the observed values N_i and the predicted values Np_i.

The number of degrees of freedom is $r = 3-1 = 2$, so that an $\alpha = 0.05$ level test gives a cutoff value of $\chi_0^2 = 5.99$. The computed value of the test statistic $X^2 = 0.3832$ is (substantially) smaller than this cutoff value, indicating that we can accept, at the $\alpha = 0.05$ level, the assumption that the underlying distribution for the number of catastrophes is binomial with $n = 5$ and $p = 0.04$.

2.5 For Further Reading

The student version of Data Desk comes with a manual/workbook and a disk of data. A lot can be learned working through the workbook and exploring the data. It is, however, software specific:

Paul Velleman, *Learning Data Analysis with Data Desk* (Revised and updated edition), W.H. Freeman and Company, New York, 1993.

A basic statistics book geared toward the concepts of statistics with few formulas is:

Freedman, Pisani, and Purves, *Statistics* (Third edition), W.W. Norton and Company, New York, 1998.

A classic work in data visualization is:

John W. Tukey, *Exploratory Data Analysis*, Addison-Wesley Publishing Company, Reading Mass., 1977.

For a good introduction to probability theory see:

Hogg and Tanis, *Probability and Statistical Inference* (Fifth edition), Prentice Hall, Englewood Cliffs, NJ, 1997.

Even though its computer illustrations are dated (punch cards), the ideas of stochastic modeling found in the next book are useful:

Francis Martin, *Computer Modeling and Simulation*, John Wiley & Sons, Inc., New York, 1968.

2.6 Exercises

1. House sales (in thousands of dollars) for the last month in a community are: 70, 65, 80, 91, 105, 143, 150, 145, 120, 92, 100, 95, 83, 139, 510, 101, 675, 84, and 580. Graphically present this data. Intuitively pick the point you consider to be the "center" of the data. Compute the mean and the median. Which is closer to your intuitive "center?"

2. This exercise uses the results from the M&M project in Section 2.1.2. If done as a class, data from everyone should be combined for best results. If not done as a class, your results will be better if you repeat the simulation at least 10 more times and preferably more.

a. Construct histograms for each time step.

b. Create side-by-side box plots with one box plot for each time step.

Notice that while the distributions are more thoroughly presented in the histograms, the side-by-side box plot conveys much of the same information, plus dramatically shows the overall trend in the data.

3. a. In a medical study, 70 guinea pigs were infected by a virus. The survival times (in days) is given in Table 2.20. Construct a frequency distribution and histogram for the guinea pig survival times.

40	45	53	56	56	57	58	62	64	73
74	79	78	80	81	81	81	82	83	83
84	85	89	90	91	92	92	97	99	99
99	99	101	102	102	102	103	104	107	108
109	113	114	118	121	123	126	128	137	138
134	144	145	147	156	162	174	174	179	184
191	198	208	214	247	249	328	383	403	511

TABLE 2.20 Guinea Pig Survival Times (in days).

b. A popular brand of cereal is sold in 16-ounce boxes. The cereal boxes are machine filled with weights around 16 ounces. Fifty boxes of cereal were weighed, and the data in Table 2.21 were obtained. Construct a frequency distribution and histogram for the weights of the cereal boxes. What distribution does the shape of the histogram suggest?

c. Create a frequency distribution and histogram for the number of viewers (in millions) for the second season of the television show *The X-files*. Season two's data is reproduced in Table 2.22.

2 Discrete Stochasticity

16.22	15.79	16.03	15.76	15.72	15.89	16.25	15.88	15.89	15.62
15.95	15.89	15.99	16.03	15.94	16.29	16.06	16.17	15.97	16.16
16.14	16.00	16.16	15.96	15.89	16.08	15.94	15.96	16.08	16.06
16.17	16.18	16.10	15.98	16.17	15.93	15.87	15.92	16.03	16.05
15.98	15.99	15.92	16.22	16.25	15.64	16.03	16.05	16.26	16.14

TABLE 2.21 Weights of Cereal Boxes.

9.8	9.3	8.7	8.2	8.5	9.2	9	9.1	8.6	9.9
8.5	9.7	8.8	10.2	10.8	9.8	10.7	9.6	10.2	9.8
7.9	8.5	8.1	9	9.6					

TABLE 2.22 The Number of Viewers (in millions) of the Second Season of Television Show *The X-files*.

7.4	6.9	6.8	5.9	6.2	5.6	5.6	6.2	6.1	5.1
6.4	6.4	6.2	6.8	7.2	6.8	5.8	7.1	7.2	7.5
8.1	7.7	7.4	8.3	9.8	9.3	8.7	8.2	8.5	9.2
9	9.1	8.6	9.9	8.5	9.7	8.8	10.2	10.8	9.8
10.7	9.6	10.2	9.8	7.9	8.5	8.1	9	9.6	

TABLE 2.23 The number of Viewers (in millions) of the First Two Seasons of Television Show *The X-files*.

d. Create a frequency distribution and histogram for the number of viewers (in millions) of first two seasons of the television show *The X-files*. The combined data is reproduced in Table 2.23.

4. Create a box plot and five-point summary for the data in Table 2.22 which represents the number of viewers in millions of the second season of television show *The X-files*.

5. Rerun the sandhill crane model with environmental stochasticity, this time assuming that the catastrophe lowers the birth rate by a random amount between 38% and 42% (uniformly distributed) and raises the death rate by a random amount between 23% and 27% (uniformly distributed). What is the effect on the web-like pattern? Repeat this exercise several times, each time altering the effect of the catastrophes on the birth and death rates. What conclusions can you draw?

6. The expected value of a discrete random variable X with probability density function (distribution) $f(x)$ was defined to be

$$E(X) = \sum_x x f(x)$$

where the sum is over all values of the random variable X. The expected value of a continuous random variable X with density function $f(x)$ was similarly defined as

$$E(X) = \int_{-\infty}^{\infty} x f(x)\, dx.$$

a. Show that the expectation of a random variable is linear; that is, $E(X)$ has the property

$$E(aX + bY) = aE(X) + bE(Y),$$

where X and Y are random variables and a and b are any real numbers.

b. The variance of a random variable was defined in the text as

$$V(X) = E((X-\mu)^2).$$

Show that the variance of a random variable can be computed as

$$V(X) = E(X^2) - (E(X))^2.$$

7. The uniform distribution is defined by

$$f(x) = \begin{cases} \frac{1}{\beta-\alpha} & \text{for } \alpha < x < \beta, \\ 0 & \text{elsewhere.} \end{cases}$$

a. Show that intervals of the same length are equally probable.

b. Show that the mean of the uniform distribution on (α, β) is the midpoint of the interval (α, β), that is $(\alpha + \beta)/2$.

c. Show that the variance of the uniform distribution is $(\beta - \alpha)^2/12$.

8. The normal distribution for a continuous random variable is defined as

$$f(x) = \frac{1}{\sigma\sqrt{2\pi}} e^{\frac{-(x-\mu)^2}{2\sigma^2}} \quad \text{for } \sigma > 0, -\infty < \mu < \infty, -\infty < x < \infty.$$

a. Show that the location parameter μ is the expected value or mean of the normal random variable X.

b. Show that the shape parameter σ^2 is the variance of the normal random variable X.

c. Show that the normal distribution is an increasing function for $x < \mu$ and is a decreasing function for $x > \mu$; hence $f(x)$ obtains a maximum at $x = \mu$.

d. Show that the normal distribution has inflection points at $x = \mu \pm \sigma$.

e. The 68 - 95 - 99 Rule. Using either a computer algebra package or a graphing calculator, verify the 68 - 95 - 99 rule. That is, verify that approximately 68% of area under the distribution occurs within one standard deviation of the mean, 95% of the area under the distribution is within two standard deviations, and 99% is within three standard deviations.

9. Let Y represent the number of successes which occur on n independent Bernoulli trials; then the distribution of Y is known as the *binomial* distribution and has the form

$$P[Y = k] = \binom{n}{k} p^k q^{n-k} \quad \text{for } k = 0, 1, \ldots, n.$$

2 Discrete Stochasticity

a. Show that a binomial random variable is a sum of n independent Bernoulli random variables.

b. Show that the mean of a binomial random variable Y is

$$E(Y) = np.$$

c. Show that the variance of a binomial random variable Y is

$$V(Y) = E(Y^2) - [E(Y)]^2 = np(1-p).$$

d. The value of a random variable with the largest probability is called the *mode* of the random variable. Show that the integer $k_{\max}$ is the mode of a binomial random variable X, where $k_{\max}$ satisfies

$$\binom{n}{k} p^k (1-p)^{n-k} \leq \binom{n}{k_{\max}} p^{k_{\max}} (1-p)^{n-k_{\max}} \text{ for } k = 0, 1, \ldots, n, k \neq k_{\max}.$$

Solve this inequality for k_{max} to obtain the mode of a binomial random variable as

$$k_{max} = \lceil (n+1)p - 1 \rceil \tag{52}$$

where $\lceil x \rceil$ is the least integer not smaller than x.

10. a. Using the ideas from Section 2.3.4, show by induction that the population size, after n time steps when $0 \leq k \leq n$ catastrophes have occurred, is

$$x(n) = (1 + r_c)^k (1 + r)^{n-k} x(0).$$

b. Using the mode of a binomial random variable, show that the most likely population size in a population with binomial catastrophes is

$$x(n) = (1 + r_c)^{\lceil (n+1)p - 1 \rceil} (1 + r)^{n - \lceil (n+1)p - 1 \rceil} x(0).$$

2.7 Projects

Project 2.1. Stochastic Bobcats

This project investigates demographic and environmental stochasticity for the bobcats considered in Project 1.2. For this project assume that bobcats have a mean birth rate of 0.4 with a standard deviation of 0.1 and a mean survival rate of 0.68 with a standard deviation of 0.07. In this model, assume that these demographic parameters follow a normal distribution.

Environmental stochasticity can take many forms. For this project it will take the form of a catastrophe which occurs an average of once every 25 years or 4% of the time. These catastrophes affect both the bobcat birth and bobcat survival rates by decreasing both rates by 30% during the year that the catastrophe occurred.

For all parts of this project, assume an initial bobcat population of 100.

1. Deterministic. Construct and run a deterministic model for the bobcats using the birth rate of 0.4 and the survival rate of 0.68, but ignoring all random factors. Graph the population over a 20-year period with these parameters and make a table of this information.

2. Demographic Stochasticity. Create a new model using only the demographic stochastic information. Make a graph and a data table with at least 50 runs for 20 years.

3. Environmental Stochasticity. Now construct a model using only environmental stochasticity. Again construct a data table and graph for at least 50 runs of this case for 20 years.

4. Take the data from the 20th year of both stochastic models and, if necessary, export it to a program that calculates statistics. Make histograms and box plots. Make reports of the following summary statistics: mean, median, 25th and 75th percentiles, standard deviation, range, inter-quartile range, minimum, and maximum. Compare the results from these two models.

5. For each model construct a graph with side-by-side box plots for the first 10 years showing the change in distribution over time. Compare both of these models with each other and with the deterministic model.

6. From the work so far, which stochastic effect is more pronounced? Either construct and analyze a model with both demographic and environmental stochastic effects, or argue that the combined effects will not be significantly different from the individual effects so there is no need to complicate the model.

Project 2.2. The Poisson Forest

In Exercise 3 of Chapter 0, you were asked to estimate the number of dots in Figure 0.2. One method you may have used to solve this problem was to count the dots in a small area and then multiply this number by the appropriate factor to estimate the whole square. In order for this method to be valid, the dots must be uniformly distributed over the square. If, for example, the dots were clustered near the center, then counting the dots in a small area near the center and multiplying would drastically overestimate the number of dots.

Ecologists are often interested in the spatial distributions of plants and animals. Aerial photographs of a forest or a large herd of animals look similar to Figure 0.2, only with trees or animals instead of dots. If the distribution of trees or animals is uniform, then an ecologist can use the method of Exercise 0.3 to estimate the size of a population. But again, the distribution must be uniform in order to use this method. A forest where the trees are uniformly distributed is called a *Poisson forest* and is related to the Poisson process considered in Chapter 6.

This project explores a method for determining whether or not the distribution of objects in a two-dimensional area is uniform. It is merely an application of the chi-square goodness-of-fit test. Partition the area into pieces or cells of equal area. The cell size should be chosen so that there are as many cells as possible subject to the condition that each cell has at least five objects in it. Count and record the number of objects in each cell. Determine the average number of objects per cell. If the distribution of objects were uniform, we would expect the average number of objects in each cell. Instead we have the observed number of objects in each cell. Recall that the X^2 statistic measures the

2 Discrete Stochasticity

FIGURE 2.18 Trees.

difference between observed and expected values over a number of categories (which are cells in this situation). It will not tell us if the distribution is uniform, but it will tell us the degree of confidence with which we can assert that the distribution is uniform.

1. Consider the "forest" shown in Figure 2.18. Break the forest into appropriately sized cells. Use the chi-square goodness-of-fit test to determine whether we are 95% confident that the distribution is uniform.

2. Find an aerial photograph of a tree farm, a crowd of people, or herd of animals (or equivalently an earth-based photograph of a large flock of birds or bats). Use the chi-square goodness-of-fit test to determine if the objects in your photograph are uniformly distributed (i.e., is your forest a Poisson forest) at a confidence level of 90% or better. If the number of objects is too large to reasonably count, test several sub-areas for uniformly distributed objects. If you find that it is reasonable to assume that the distribution is uniform, estimate the number of objects in the photograph.

3. Find a photograph of the moon which clearly shows the craters. Use the ideas of this project to determine if the craters are uniformly distributed to a reasonable level of confidence. You should try this at least twice—once with a view from above the equator, and again with a view looking down on a pole.

Project 2.3. Black-Capped Chickadees

In the article *Dynamics of a Black-Capped Chickadee Population 1958–1983* by G. Loery and J. Nichols [39], the wintering population of black-capped chickadees (*Parus atricapillus*) in northwestern Connecticut was studied. This article studies many factors including the effects on population dynamics by clear-cutting portions of nearby red pine plantations, the natural introduction of a competing species (tufted titmice), and winter temperature variations. We use some of the information in this article to produce a simplified chickadee model. The interested reader is referred to the original article for more information.

Year	Recruitment	Survival Rate	Actual Population
1958	–	.90	–
1959	-47.8	.64	273.8
1960	70.3	.85	126.8
1961	100.6	.57	178.1
1962	35.3	.58	201.9
1963	62.0	.64	151.4
1964	89.3	.59	159.0
1965	34.3	.51	183.5
1966	57.8	.69	126.5
1967	13.0	.49	145.2
1968	112.8	.61	85.0
1969	239.4	.54	164.3
1970	-9.3	.50	328.1
1971	61.7	.43	155.7
1972	99.6	.75	127.4
1973	57.9	.54	194.2
1974	49.2	.50	161.2
1975	48.9	.50	129.0
1976	61.1	.70	112.8
1977	73.4	.54	137.9
1978	43.6	.48	147.0
1979	87.7	.63	113.2
1980	49.2	.55	157.5
1981	72.1	.47	135.7
1982	–	–	134.2

TABLE 2.24 Birth and Death Rates and Population Counts for the Black-Capped Chickadee from *Dynamics of a Black-Capped Chickadee Population 1958–1983* by G. Loery and J. Nichols, used by permission of The Ecological Society of America.

Table 2.24 summarizes some of this information about survival rates and recruitment (births and immigration) for the black-capped chickadee for the years 1958–1981. Note that survival is a *rate* (fraction of the existing population), while recruitment is a *number* of birds. The fact that there are fractional birds is due to using sampling (capture-recapture and capture-resighting) methods to estimate the whole population and recruitment rates. Loery and Nichols note that negative recruitments have confidence intervals which include zero and should probably be considered to be zero, but are left negative for purposes of computing the mean and other statistics.

These data will be used for another project later in this book. For now we use them to construct a stochastic model.

1. Find the mean and standard deviations for both recruitment and the survival rates. These sample means and sample standard deviations will be used to estimate the population means and standard deviations.

2 Discrete Stochasticity

2. Using the means for the recruitment and death rates (1 minus survival rate) in part **1**, construct a deterministic model. This should be an affine model. Analytically solve the affine equation and numerically graph a solution curve for 1959 to 1982. Compare its output to the observed data.

3. Construct a model for the chickadees with demographic stochasticity. Assume death rates and recruitment follow normal distributions with means and standard deviations approximately those found in part **1**. Run at least 50 simulations for the years 1959 to 1982. Display these results using multiple line graphs and side-by-side box plots. Compare to the observed data. Compare the results of the stochastic and the deterministic model.

4. In part **3** a standard assumption of normality was made. A normal probability plot (or quantile-quantile [q-q] plot) can be used to test this assumption. Intuitively it compares quantiles (percentiles) of the data with quantiles of a true normal. More precisely it first transforms the data by subtracting the mean and dividing by the standard deviation and then compares the quantiles of this data with the quantiles of a standard normal $N(0, 1)$. Perfectly normal data will lie on a line. Deviations indicate some degree of non-normality. If you have easy access to software that generates a normal probability plot, create one for the birth rates and one for the death rates. Comment on the assumption of normality.

Project 2.4. More M&M's

In the beginning of this chapter, we performed a physical simulation using M&M's. During time step, each M&M has a 50% chance of "dying" (landing letter side up). This project involves the construction of a computer model to simulate the same behavior. If initially there are 100 M&M's, the deterministic model is simple; the population will be 100, 50, 25, 12.5, 6.25, etc. The point of this project is the model construction, since we already know the expected output.

1. Individual-Based Model. Construct a computer model which has 100 M&M's (i.e., has 100 compartments) each of which live or die based on a random number generation. At each time step, compute the number of M&M's still "alive." This is called an *individual-based model* since each member of the population is individually modeled. This type of model is useful when individuals have important characteristics, such as tracking genetic factors.

2. Population-Based Model. The model in part **1** can be thought of as performing n Bernoulli trials at each step (where n is the number of members still alive). Recall from the text that the sum of Bernoulli trials has a binomial distribution. This time, model the 100 M&M's as an entire population (a one compartment model), using a random number from a binomial distribution to determine the number that survive each time step.

3. Using a Normal Approximation. If n is large and p, the probability of "success," is not near either 0 or 1, a binomial distribution is frequently approximated by a normal with mean np and standard deviation $np(1 - p)$. Repeat part **2** with this approximation and discuss advantages and disadvantages of using this approximation.

Project 2.5. Capture-Recapture

Note: This project is suitable for students with some background in probability. It requires knowledge of hypergeometric distributions and confidence intervals.

The actual population size N of a specific species of animals in their natural environment is often unknown. It is of practical interest to estimate N without counting every individual. Capture-recapture is one technique of doing this.

In capture-recapture, biologists capture and mark (in some fashion) M individuals from the population. These M individuals are then released back into the population. Later, after a sufficient amount of time has passed to allow the marked individuals to mix with the unmarked individuals, a sample of size n is taken from the population and the number X of marked individuals is noted. The problem is to estimate the population size based upon the number of marked individuals in the sample of size n. Note that X is a discrete random variable since we can not predict with certainty how many marked individuals will be recaptured.

If we assume that a sufficient amount of time has passed to allow the population of marked and unmarked individuals to be thoroughly mixed, the proportion $\frac{M}{N}$ of marked individuals in the population should be (approximately) the same as the proportion, $\frac{X}{n}$ of marked individuals in the sample. Equating these two proportions

$$\frac{M}{N} = \frac{X}{n}$$

and solving for the unknown population size N gives

$$N = \frac{Mn}{X}.$$

This equation can be used to estimate the unknown population size N. Since this is an estimate of the unknown population size N, we will denote it (as usual) by $\hat{N}$. That is,

$$\hat{N} = \frac{Mn}{X}.$$

Properties of the estimate $\hat{N}$ may be deduced from the distribution of the random variable X. We divide the population into two types. We consider marked individuals to be of type 1 and the unmarked individuals to be of type 2.

The distribution for X the numbers of type 1 (marked) individuals in a sample of n individuals drawn *without* replacement from the population is called the *hypergeometric* and can be found in any probability and statistics text (Hogg and Tanis [26] for example). The probability density function for a hypergeometric distribution is given by

$$P(X = k) = \frac{\binom{M}{k}\binom{N-M}{n-k}}{\binom{N}{n}} \text{ for } k = 0, 1, \ldots, n.$$

Suppose that X is observed to equal k. Let $P_k(N)$ represent the probability of observing k individuals (the observed event) when there are actually N animals in the

2 Discrete Stochasticity

population. It would appear reasonable that an estimate of N would be the value of N that maximizes $P_k(N)$. Such an estimate is called a *maximum-likelihood* estimate.

1. Show that
$$\hat{N} = \frac{Mn}{X}$$
is a maximum-likelihood estimator for N by maximizing $P_k(N)$. This is best shown by considering when
$$P_k(N) \geq P_k(N-1).$$

2. In situations of practical interest, N will be much larger than both M and n so that the sampling of the population may be considered as sampling *with* replacement. Show that in this situation, the distribution for X is approximately binomial with parameters n and $p = \frac{M}{N}$.

3. Binomial distributions are approximated fairly well by normal distributions when the sample size n is large. Using part **2**, show that the distribution of X is approximately normal with mean $\frac{nM}{\hat{N}}$ and variance $(nM/\hat{N})(1 - M/\hat{N})$.

4. Using part **3**, derive the following $100 \times (1-\alpha)\%$ confidence interval for the estimate $\hat{N}$.
$$P\left[\frac{nM}{k + z_{\alpha/2}\sqrt{k(1-\frac{k}{n})}} < \hat{N} < \frac{nM}{k - z_{\alpha/2}\sqrt{k(1-\frac{k}{n})}}\right] = 1 - \alpha,$$
where k is the number of recaptured individuals in the sample of size n and
$$P(Z > z_{\alpha/2}) = \frac{\alpha}{2}$$
with Z the standard normal random variable.

5. Simulation of capture-recapture. These results can be empirically verified through numerical simulation. One approach to this simulation is to create an array of length 500. Initially, the entries in the array will be zero. To "capture the animals," generate 100 random integers between 1 and 500. For each of these random integers, change the entry in this position of the array to a 1. This will represent a tagged animal. (Care must be taken not to tag an animal twice.) Once this step is completed, the "recapturing" will occur from generating another 100 random integers and checking the array position for these integers. The number of 1's will represent the number of animals recaptured. Using the formula obtained in part **1**, how well does your number predict the population size of 500? Perform this simulation 100 times. Does the average of the 100 trials predict the population size better? Using these 100 simulations, construct an empirical 95% confidence interval. Compare this confidence interval with that obtained in part **4**.

MORTON COLLEGE
LEARNING RESOURCES CENTER
CICERO, ILL.

3

Stages, States, and Classes

In the first two chapters, we looked at models involving growth in which the entity that was growing was considered as an aggregate. We used single growth and decay parameters which described on average the behavior of the whole. In many real situations, the whole can be divided into parts which behave differently. In populations, the young have different survival and birth rates than the old, and males and females may have different average lifetimes. In personal finance, your savings may be invested in several places, each with its own interest rate, and each loan you obtain probably has a different rate. Further, an investment plan that puts money into an account at regular intervals, say monthly, has money earning different rates, depending on how long it is in an account. It is obvious that using one rate for growth and decay (or interest earned and interest paid) for these situations is only an approximation of the true behavior. The purpose of the model should help you decide if the aggregate approach is acceptable or if you need to increase the resolution of your model by incorporating this additional structure. This chapter assumes that your purposes require this degree of model refinement. In this chapter, we discuss a method for diagramming these structured models (state diagrams) and for converting the diagrams into equations. While the resulting equations can be numerically investigated, they can also be expressed in matrix notation, and we explore methods of analyzing matrix equations. The first part of this chapter explores deterministic models; the second part deals with a special type of stochastic model called a Markov chain. The last section presents a bookkeeping device used in population modeling called a life table. The models in this chapter have parameters which are constants. While this is necessary for some of the mathematical analysis, it can be relaxed for models which are numerically investigated.

3.1 A Human Population Model

Based on our everyday experiences, there are several facts about human biology that should be evident. The first is that only females give birth. It may be a hard blow to the male ego, but if there are enough men to go around in some sense, the number of births is determined only by the number of females. Many population models consider only the female portion of the population, and we do likewise in this example. The second observation is that the probability of a woman either giving birth or dying during the course of a year changes throughout her lifetime. Thus a high resolution model must have different birth and survival rates for the different stages of life. After menopause, a woman no longer contributes to bringing new children into the population. Because of this, many population models consider only females of child-bearing age.

There is extensive data on survival and reproduction rates, and they are computed year by year for 102 years of a woman's life. Moreover, survival rates are computed daily for the first 28 days of life and monthly for the first year. Why? Insurance companies make their profit from this information. A table with the information just stated can be found in the book *Introduction to the Mathematics of Demography* by R. L. Brown [7], a book written for actuary students.

Needless to say there is plenty of data for modeling a human population with multiple classes. While an actuary may be interested in the resolution given by 102 age classes, we would like to illustrate these methods without this degree of complication. Frequently the lifetime of an individual is split into several age classes of equal length. We consider only women of child-bearing age in our model and make a simplifying assumption that child bearing stops after the age of 45. Secondly we split the 45 years into three equal size classes, namely 0–15, 15–30, and 30–45. Based on data collected by insurance companies as reported in Brown's book and reprinted here by permission, we know the birth rates and survival rates for each age class. We let b_i indicate the birth rate of the ith age class and s_{ij} denote the survival rate from the ith class to the jth class. Here $b_1 = 0.4271$, $b_2 = 0.8498$, $b_3 = 0.1273$, $s_{12} = 0.9924$, and $s_{23} = 0.9826$.

We let $x_i(n)$ denote the population of the ith age class at the nth time step. One word of **CAUTION**. Since the population has been grouped into age classes of length 15 years, one time step is 15 years. Thus $n = 1$ corresponds to $t = 15$, $n = 2$ corresponds to $t = 30$, and so on. The equations are easy to set up. How does one become a $0 - 15$-year-old in the next time step? By being born during this time step to a mother in any age class. How does one enter the second age class? Simply by surviving the first age class. Finally in order to be in the third age class, one must survive the second age class. From this we set up the following equations:

$$x_1(n+1) = 0.4271 x_1(n) + 0.8498 x_2(n) + 0.1273 x_3(n) \tag{53}$$

$$x_2(n+1) = 0.9924 x_1(n) \tag{54}$$

$$x_3(n+1) = 0.9826 x_2(n). \tag{55}$$

A spreadsheet is an ideal tool for analyzing equations in this form. Figure 3.1 tracks an initial population of 30 million women in each class through 10 time steps (150 years). A graph of these results is presented in Figure 3.2. Several trends should be evident. First, even though the population is broken into classes, each class is eventually growing exponentially, and the overall population is also eventually growing exponentially. This suggests that there is one single growth rate that averages all the individual birth and

3 Stages, States, and Classes

Year	Age 0-15	Age 15-30	Age 30-45	Total
0	30	30	30	90
15	42.126	29.772	29.478	101.376
30	47.0448096	41.8058424	29.2539672	118.104619
45	59.3434731	46.687269	41.0784207	147.109163
60	70.2497215	58.8924627	45.8749106	175.017095
75	85.890347	69.7158237	57.8677338	213.473904
90	103.294837	85.2375803	68.5027683	257.035185
105	125.272523	102.509796	83.7544464	311.536765
120	151.27866	124.320452	100.726125	376.325237
135	183.081071	150.128942	122.157276	455.36729
150	221.324122	181.689655	147.516699	550.530476

FIGURE 3.1 Spreadsheet For Human Female Population Model.

survival rates. On second thought, this is no surprise; that is what we did in Chapter 1. Secondly, even though the age classes started out with one-third of the population in each class, they redistributed themselves into some new proportions which stabilized at 40.2% in first class, 33.0% in second class, and 26.8% in the third class. While this is not obvious from the spreadsheet analysis, the mathematics of this chapter will show that this is not an accidental occurrence. The overall growth rate is called an eigenvalue as in Chapter 1 and the stabilizing proportions are called an eigenvector which is discussed below.

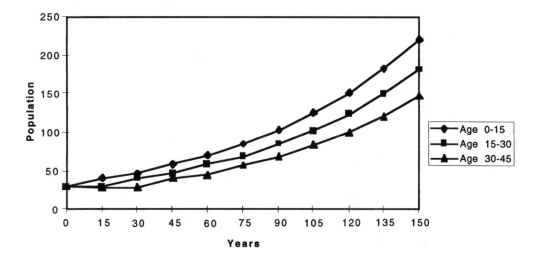

FIGURE 3.2 Graph of Output From Human Female Population Model.

3.2 State Diagrams

This chapter deals with models that have structure in the form of compartments or age classes or stages. Frequently we talk about a *state* or *state vector* or *state distribution* which are the values of all the classes at a given time. Actually the word *state* has several meanings. Sometimes a state is the value of one compartment, and a state vector is the value of all compartments; sometimes a state is synonymous with state vector and means the value of all compartments. Context makes clear which is meant. Usually we know one state and wish to know the state for the next time increment (or many time increments later). In our human population example, the original state was 30 million women in each class. Each row of the spreadsheet in Figure 3.1 is a subsequent state. While compartmental diagrams seems to be ideal for a model with classes, they are awkward if we are interested in states. There is a related type of diagram that goes by a variety of names, which we call a *state diagram*, which works much better for our current purposes.

A state diagram, like a compartmental diagram, lists the states or stages or age classes or compartments, or whatever else the structures may be called, as circles with arrows drawn between the circles representing states that interact in some way. Numbers or symbols adjacent to the arrows are called *weights*. We adopt four conventions to make the modeling process uniform across applications. First, all such diagrams are called state diagrams regardless of whether a particular application is calling them states or something else. Similarly we refer to an age class or other structure as a state if it is convenient to do so. Second, the arrows always refer to what is entering a state, not leaving. It is assumed that between every time step the state completely clears and then refills. Third, the number over an arrow indicates the fraction of the contents of the state at the arrow's tail that flows into the state at the arrow's head. This fraction may be greater than one, as in the case of births resulting from a particular age class. Arrows, then, should not always be interpreted to be flows, although sometimes this is a legitimate interpretation. Finally, implicit in the third assumption is the assumption that the amount that flows into a state during a time step depends only on the contents of the states during the immediately preceding time step and not on the condition of the states at any previous time. It also makes no difference how the contents of a state came into being.

For the purposes of this book, state diagrams always involve circles and arrows, while compartmental diagrams involve rectangles and arrows.

Consider the following examples:

3.2.1 Human Population

The states represent the age classes. The numbers over the arrows are the birth and survival rates. The arrows indicate how many women move into each age class during each time step. Notice that it does not indicate how many flow out. The diagram assumption is that all of the population moves out. Some of it flows to the next age class as indicated, and some it to a "death state" which is not part of this model.

3.2.2 Money in a bank account I

This is a model that is probably more naturally set up as a compartmental diagram, but it makes a good example of a state diagram. Here the amount of money in the bank account

3 Stages, States, and Classes

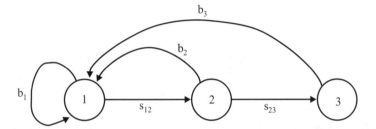

FIGURE 3.3 State Diagram For the Female Human Population Model.

is the state. The loop arrow indicates the earned interest being added to the account. The number over the arrow is not the interest rate, but rather one plus the interest rate since the diagram assumption assumes the state empties each time step and we need to refill the old balance.

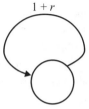

FIGURE 3.4 State Diagram for a Bank Account.

3.2.3 Money in a bank account II

This time we add a regular investment to the previous bank account model. There are two states: one for the investment and one for the bank account. The loop arrow on the investment state indicates that it gets completely refilled each time step. The arrow from the investment state to the account state indicates that the amount in the investment state gets added to the account each time step.

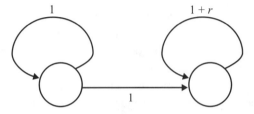

FIGURE 3.5 State Diagram Bank Account With Regular Investing.

3.3 Equations From State Diagrams

With the assumptions we made in drawing state diagrams, it is easy to construct the equations for the system. We begin by numbering the states, if possible, by a scheme that has some physical meaning. We use the symbol $x_i(t)$ to refer to the contents of the ith state at time t, and, when the time is unimportant or understood, we just write x_i. The symbol w_{ij} is the number or weight associated with the arrow from the ith state to jth state. If there is no arrow, w_{ij} is assumed to be zero. Since the diagram assumption is that the arrow heads indicate where things are flowing, the value of $x_i(t+1)$ is just the sum of each $x_j(t)$ times the weight w_{ji} from the jth state to the ith state. Thus

$$x_i(t+1) = w_{1i}x_1(t) + w_{2i}x_2(t) + \cdots + w_{ni}x_n(t)$$

or

$$x_i(t+1) = \sum_{j=1}^{n} w_{ji}x_j(t).$$

What should be obvious here is that state diagrams are useful in setting up recurrence relations; compartmental diagrams are useful in setting up difference equations.

The $x_i(t)$ are useful to us in a number of ways. Clearly they tell us the size of the ith class at time t. In the case of populations, their sum gives us the total population. For population and Markov models (discussed below) the collection $X = \{x_1, \ldots, x_n\}$ gives the distribution of the states or state vector.

Using the state diagram for the human female population diagram shown in Figure 3.3, we construct the equations

$$x_1(n+1) = b_1 x_1(n) + b_2 x_2(n) + b_3 x_3(n) \tag{56}$$

$$x_2(n+1) = s_{12} x_1(n) \tag{57}$$

$$x_3(n+1) = s_{23} x_2(n). \tag{58}$$

If we put in the values for the b's and s's, these are the equations we found before (Equations (53), (54), and (55)).

3.4 A Primer of Matrix Algebra

Despite their diversity, the equations involved in all the models in this chapter have the same form. In particular these models have some number of states, say n, and a variable, $x_i(t)$, for each state. The equation for $x_i(t+1)$ is the sum of all the products of a constant with each $x_j(t)$, i.e., $x_i(t+1) = a_{1i}x_1(t) + a_{2i}x_2(t) + \cdots + a_{ni}x_n(t)$. The frequency and importance of models with equations of this form resulted in the development of mathematics that deals explicitly with systems of equations of this type. This mathematics is called linear algebra, and we are primarily interested in a subset of linear algebra called matrix algebra. This section is a concise introduction to matrix algebra. We begin by assuming no prior knowledge of matrices and invite readers with some background in linear algebra to skip ahead.

3 Stages, States, and Classes

Simply put, a *matrix* is a rectangular array of numbers. The horizontal lists are called rows, and the vertical lists are called columns. For example, let

$$A = \begin{pmatrix} 2 & 3 & 4 \\ 0 & 1 & -2 \end{pmatrix}.$$

This matrix has two rows: $\begin{pmatrix} 2 & 3 & 4 \end{pmatrix}$ and $\begin{pmatrix} 0 & 1 & -2 \end{pmatrix}$. It has three columns: $\begin{pmatrix} 2 \\ 0 \end{pmatrix}, \begin{pmatrix} 3 \\ 1 \end{pmatrix}$, and $\begin{pmatrix} 4 \\ -2 \end{pmatrix}$.

We refer to the size of a matrix as its number of rows by its number of columns. A 2×3 matrix (read "two by three matrix") is a matrix with two rows and three columns. A *vector* is a matrix with either one row or one column; sometimes a vector with only one row is called a *row vector*, and a vector with only one column is called a *column vector*. In the example above of the sandhill crane, the distribution of age classes is a vector. We frequently need to refer to specific elements of a matrix, and we use the row and column information to provide an address for each element. For example, the ijth element is the element in the ith row and the jth column (rows are always listed before columns). Thus we denote a generic matrix by

$$A = \begin{pmatrix} a_{11} & a_{12} & \cdots & a_{1n} \\ a_{21} & a_{22} & \cdots & a_{2n} \\ \vdots & \vdots & \ddots & \vdots \\ a_{m1} & a_{m2} & \cdots & a_{mn} \end{pmatrix}$$

or more succinctly by $A = (a_{ij})_{m \times n}$.

Matrices can be considered to be a number system much as the integers, the real numbers, or the complex numbers are number systems. As with other number systems we define ways to add, subtract, multiply, and divide matrices. Thus we construct an algebra for matrices which we call, appropriately enough, matrix algebra. Two matrices can be added provided that they have the same size; thus two 2×3 matrices can be added together, but a 2×3 cannot be added to a 3×5 matrix. Addition and subtraction are defined elementwise. Thus $A + B = C$ means that for all i and j, $c_{ij} = a_{ij} + b_{ij}$. Similarly $A - B = C$ means $c_{ij} = a_{ij} - b_{ij}$. For example

$$\begin{pmatrix} a & b \\ c & d \end{pmatrix} + \begin{pmatrix} e & f \\ g & h \end{pmatrix} = \begin{pmatrix} a+e & b+f \\ c+g & d+h \end{pmatrix},$$

$$\begin{pmatrix} a & b \\ c & d \end{pmatrix} - \begin{pmatrix} e & f \\ g & h \end{pmatrix} = \begin{pmatrix} a-e & b-f \\ c-g & d-h \end{pmatrix}.$$

In algebra on the real numbers, 0 is a distinguished number called the additive identity. This means that for any real a, $a + 0 = a$ and $0 + a = a$. There is a matrix (for each size of matrix) playing this same role for matrix arithmetic. It is called the *zero matrix* and is the matrix consisting of all zeros, and denoted by **0**. It is easy to verify that $A + \mathbf{0} = \mathbf{0} + A = A$.

There are actually two number systems involved in matrix algebra: the matrices themselves and the elements which make up the matrices. We distinguish these by calling the former matrices and the latter *scalars*. Technically the 1×1 matrix (a) is a matrix, while a is a scalar. Whenever feasible, matrices are denoted by capital letters and scalars are denoted by lower case letters. We occasionally need to multiply scalars and matrices together. If s is a scalar and A is a matrix, we define a new matrix sA to be the matrix which results from multiplying all the entries of A by s. Thus if $B = sA$, then $b_{ij} = sa_{ij}$. For example,

$$s \begin{pmatrix} a & b \\ c & d \end{pmatrix} = \begin{pmatrix} sa & sb \\ sc & sd \end{pmatrix}.$$

Multiplication in matrix algebra is in general much stranger than multiplication in the real numbers. While the multiplication by a scalar that we just defined is similar to addition and subtraction of matrices, there is a major conceptual difference. Multiplication of a matrix by a scalar is the multiplication of two different types of objects, namely scalars and matrices. It is useful to define a multiplication between two matrices. The motivation for the definition is perhaps not clear, but this definition is formed out of naturally arising patterns. We will soon see that it is extremely useful. Let $A = (a_{ij})_{n \times r}$ and $B = (b_{ij})_{r \times s}$. The product $C = AB$ is defined by $c_{ij} = a_{i1}b_{1j} + a_{i2}b_{2j} + \cdots + a_{ir}b_{rj}$. In a sense what we have done is multiply the ith row of A by the jth column of B. Here are a few examples.

$$\begin{pmatrix} a & b \\ c & d \end{pmatrix} \begin{pmatrix} e & f \\ g & h \end{pmatrix} = \begin{pmatrix} ae+bg & af+bh \\ ce+dg & cf+dh \end{pmatrix}$$

$$\begin{pmatrix} 1 & 2 \\ 3 & 4 \end{pmatrix} \begin{pmatrix} 0 & 1 \\ 1 & 0 \end{pmatrix} = \begin{pmatrix} 2 & 1 \\ 4 & 3 \end{pmatrix}$$

$$\begin{pmatrix} 0 & 1 \\ 1 & 0 \end{pmatrix} \begin{pmatrix} 1 & 2 \\ 3 & 4 \end{pmatrix} = \begin{pmatrix} 3 & 4 \\ 1 & 2 \end{pmatrix}$$

$$\begin{pmatrix} 1 & 3 \\ 2 & 1 \end{pmatrix} \begin{pmatrix} -1 & 3 & 2 \\ 2 & -3 & 4 \end{pmatrix} = \begin{pmatrix} 5 & -6 & 14 \\ 0 & 3 & 8 \end{pmatrix}.$$

From these examples several things are evident: 1) If you multiply an $m \times r$ matrix by an $r \times n$ matrix, the result is an $m \times n$ matrix. 2) You can multiply matrices of different sizes, but the number of columns of the matrix on the left must equal the number of columns of the matrix on the right. 3) In general AB and BA are different matrices if, indeed, they both exist. This is often phrased as "matrix multiplication is not commutative," which is different from multiplication of the real numbers so beginners should always use great care when manipulating matrices.

After you have practiced multiplying some matrices, you should become convinced that, while more complicated than real number multiplication, matrix multiplication is not very hard to do. Despite being not very hard, it is cumbersome and not very pleasant if the matrices are large and many need to be multiplied together. These two statements suggest that matrix multiplication would be ideal for a calculator or computer. Indeed many calculators and software packages are able to perform matrix multiplication, and we assume that you have access to one and have mastered its use from this point on.

In what follows, we frequently need to multiply a matrix by itself repeatedly. The analogous notation from real number multiplication carries through. The product AA is written A^2, AAA is A^3, and in general the product of n copies of A is A^n.

While dealing with notation for powers of matrices, we should mention the transpose operation. The *transpose* of a matrix is a new matrix which is the same as the old except the rows and columns have been switched. The transpose of an $n \times m$ matrix is then an $m \times n$ matrix. The notation for the transpose of A is A^T. We always use a capital T for transpose. Later on we have occasion to write A^t which is A multiplied by itself t times. Our main use of the transpose is to simplify the typesetting. When referring to a column vector, we usually write it in transposed form $(x_1, \ldots, x_n)^T$.

The last operation we consider at this time is division. Strictly speaking, division is undefined, but we define an equivalent. When working with reals we have both 2 and $1/2$. These numbers are called multiplicative inverses of each other because $2 \times 1/2 = 1$ and $1/2 \times 2 = 1$. In an abstract algebra course, we express the fact that $1/2$ is the multiplicative inverse of 2 by writing it as 2^{-1}. Similarly $2 = (1/2)^{-1}$. Here the symbol "-1" can be interpreted either as physically inverting the number, or as the abstract inverse, i.e., 2^{-1} is the number which when multiplied by 2 gives 1. If you have not thought about it before, notice that all division can be expressed as multiplying by a multiplicative inverse. This is the idea that is carried over into matrix multiplication. The symbol I will denote any square matrix (same number of rows as columns) with ones on the diagonal and zeros elsewhere and is called an *identity matrix*. For example the 3×3 identity matrix is the matrix

$$I = \begin{pmatrix} 1 & 0 & 0 \\ 0 & 1 & 0 \\ 0 & 0 & 1 \end{pmatrix}.$$

This matrix plays the role of "1" in matrix multiplication. You should verify that $AI = A$ and $IA = A$. Because of this, the identity matrix is frequently called the multiplicative identity. The *inverse of a matrix* A is denoted by A^{-1}. The matrix A^{-1} is defined to be the matrix such that $AA^{-1} = I$ and $A^{-1}A = I$ if it exists. If you followed the discussion about 2 and $1/2$, then you should see this as exactly analogous. The existence of A^{-1} is an issue. By the way A^{-1} is defined, it needs to multiply A on both sides, so A has to be a square matrix for A^{-1} to exist. Further there is a number associated with a matrix called its *determinant*. Finding a determinant of a large matrix by hand is computationally intensive. Fortunately most calculators and software that deal with matrices have the capability of evaluating a determinant. A brief discussion of how to find a determinant by hand is found in Exercise 3. A matrix has an inverse if and only if the value of its determinant is non-zero. You can think of a matrix with a zero determinant as being kin to the real number zero which also has no multiplicative inverse.

Finding the inverse of a matrix is a non-trivial, though completely algorithmic, process. We do not discuss the technique for finding one by hand and refer the interested reader to any elementary linear algebra text. As with matrix multiplication, computing inverses is rarely done by hand, and we assume that the reader has access to software or a calculator which is capable of doing these computations.

The idea of an *inner product* or *scalar product* is used briefly in Chapter 4. The inner product is a method for multiplying two vectors. If U and V are vectors of length n, then the inner product (denoted by angle brackets) is

$$<U,V> = \sum_{i=1}^{n} u_i v_i.$$

If U and V are thought of as row vectors, then the inner product can be expressed in terms of matrix multiplication

$$<U,V> = UV^T.$$

Notice that the result of multiplying two vectors together in this manner is a scalar, not a vector or matrix. For this reason, it is sometimes called a scalar product.

At the beginning of this section, we mentioned that state models all have similar equations. We conclude this section by describing how such models are formulated using matrices. Recall that these models have equations of the form $x_i(t+1) = a_{i1}x_1(t) + a_{i2}x_2(t) + \cdots + a_{in}x_n(t)$ for each of the n states. Let $X(t)$ be the column vector $(x_1, x_2, \ldots, x_n)^T$. Let

$$A = \begin{pmatrix} a_{11} & a_{12} & \cdots & a_{1n} \\ a_{21} & a_{22} & \cdots & a_{2n} \\ \vdots & & & \vdots \\ a_{n1} & a_{n2} & \cdots & a_{nn} \end{pmatrix}.$$

We leave it as an exercise in matrix multiplication to show that $X(t+1) = AX(t)$ gives the system of equations we desire. This illustrates the power of matrices to express cumbersome equations with just a few symbols.

3.5 Applying Matrices to State Models

This section applies matrix equations to the human population model. From the state diagram (Figure 3.2), we have three states, so the state vector is

$$X(n) = (x_1(n), x_2(n), x_3(n)).$$

We also call this the age distribution vector. Strictly speaking we should write this as a column vector

$$X(n) = \begin{pmatrix} x_1(n) \\ x_2(n) \\ x_3(n) \end{pmatrix}$$

in order for the matrix multiplication to work correctly, but we write it in its transposed form in the body of the text to save space. We also need an initial age distribution $X(0)$. Recall that we started with a population of 90 million females uniformly distributed among the three age classes. Thus

$$X(0) = (30, 30, 30)^T$$

3 Stages, States, and Classes

Recall that the following system of equations described our model.

$$x_1(n+1) = 0.4271 x_1(n) + 0.8498 x_2(n) + 0.1273 x_3(n) \tag{59}$$

$$x_2(n+1) = 0.9924 x_1(n) \tag{60}$$

$$x_3(n+1) = 0.9826 x_2(n) \tag{61}$$

This is exactly the form discussed at the beginning and end of the Primer of Matrix Algebra section. In particular, if we define A to be the matrix

$$\begin{pmatrix} 0.4271 & 0.8498 & 0.1273 \\ 0.9924 & 0.0 & 0.0 \\ 0.0 & 0.9826 & 0.0 \end{pmatrix},$$

then we can write the system of equations very succinctly as the single matrix equation $X(n+1) = AX(n)$. Notice that this has nearly the same form as the first difference equation we studied in Chapter 1, namely $x(n+1) = ax(n)$. The difference now is that the capital letters represent matrices. The similarity in form suggests that we might approach the analysis of this matrix equation in much the same way that we analyzed the difference equation. This is examined in great detail in the next section which is on eigenvector and eigenvalue analysis. Right now we recall that if $x(n+1) = ax(n)$, then $x(n+1) = a^{n+1}x(0)$. The same analysis works here using matrices and matrix algebra. We know that $X(1) = AX(0)$ and $X(2) = AX(1)$, so $X(2) = AAX(0) = A^2 X(0)$. Thus if $X(n) = AX(n-1)$ for any $n > 0$, it follows that $X(n+1) = AX(n) = AA^{n-1}X(0) = A^n X(0)$. So by induction $X(n) = A^n X(0)$ for all non-negative integers n.

The form of the matrix A is, of course, independent of the particulars of humans. This form is observed in any population divided into equal age classes, where the time step is the length of the age class and we know survival and reproduction rates for each age class. In general if there are n age classes, survival probabilities $s_{i,i+1}$ for the ith class surviving to the $(i+1)$th age class, and birth rates b_i giving the reproduction rate for each age class, then we form the matrix

$$A = \begin{pmatrix} b_1 & b_2 & b_3 & \ldots & b_{n-1} & b_n \\ s_{12} & 0 & 0 & \ldots & 0 & 0 \\ 0 & s_{23} & 0 & \ldots & 0 & 0 \\ \vdots & & & & & \vdots \\ 0 & 0 & 0 & \ldots & s_{n-1,n} & 0 \end{pmatrix}.$$

Matrices of this form are called Leslie matrices after P. Leslie who introduced them in a landmark paper in 1945 [37].

Also note that we can formulate the transition matrix directly from the state diagram without explicitly writing down the system of linear equations. Simply put, the s_{ij} entry is the weight assigned to the arrow from state i to state j.

In the next section, we analytically study the matrix equation $X(n) = A^n X(0)$. We could analyze this numerically using computer software for the human female data, but the results would be identical to the results we obtained by using the spreadsheet. We forego this at this time, but do an analysis of this type in the teasel example below.

3.6 Eigenvector and Eigenvalue Analysis

As mentioned in the preceding section, the matrix equation $X(n+1) = AX(n)$ has the same form as the difference equation $x(n+1) = ax(n)$ studied in Chapter 1. There we found that we could solve this equation to get $x(n) = a^n x(0)$. At that time, we called the letter a an eigenvalue or growth constant. We then proceeded to give a scheme for analyzing linear difference equations. This resulted in multiple growth rates or eigenvalues, and we discussed short-term and long-term effects of these; in particular, the long-term growth rate was determined by the dominant eigenvalue. We follow a very similar analysis here, only now working with the matrix equation $X(t+1) = AX(t)$. We will do a similar analysis again when we study continuous models needing differential equations in Chapter 5. One of the objectives of this book is to identify mathematical threads which cross back and forth between fields. Solving linear equations of various types is one such thread. Typically the solution to the difference equations would be seen in a course in either difference equations or discrete dynamical systems. What we are about to explain would be seen in a course in linear algebra, and the solution of systems of linear differential equations would be seen, naturally, in a course in differential equations.

As in Chapter 1, we need to find eigenvalues. Unlike Chapter 1, we need to define a new object, called an eigenvector. An *eigenvector* of a matrix A is a vector V for which there is a scalar λ such that the equation $AV = \lambda V$ is true. The scalar λ is called an eigenvalue associated with the eigenvector V. While this equation gives us the definition of an eigenvalue and its associated eigenvectors, it may not be clear what eigenvectors and eigenvalues represent. The ideas of population models will provide a context for understanding eigenvectors and eigenvalues. If A is a matrix describing how a population changes from year to year and V is an age distribution vector, we see that an eigenvector is a stable population distribution and that the eigenvalue is the growth rate for the population. In other words, if the population is currently distributed according to the proportions given by an eigenvector V, then AV, the population next year is λV. This has the same proportions, but all the age classes are multiplied by a factor λ. The reader should verify that, if $X(0) = V$, then $X(t) = \lambda^t X(0)$. Consequently if we start out with an eigenvector to begin with, the population remains in the same proportions but grows with a growth rate λ. We recall that we observed this behavior with the human population model. At first the age classes grew erratically, but eventually they stabilized and remained in the same proportions, namely eigenvector proportions (see Figure 3.1).

Finding eigenvectors and eigenvalues is a matter of solving the equation $AV = \lambda V$. We outline the procedure, but will not actually do it by hand. This is a procedure at which computers and calculators are fairly competent. We subtract λV from both sides to get $AV - \lambda V = \mathbf{0}$. Next we factor out V which gives $(A - \lambda I)V = \mathbf{0}$. The I is a necessary part of the factoring if the matrix multiplications are to make sense. We can solve this by inverting the matrix $A - \lambda I$ and get $V = \mathbf{0}$ which is not an interesting solution. The only way to get non-zero solutions for V is to force $A - \lambda I$ not to have an inverse, which is the case when the value of the determinant of this matrix is equal to zero. Setting the determinant equal to zero gives us a polynomial equation with λ as its variable. We then solve for λ (usually getting more than one value). After finding the λ, we substitute them, one at a time, into the equation $(A - \lambda I)V = \mathbf{0}$ to find V. Thus

3 Stages, States, and Classes

eigenvectors are associated with eigenvalues. This process is taught in a beginning linear algebra class. It can be done by hand easily for 2×2 and occasionally for 3×3 matrices. Implicit in the procedure of finding the eigenvectors of an $n \times n$ matrix is solving an nth degree polynomial which is hard or impossible for n greater than 3. So while it may be advantageous to understand what is going on in the eigenvector-eigenvalue solving procedure, in practice it is rare to find them without using computers or calculators.

An important point is worth mentioning. If V is an eigenvector of λ, then so is any scalar times V. Further if $V_1, \ldots, V_n$ are eigenvectors associated with λ, then so is any linear combination of these vectors. If these facts are not obvious, the reader should take a few moments to figure out their simple proofs. So to be precise, we should never say "the" eigenvector of A associated with λ, rather "an" eigenvector of A associated with λ.

Throughout mathematics there are ideas which are generally true, but which fail in specific cases. We make some assumptions here which are valid for most matrices one would run across in modeling situations, but which do not hold for all matrices. The reader should consult a text in linear algebra for details on working with more pathological matrices. The primary assumption is that all $n \times n$ matrices under consideration have n distinct eigenvectors. If this is the case, any state vector can be expressed as a linear combination of these n eigenvectors. (A linear combination is the sum of each vector times a scalar). The initial population vector is expressible in terms of the eigenvectors as

$$X(0) = c_1 V_1 + c_2 V_2 + \cdots + c_n V_n.$$

Since V_i is an eigenvector, $AV_i = \lambda_i V_i$ and, as we saw above, $A^n V_i = \lambda_i^n V_i$. Thus $X(t) = A^t n X(0)$, implying that

$$X(t) = c_1 \lambda_1^t V_1 + c_2 \lambda_2^t V_2 + \cdots + c_n \lambda^t V_n.$$

Compare this to the solution of a linear difference equation in Chapter 1. The only difference is the introduction of the eigenvectors, which makes sense as we are now in a matrix setting.

The effect of each eigenvector on the whole is determined by its eigenvalue. If $\lambda_i > 1$, then λ_i^t grows exponentially. If $\lambda_i = 1$, then $\lambda_i^t = 1$. If $0 < \lambda_i < 1$, then λ_i^t decays exponentially. If $\lambda_i < 0$, then λ_i^t oscillates between positive and negative values, but grows if $\lambda_i < -1$ (undamped oscillations), oscillates between 1 and -1 if $\lambda_i = -1$, and decays if $0 > \lambda_i > -1$ (damped oscillations). If λ_i is complex, then λ_i^t oscillates and grows if the modulus of λ_i is greater than one and decays if the modulus of λ_i is less than one.

The key then is $|\lambda_i|$. If $|\lambda_i| > 1$, there is exponential growth. If $|\lambda_i| < 1$, there is exponential decay.

We turn now to the long-term or asymptotic behavior of $X(t)$. We have seen that we can decompose $X(t)$ into pieces determined by the eigenvectors, namely $X(t) = c_1 \lambda_1^t V_1 + c_2 \lambda_2^t V_2 + \cdots + c_n \lambda_n^t V_n$. We studied how each piece behaves. Given two equations $x(t) = a^t$ and $y(t) = b^t$, we learn in calculus that, if $|a| \geq |b|$, then the $x(t)$ graph eventually dominates the $y(t)$ graph and the domination is so complete that the function $f(t) = a^t + b^t$ becomes indistinguishable from the graph of $x(t)$. Our situation is the same, though with, perhaps, more exponential terms in the sum. Still the dominant growth rate (largest eigenvalue in absolute value or modulus) λ_d wins out eventually. This also implies that the vector $X(t)/\lambda_d^t$ converges to the eigenvector associated with λ_d. The existence of dominant eigenvalues is beyond our scope, and the interested reader is invited to investigate the Perron-Frobenius theorem which can be found in many sources, one of which is *Matrix Population Models* by Caswell [9].

We conclude this section by applying the ideas to the human female population model. Recall that the Leslie matrix A for this model is

$$\begin{pmatrix} 0.4271 & 0.8498 & 0.1273 \\ 0.9924 & 0.0 & 0.0 \\ 0.0 & 0.9826 & 0.0 \end{pmatrix}.$$

The eigenvalues of A are $\lambda_1 = 1.20934$, $\lambda_2 = -0.615455$, and $\lambda_3 = -0.166782$. The dominant eigenvalue is 1.2093 which says that after the population distribution stabilizes, the system grows at a rate of 0.2093. An eigenvector associated with this eigenvalue is

$$V = \{-0.6871, -0.5639, -0.4581\}$$

which normalized (scaled so the components sum to one) gives

$$\{0.4020, 0.3299, 0.2681\}.$$

Note that these are the proportions found by the spreadsheet analysis in Section 3.1.

Example 3.1. The Florida sandhill crane In the first two chapters, we built models based on demographic data for the Florida sandhill crane. In both chapters, the survival and birth rates were based on an average for the population as a whole. In this section, we extend this example based on the additional data presented in Table 3.1 (from Cox et al. [13]) and the ideas presented above.

We will be looking at only the best-case data. The worst-case data can be investigated on your own.

There are several points to notice in this data. As with the human population model, the birth rates are for the female portion of the population only. This makes sense considering that only the females give birth. The model we construct is for only the female portion of the population which is why the birth rates are roughly half of the aggregate rate used in the first two chapters. We assume equal sex ratios at birth and that male and female survival rates are the same. With these assumptions, we get the whole population by multiplying the female population by two. If these assumptions are incorrect, we need a separate submodel for the males. Secondly, there are 20 age classes, each of one year's duration. The data provides a survival rate and birth rate for each of these age classes. As one might expect, the first year of life is hard, with both reproduction and survival low. After that there is a general increase in reproduction rate to a certain level. After a jump in the second year, there is a general decline in survival rates. The average generation length for cranes is eight years, so 20 age classes allows for individuals to grow old, but we assume no individual sandhill crane lives past the age of 20.

3 Stages, States, and Classes

Age Class	Survival Worst Case	Survival Best Case	Reproduction Worst Case	Reproduction Best Case
1	0.60	0.63	0.00	0.00
2	0.88	0.93	0.00	0.00
3	0.93	0.98	0.01	0.01
4	0.93	0.98	0.04	0.04
5	0.92	0.97	0.16	0.17
6	0.92	0.96	0.26	0.27
7	0.91	0.95	0.26	0.27
8	0.90	0.94	0.26	0.27
9	0.89	0.91	0.26	0.27
10	0.86	0.89	0.26	0.27
11	0.84	0.89	0.26	0.27
12	0.84	0.89	0.26	0.27
13	0.84	0.87	0.26	0.27
14	0.77	0.81	0.26	0.27
15	0.72	0.76	0.26	0.27
16	0.65	0.68	0.26	0.27
17	0.48	0.51	0.26	0.27
18	0.47	0.50	0.26	0.27
19	0.47	0.50	0.26	0.27
20	0.47	0.50	0.26	0.27

TABLE 3.1 Age Structured Parameters for the Sandhill Crane. Reproduction Reflects the Number of Females Born on Average to Each Adult Female Each Year.

A state diagram for the sandhill crane is shown in Figure 3.6. Following the discussion for converting diagrams to equations, we get the following system of equations.

$$x_1(t+1) = 0.01x_3(t) + 0.04x_4(t) + 0.17x_5(t) + 0.27x_6(t) + 0.27x_7(t)$$
$$+ 0.27x_8(t) + 0.27x_9(t) + 0.27x_{10}(t) + 0.27x_{11}(t) + 0.27x_{12}(t)$$
$$+ 0.27x_{13}(t) + 0.27x_{14}(t) + 0.27x_{15}(t) + 0.27x_{16}(t)$$
$$+ 0.27x_{17} + 0.27x_{18}(t)(t) + 0.27x_{19}(t) + 0.27x_{20}(t)$$

$$x_2(t+1) = 0.63x_1(t)$$
$$x_3(t+1) = 0.93x_2(t)$$
$$x_4(t+1) = 0.98x_3(t)$$
$$x_5(t+1) = 0.98x_4(t)$$
$$\vdots$$
$$x_{17}(t+1) = 0.68x_{16}(t)$$
$$x_{18}(t+1) = 0.51x_{17}(t)$$
$$x_{19}(t+1) = 0.5x_{18}(t)$$
$$x_{20}(t+1) = 0.5x_{19}(t).$$

3.6 Eigenvector and Eigenvalue Analysis

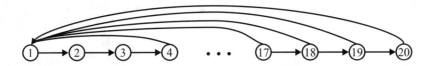

FIGURE 3.6 State Diagram For the Florida Sandhill Crane.

Before these equations are used, we need an initial population distribution. We start with a population of 40 females uniformly distributed among the 20 states, thus

$$X(0) = \{2,2,2,2,2,2,2,2,2,2,2,2,2,2,2,2,2,2,2,2\}.$$

A spreadsheet projection is shown in Figures 3.7 and 3.8. The graphs of the total population are shown in Figures 3.9 and 3.10. Notice the initial population decline followed by the exponential growth. This type of behavior is impossible to observe in a model which lumps all the age classes together. We have also included a row listing the percentage growth rate. After approximately 30 years this growth rate stabilizes at 2.68%.

Florida Sandhill Crane Demographic Data											
Age Class	Birth Rate	Survival Rate	t	0	1	2	3	4	5	6	7
1	0.00	0.63	2.00	8.54	6.93	6.03	5.49	5.14	5.28	5.70	
2	0.00	0.93	2.00	1.26	5.38	4.36	3.80	3.46	3.24	3.33	
3	0.01	0.98	2.00	1.86	1.17	5.00	4.06	3.53	3.21	3.01	
4	0.04	0.98	2.00	1.96	1.82	1.15	4.90	3.98	3.46	3.15	
5	0.17	0.97	2.00	1.96	1.92	1.79	1.13	4.81	3.90	3.39	
6	0.27	0.96	2.00	1.94	1.90	1.86	1.73	1.09	4.66	3.78	
7	0.27	0.95	2.00	1.92	1.86	1.83	1.79	1.66	1.05	4.47	
8	0.27	0.94	2.00	1.90	1.82	1.77	1.73	1.70	1.58	1.00	
9	0.27	0.91	2.00	1.88	1.79	1.71	1.66	1.63	1.60	1.49	
10	0.27	0.89	2.00	1.82	1.71	1.63	1.56	1.51	1.48	1.45	
11	0.27	0.89	2.00	1.78	1.62	1.52	1.45	1.39	1.35	1.32	
12	0.27	0.89	2.00	1.78	1.58	1.44	1.36	1.29	1.24	1.20	
13	0.27	0.87	2.00	1.78	1.58	1.41	1.28	1.21	1.15	1.10	
14	0.27	0.81	2.00	1.74	1.55	1.38	1.23	1.12	1.05	1.00	
15	0.27	0.76	2.00	1.62	1.41	1.25	1.12	0.99	0.90	0.85	
16	0.27	0.68	2.00	1.52	1.23	1.07	0.95	0.85	0.76	0.69	
17	0.27	0.51	2.00	1.36	1.03	0.84	0.73	0.65	0.58	0.51	
18	0.27	0.5	2.00	1.02	0.69	0.53	0.43	0.37	0.33	0.29	
19	0.27	0.5	2.00	1.00	0.51	0.35	0.26	0.21	0.19	0.17	
20	0.27			2.00	1.00	0.50	0.26	0.17	0.13	0.11	0.09
		Year		0	1	2	3	4	5	6	7
		Total		40.00	39.64	38.02	37.17	36.82	36.71	37.10	37.98
		Pecent Growth Rate			-0.90%	-4.08%	-2.24%	-0.93%	-0.30%	1.04%	2.39%

FIGURE 3.7 Spreadsheet for the Florida Sandhill Crane.

3 Stages, States, and Classes

8	9	10	11	12	13	14	15	16	17	18	19	20	21
5.97	6.17	6.33	6.47	6.64	6.84	7.06	7.29	7.51	7.71	7.90	8.07	8.25	8.47
3.59	3.76	3.89	3.99	4.08	4.18	4.31	4.45	4.59	4.73	4.86	4.98	5.08	5.20
3.09	3.34	3.50	3.62	3.71	3.79	3.89	4.00	4.14	4.27	4.40	4.52	4.63	4.73
2.95	3.03	3.27	3.43	3.54	3.63	3.72	3.81	3.92	4.05	4.19	4.31	4.43	4.54
3.09	2.89	2.97	3.20	3.36	3.47	3.56	3.64	3.73	3.85	3.97	4.10	4.22	4.34
3.29	2.99	2.81	2.88	3.11	3.26	3.37	3.45	3.53	3.62	3.73	3.85	3.98	4.10
3.63	3.16	2.88	2.69	2.77	2.98	3.13	3.23	3.32	3.39	3.48	3.58	3.70	3.82
4.25	3.45	3.00	2.73	2.56	2.63	2.83	2.97	3.07	3.15	3.22	3.30	3.40	3.51
0.94	4.00	3.24	2.82	2.57	2.41	2.47	2.66	2.79	2.89	2.96	3.03	3.11	3.20
1.35	0.85	3.64	2.95	2.57	2.34	2.19	2.25	2.42	2.54	2.63	2.69	2.76	2.83
1.29	1.20	0.76	3.24	2.62	2.28	2.08	1.95	2.00	2.16	2.26	2.34	2.40	2.45
1.17	1.15	1.07	0.67	2.88	2.34	2.03	1.85	1.73	1.78	1.92	2.01	2.08	2.13
1.07	1.05	1.02	0.95	0.60	2.56	2.08	1.81	1.65	1.54	1.58	1.71	1.79	1.85
0.96	0.93	0.91	0.89	0.83	0.52	2.23	1.81	1.57	1.43	1.34	1.38	1.49	1.56
0.81	0.78	0.75	0.74	0.72	0.67	0.42	1.81	1.47	1.27	1.16	1.09	1.12	1.20
0.65	0.61	0.59	0.57	0.56	0.55	0.51	0.32	1.37	1.11	0.97	0.88	0.83	0.85
0.47	0.44	0.42	0.40	0.39	0.38	0.37	0.35	0.22	0.93	0.76	0.66	0.60	0.56
0.26	0.24	0.22	0.21	0.20	0.20	0.19	0.19	0.18	0.11	0.48	0.39	0.34	0.31
0.15	0.13	0.12	0.11	0.11	0.10	0.10	0.10	0.10	0.09	0.06	0.24	0.19	0.17
0.08	0.07	0.07	0.06	0.06	0.05	0.05	0.05	0.05	0.05	0.04	0.03	0.12	0.10
8	9	10	11	12	13	14	15	16	17	18	19	20	21
39.05	40.24	41.44	42.63	43.87	45.19	46.59	48.00	49.37	50.69	51.91	53.16	54.51	55.92
2.82%	3.04%	2.99%	2.87%	2.89%	3.02%	3.11%	3.01%	2.86%	2.68%	2.40%	2.41%	2.54%	2.58%

22	23	24	25	26	27	28	29	30	31	32	33	34	35
8.70	8.95	9.20	9.45	9.71	9.97	10.24	10.51	10.79	11.08	11.38	11.68	12.00	12.32
5.34	5.48	5.64	5.80	5.96	6.12	6.28	6.45	6.62	6.80	6.98	7.17	7.36	7.56
4.84	4.96	5.10	5.24	5.39	5.54	5.69	5.84	6.00	6.16	6.32	6.49	6.67	6.85
4.63	4.74	4.86	5.00	5.14	5.28	5.43	5.58	5.72	5.88	6.03	6.20	6.36	6.53
4.45	4.54	4.64	4.77	4.90	5.03	5.18	5.32	5.46	5.61	5.76	5.91	6.07	6.24
4.21	4.31	4.40	4.51	4.62	4.75	4.88	5.02	5.16	5.30	5.44	5.59	5.74	5.89
3.93	4.04	4.14	4.23	4.33	4.44	4.56	4.69	4.82	4.95	5.09	5.22	5.36	5.51
3.63	3.74	3.84	3.93	4.02	4.11	4.22	4.33	4.45	4.58	4.71	4.83	4.96	5.09
3.30	3.41	3.51	3.61	3.70	3.77	3.86	3.96	4.07	4.19	4.31	4.42	4.54	4.66
2.91	3.01	3.10	3.20	3.28	3.36	3.43	3.51	3.61	3.70	3.81	3.92	4.03	4.13
2.52	2.59	2.67	2.76	2.85	2.92	2.99	3.06	3.13	3.21	3.30	3.39	3.49	3.58
2.18	2.24	2.31	2.38	2.46	2.53	2.60	2.66	2.72	2.78	2.86	2.93	3.02	3.10
1.90	1.94	1.99	2.05	2.12	2.19	2.25	2.31	2.37	2.42	2.48	2.54	2.61	2.69
1.61	1.65	1.69	1.73	1.79	1.84	1.90	1.96	2.01	2.06	2.11	2.16	2.21	2.27
1.26	1.31	1.34	1.37	1.40	1.45	1.49	1.54	1.59	1.63	1.67	1.71	1.75	1.79
0.92	0.96	0.99	1.02	1.04	1.07	1.10	1.13	1.17	1.21	1.24	1.27	1.30	1.33
0.58	0.62	0.65	0.67	0.69	0.71	0.73	0.75	0.77	0.80	0.82	0.84	0.86	0.88
0.29	0.29	0.32	0.33	0.34	0.35	0.36	0.37	0.38	0.39	0.41	0.42	0.43	0.44
0.15	0.14	0.15	0.16	0.17	0.17	0.18	0.18	0.19	0.19	0.20	0.20	0.21	0.21
0.08	0.08	0.07	0.07	0.08	0.08	0.09	0.09	0.09	0.09	0.10	0.10	0.10	0.10
22	23	24	25	26	27	28	29	30	31	32	33	34	35
57.43	59.01	60.63	62.28	63.97	65.69	67.46	69.27	71.13	73.04	75.00	77.01	79.07	81.19
2.70%	2.75%	2.75%	2.73%	2.71%	2.70%	2.69%	2.69%	2.69%	2.68%	2.68%	2.68%	2.68%	2.68%

FIGURE 3.8 Spreadsheet for the Florida Sandhill Crane (continued).

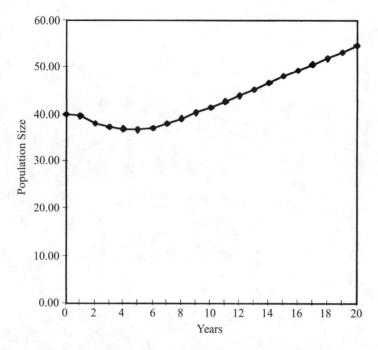

FIGURE 3.9 Graphs of Florida Sandhill Crane Projections for 20 Years.

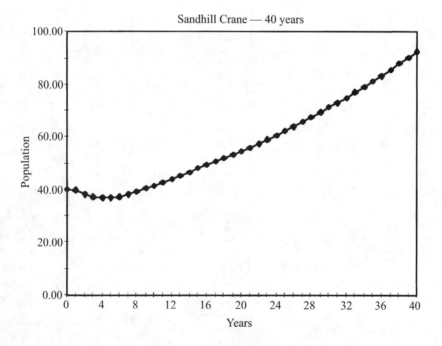

FIGURE 3.10 Graphs of Florida Sandhill Crane Projections for 40 Years.

3 Stages, States, and Classes

Next we look at some of the matrix analysis we discussed. The transition matrix for the cranes is

$$\begin{pmatrix} 0.0 & 0.0 & 0.01 & 0.04 & \ldots & 0.27 & 0.27 \\ 0.63 & 0.0 & 0.0 & 0.0 & \ldots & 0.0 & 0.0 \\ 0.0 & 0.93 & 0.0 & 0.0 & \ldots & 0.0 & 0.0 \\ 0.0 & 0.0 & 0.98 & 0.0 & \ldots & 0.0 & 0.0 \\ \vdots & & & & & & \vdots \\ 0.0 & 0.0 & 0.0 & 0.0 & \ldots & 0.5 & 0.0 \end{pmatrix}.$$

The eigenvalues of this matrix are shown in Table 3.2.

Examining the eigenvalues in Table 3.2, we see that all but two are complex. The dominant eigenvalue in this case is the positive real one, so $\lambda_d = 1.02686$. This implies that the population as a whole is growing and that the growth rate is 2.686%. The full list of eigenvectors fills up a lot of space so we report only the dominant eigenvector which is

$$V = \begin{pmatrix} 0.546 & 0.335 & 0.304 & 0.290 & 0.277 & 0.262 & 0.245 & 0.227 & 0.207 & 0.183 \\ 0.159 & 0.138 & 0.119 & 0.101 & 0.080 & 0.059 & 0.039 & 0.019 & 0.009 & 0.005 \end{pmatrix}.$$

These are the ratios into which the various age classes eventually stabilize.

We direct the reader's attention back to the spreadsheet analysis of the sandhill crane. There we computed the growth rate for the population as a whole. Notice that it is converging to 2.68% which is the dominant eigenvalue (minus 1 of course). Thus by computing the eigenvalues, we predict the overall growth rate without actually tracking the population year by year. Also notice in the spreadsheet that the age class distribution is approaching the distribution given by the eigenvector associated with the dominant eigenvalue. To see this, both vectors must be standardized. The most common way to do this is to divide each vector by the sum of its entries. The result is a vector which gives the proportion of the total in each age class. If we look at the graph after running the model, we see that the long-term behavior is determined by the dominant eigenvalue; however, the curve is not a smooth exponential to begin with, due to the effects of all the

Eigenvalue	Eigenvalue
1.02686	0.745798 + 0.377527 i
0.745798 - 0.377527 i	0.560372 + 0.589528 i
0.560372 - 0.589528 i	-0.456006 + 0.590951 i
-0.456006 - 0.590951 i	-0.250262 + 0.701766 i
-0.250262 - 0.701766 i	-0.603457 + 0.409628 i
-0.603457 - 0.409628 i	0.340707 + 0.642955 i
0.340707 - 0.642955 i	-0.0113737 + 0.720982 i
-0.0113737 - 0.720982 i	-0.685756 + 0.204179 i
-0.685756 - 0.204179 i	-0.71228
0.202688 + 0.662535 i	0.202688 - 0.662535 i

TABLE 3.2 Eigenvalues for the Florida Sandhill Crane.

3.7 A Staged Example

Teasel *Dipsacus sylvestris* is a perennial plant which has six stages: x_1=first-year dormant seeds, x_2= second-year dormant seeds, x_3=small rosettes, x_4=medium rosettes, x_5=large rosettes, and x_6=flowering plants. A rosette is a nonflowering vegetative stage which gets its name from the leaves extending out from a central point. A state diagram for teasel is shown in Figure 3.11.

The transition matrix for this diagram is

$$A = \begin{pmatrix} 0 & 0 & 0 & 0 & 0 & 322.38 \\ 0.966 & 0 & 0 & 0 & 0 & 0 \\ 0.013 & 0.010 & 0.125 & 0 & 0 & 3.448 \\ 0.007 & 0 & 0.125 & 0.238 & 0 & 30.170 \\ 0.008 & 0 & 0 & 0.245 & 0.167 & 0.862 \\ 0 & 0 & 0 & 0.023 & 0.750 & 0 \end{pmatrix}.$$

It is hard to work with a matrix of this size by hand. The rest of this analysis was made using *Mathematica*. The notebook performing these calculations is in the *Mathematica* Appendix.

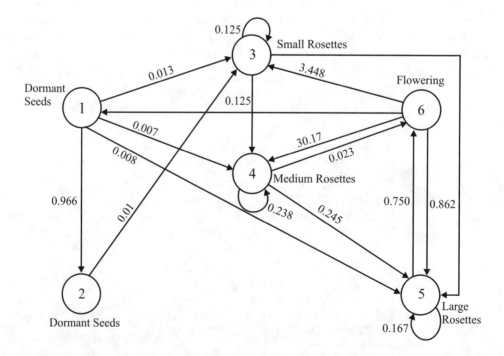

FIGURE 3.11 State diagram for teasel.

3 Stages, States, and Classes

The eigenvalues of this matrix are

$$\{2.3219, -0.9574 + 1.4886\,i, -0.9574 - 1.4886\,i,$$
$$0.1424 + 0.1980\,i, 0.1424 - 0.1980\,i, -0.1619\}.$$

To find the dominant eigenvalue, we find the modulus of each eigenvalue. The moduli of these eigenvalues are

$$\{2.3219, 1.7699, 1.7699, 0.2439, 0.2439, 0.1619\}.$$

Thus the dominant eigenvalue is the positive real one, namely 2.3219. This tells us that the teasel, in all its states (including seeds), is growing at a rate of 132% (i.e., $(2.3219 - 1)100\%$).

The eigenvectors take up considerable space, so we report only the eigenvector associated with the dominant eigenvalue.

$$X = (0.9183, 0.3821, 0.01755, 0.09990, 0.01741, 0.006614).$$

This gives the proportions of the various stages when the model stabilizes. It is easier to understand when normalized (divided by the sum of the entries so that the sum of the resulting vector is one).

$$X = (0.6369, 0.2650, 0.01217, 0.06928, 0.01208, 0.004587).$$

This tells us that eventually 64% of the teasel will be first-year dormant seeds, 26% will be second-year dormant seeds, 1.2% will be small rosettes, 6.9% will be medium rosettes, 1.2% will be large rosettes, and 0.4% will be in a flowering state.

While we know this to be true, we conclude by verifying it. We start with the various stages uniformly distributed (i.e., $X(0) = (10, 10, 10, 10, 10, 10)$), and look at the distribution after a long time, say 100 years. These numbers are big, as predicted by the exponential nature of the matrix equation, and presumably the modeling assumptions no longer hold; but remember we are doing this analysis just to verify the mathematics.

$$X(100) = (2.5790\,10^{39}, 1.0730\,10^{39}, 4.9299\,10^{37}, 2.8054\,10^{38}, 4.8901\,10^{37}, 1.8575\,10^{37}).$$

Normalizing, i.e., computing,

$$\frac{X(100)}{\sum_{i=1}^{6} X_i(100)}$$

gives

$$(0.6369, 0.2650, 0.01217, 0.06928, 0.01208, 0.004587)$$

which agrees with the predicted result to the number of decimal places reported. (In the *Mathematica* Appendix these results are presented to more decimal places, and the results are seen to be in close agreement though they are not identical. Again we expect some deviation in the answer since we are using $X(100)$ and not $X(\infty) = \lim_{n \to \infty} X(n)$.)

3.8 Fundamentals of Markov Chains

A Markov chain is a special class of state model. As with earlier state models, it consists of a collection of states, only now we are modeling probabilities of transitions between states. The weight assigned to each arrow is now interpreted as either the probability that something in the state at the arrow's tail moves to the state at the arrow's head, or the percentage of things at the arrow's tail which move to the state at the arrow's head. At each time step, something in one state must either remain where it is or move to another state. Thus the sum of the arrows into (or out of) a state must be one. The state vector $X(t)$ in a Markov model traditionally lists either the probability that a system is in a particular state at a particular time, or the percentage of the system which is in each state at a given time. Thus $X(t)$ is a probability distribution vector and must sum to one. We have occasionally mentioned such vectors in what we have done before, but when dealing with a Markov model we deal with probability distribution vectors exclusively. Recapping, there are three properties which identify a state model as being a Markov model: 1) The Markov assumption: the probability of one's moving from state i to state j is independent of what happened before moving to state j and of how one got to state i. 2) Conservation: the sum of the probabilities out of a state must be one. 3) The vector $X(t)$ is a probability distribution vector which describes the probability of the system's being in each of the states at time n.

In some sense, we have been assuming the Markov assumption all along. By this we mean that we have been assuming that the number being assigned to a state during a time step depends only on the way things were distributed during the prior time step and not any further back than that. This was the fourth convention we made when defining state diagrams. Essentially it says that we are considering only first-order recurrence relations. Strictly speaking the Markov assumption refers to only probabilities, but we used equivalents of it with birth rates that were greater than one. When discussing the probabilities associated with a Markov chain, the term conditional probability is often used. Conditional probability means just the probability of something's happening given that something else has already happened. In our case the probability of moving from state i to state j assumes we were in state i to begin with, so, technically, this is a conditional probability.

The transition matrix for a Markov chain is then a matrix of probabilities (conditional probabilities if we are perfectly correct) of moving from one state to another. Thus

$$T = \begin{pmatrix} p_{11} & p_{21} & \cdots & p_{n1} \\ p_{12} & p_{22} & \cdots & p_{n2} \\ \vdots & \vdots & \ddots & \vdots \\ p_{1n} & p_{2n} & \cdots & p_{nn} \end{pmatrix}.$$

We also require that the each column sums to one in order to satisfy the conservation property. The system moves from states given by column indices to states given by row indices. For example, p_{21} is the probability of the system's moving from state 2 to state 1.

We can represent a Markov chain using a state diagram (Figure 3.12).

The transition probabilities p_{ij} are shown as the flows between states.

3 Stages, States, and Classes

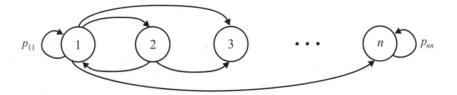

FIGURE 3.12 General State diagram of a Markov Chain.

Consider the following transition matrix for a Markov chain:

$$T = \begin{pmatrix} 0.50 & 0.30 & 0.10 \\ 0.30 & 0.60 & 0.10 \\ 0.20 & 0.10 & 0.80 \end{pmatrix}.$$

There are three states for this chain, which we label $i = 1, 2, 3$. The state diagram for this chain is shown in Figure 3.13.

Unlike models discussed earlier, the vector $X(t)$ does not give the number of individuals in each state at time t; rather it gives the probability that the system is in each state at time t. It is conventional with Markov chains to denote $X(t)$ as X_t. An initial distribution X_0, is a distribution for the chance that the system is initially in each of the states. For instance, suppose

$$X_0 = \begin{pmatrix} 0.50 \\ 0.30 \\ 0.20 \end{pmatrix}.$$

The interpretation X_0 is that there is a 50% chance the system is initially in state 1, 30% chance it is in state 2, and a 20% chance it is in state 3.

In this context, matrix multiplication gives the probability distribution one time step later. That is,

$$X_1 = TX_0,$$

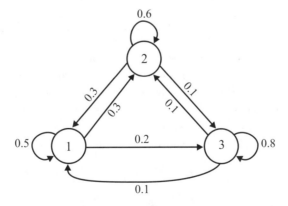

FIGURE 3.13 State diagram for the Markov Chain.

where X_0 is an initial distribution. Using the transition matrix and initial distribution from above, we have

$$X_1 = TX_0 = \begin{pmatrix} 0.50 & 0.30 & 0.10 \\ 0.30 & 0.60 & 0.10 \\ 0.20 & 0.10 & 0.80 \end{pmatrix} \begin{pmatrix} 0.50 \\ 0.30 \\ 0.20 \end{pmatrix}$$

$$= \begin{pmatrix} 0.36 \\ 0.35 \\ 0.29 \end{pmatrix},$$

so that after one time step, there is a 36% chance of the system's being in state 1, and 35% and 29% chances of being in states 2 and 3 respectively. Using this notation, the distribution after $t = n$ time steps is given by

$$X_n = T^n X_0. \tag{62}$$

An important idea, which we make use of in the next two sections, is whether the sequence of column vectors $X_n, n \geq 1$ converges to a steady-state (unchanging from time step to time step) column vector $\overline{X}$. Determining $\overline{X}$ allows us to answer long-term behavior questions we may pose.

We observe here that if $X_n \to \overline{X}$ as $n \to \infty$, we must have the matrix T^n approaching some fixed matrix L, that is

$$\begin{array}{rcl} X_n & = & T^n X_0 \\ \downarrow & & \downarrow \\ \overline{X} & = & LX_0. \end{array} \tag{63}$$

The matrix L, if it exists, is referred to as the "steady-state" matrix. The convergence of the matrix T^n to the steady-state matrix L is independent of the initial distribution X_0, as equation (63) shows. The steady-state distribution $\overline{X}$ and the steady-state matrix L can be shown to exist, provided that the transition matrix T satisfies the property that some power has all positive entries. Matrices satisfying this condition are called *regular*.

If T is regular, we find the steady-state distribution $\overline{X}$ by solving the set of equations

$$\overline{X} = T\overline{X} \tag{64}$$

for $\overline{X}$, along with the condition that the sum of the entries in $\overline{X}$ must be one. The matrix equation (64) clearly conveys the idea that the steady-state distribution $\overline{X}$ is a fixed point of the system of equations (62). Equivalently, $\overline{X}$ is an eigenvector with eigenvalue one.

An intuitively appealing method for determining the steady-state distribution $\overline{X}$, is to compute (or approximate) the steady-state matrix L. Traditionally, this is done analytically using a method called matrix diagonalization. Since

$$T^n \to L \text{ as } n \to \infty,$$

we approximate L by computing T^n for large values of n. This is easily done using a software package, or, if the number of states is small, a calculator with matrix capabilities.

3 Stages, States, and Classes

For our example, the steady-state matrix is approximately

$$L \approx T^{100} = \begin{pmatrix} 0.2692 & 0.2692 & 0.2692 \\ 0.3077 & 0.3077 & 0.3077 \\ 0.4231 & 0.4231 & 0.4231 \end{pmatrix}. \tag{65}$$

The form of the matrix in equation (65) might at first glance appear surprising. If the steady-state matrix L exists, it has the form given in (65) where each of the columns are identical. This fact follows from the equation $\overline{X} = LX_0$ and recalling that the sum of the entries in the column vector X_0 is one. This equation also demonstrates that each column of L ($\approx T^n$, for n large) is $\overline{X}$. Thus, for our example, we have the steady-state distribution

$$\overline{X} = \begin{pmatrix} 0.2692 \\ 0.3077 \\ 0.4231 \end{pmatrix}.$$

There is another class of Markov chains which have important modeling properties. Consider the following example of a transition matrix.

$$T = \begin{pmatrix} 0.20 & 0 & 0 & 0 & 0 \\ 0.20 & 0.40 & 0.25 & 0 & 0 \\ 0.20 & 0.40 & 0.25 & 0 & 0 \\ 0.20 & 0.20 & 0.25 & 1 & 0 \\ 0.20 & 0 & 0.25 & 0 & 1 \end{pmatrix}. \tag{66}$$

The state diagram for this transition matrix is shown in Figure 3.14.

This system has some important features. States 4 and 5 are called "absorbing" states. Once the system enters an absorbing state, the system remains in that state from that time on. The absorbing states are easily identified from the state diagram in that they have loops with weight one. States 4 and 5 both have loops with weight one.

Absorbing Markov chains are different in structure than those we have previously considered. An absorbing state precludes the transition matrix from being regular. The assumption that the transition matrix is regular is enough to ensure the existence of a steady-state matrix, but is not a characterization. Steady-state matrices exist for absorbing

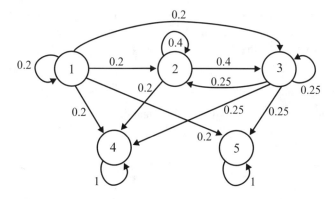

FIGURE 3.14 State diagram for the Absorbing Markov Chain.

Markov chains, and the additional structure of the absorbing chains provides useful information. The project section considers an example of a nonabsorbing, non-regular transition matrix for which a steady-state matrix can be calculated.

If we compute the steady-state matrix for the above absorbing chain, we obtain

$$L \approx T^{500} = \begin{pmatrix} 0 & 0 & 0 & 0 & 0 \\ 0 & 0 & 0 & 0 & 0 \\ 0 & 0 & 0 & 0 & 0 \\ 0.5714 & 0.7143 & 0.5714 & 1 & 0 \\ 0.4286 & 0.2857 & 0.4286 & 0 & 1 \end{pmatrix}. \tag{67}$$

This matrix exhibits several properties that we need later on. Examining the structure of the transition matrix T in equation (66), we see that it can be decomposed into blocks of the form

$$\begin{pmatrix} A_{3\times 3} & O_{3\times 2} \\ B_{2\times 3} & I_{2\times 2} \end{pmatrix}, \tag{68}$$

where $A_{3\times 3}$ is the submatrix given by

$$A_{3\times 3} = \begin{pmatrix} 0.20 & 0 & 0 \\ 0.20 & 0.40 & 0.25 \\ 0.20 & 0.40 & 0.25 \end{pmatrix},$$

the submatrix $B_{2\times 3}$ is

$$B_{2\times 3} = \begin{pmatrix} 0.20 & 0.20 & 0.25 \\ 0.20 & 0 & 0.25 \end{pmatrix},$$

and $O_{3\times 2}$ is a matrix of all zeros,

$$O_{3\times 2} = \begin{pmatrix} 0 & 0 \\ 0 & 0 \\ 0 & 0 \end{pmatrix}.$$

The matrix $I_{2\times 2}$ is just the 2×2 identity matrix, and if it was not obvious, we formed the blocks around the identity matrix block. This decomposition is always possible for absorbing Markov chains, though we may need to re-label the states so that the absorbing states are listed last (so the identity matrix is in the proper position). In general, if a Markov chain has a absorbing states and b nonabsorbing states, we can arrange the transition matrix to have the form

$$\begin{pmatrix} A_{b\times b} & O_{b\times a} \\ B_{a\times b} & I_{a\times a} \end{pmatrix}. \tag{69}$$

This block decomposition gives useful information about the absorbing Markov chain. The steady-state matrix L has the form

$$L = \begin{pmatrix} O_{b\times b} & O_{b\times a} \\ B_{a\times b}(I_{b\times b} - A_{b\times b})^{-1} & I_{a\times a} \end{pmatrix}.$$

The entries in the matrix $B_{a\times b}(I_{b\times b} - A_{b\times b})^{-1}$ represent the probability of being absorbed in ith absorbing state if the system was initially in the jth nonabsorbing state. In the

3 Stages, States, and Classes

example,

$$B_{2\times 3}(I_{3\times 3} - A_{3\times 3})^{-1} = \begin{pmatrix} 0.5714 & 0.7143 & 0.5715 \\ 0.4286 & 0.2857 & 0.4286 \end{pmatrix}.$$

These entries are viewed as "absorption" probabilities. For example, there is a 71.43% chance that the system will be absorbed in state 4, given that it initially started in state 2. To understand which state is which, refer back to the columns and rows of (67). The other entries have a similar interpretation.

Further information is obtained from the *fundamental* matrix $F_{b\times b} = (I_{b\times b} - A_{b\times b})^{-1}$. The entries $f_{i,j}$ of this matrix are the average number of times the process is in state j, given that it began in state i. A proof of this result is in Olinick [48]. For our example, the fundamental matrix is

$$F = (I - A)^{-1} = \begin{pmatrix} 1.2500 & 0 & 0 \\ 0.7143 & 2.1429 & 0.7143 \\ 0.7143 & 1.1429 & 1.7143 \end{pmatrix}. \tag{70}$$

Recalling the block form of the transition matrix (68), the position of the submatrix A indicates that i and j have values 1, 2, or 3, so that $f_{1,1} = 1.25$ is the average number of time steps that the system is in state 1, given that it was initially in state 1. The other entries have analogous interpretations. The sum of the entries of the jth column of the fundamental matrix F is the average number of time steps for a process initially in state j to be absorbed. For example, if the system is initially in state 1, it takes an average of $1.25 + 0.7143 + 0.7143 = 2.6786$ time steps before the system enters an absorbing state.

The next two sections present models based upon Markov chains and use the above analysis. The project section also contains some interesting Markov models, as well as some further points of the theory of Markov chains.

3.9 Markovian Squirrels

The American gray squirrel (*Sciurus carolinensis Gmelin*) was introduced in Great Britain by a series of releases from various sites starting in the late nineteenth century. In 1876, the first gray squirrels were imported from North America, and have subsequently spread throughout England and Wales, as well as parts of Scotland and Ireland.

Simultaneously, the native red squirrel (*Sciurus vulgaris L.*), considered the endemic subspecies, has disappeared from most of the areas colonized by gray squirrels. Originally, the red squirrel was distributed throughout Europe and eastward to northern China, Korea, and parts of the Japanese archipelago. During the last century, the red squirrel has consistently declined, becoming extinct in many areas of England and Wales, so that it is now confined almost solely to Northern England and Scotland. A few isolated red squirrel populations exist on offshore islands in southern England and mountainous Wales.

The introduction of the American gray squirrel continued until the early 1920s, by which time the gray squirrels had rapidly spread throughout England. By 1930 it was apparent that the gray squirrel was a pest in deciduous forests, and control measures were attempted. Once the pest status of the gray squirrel was recognized, national distribution surveys were undertaken. The resulting distribution maps clearly showed the tendency for the red squirrel to be lost from areas that had been colonized by the gray squirrel during the preceding 15 to 20 years.

Since 1973, an annual questionnaire has been circulated to foresters by the British Forestry Commission. The questionnaire concerns the presence or absence of the two squirrel species. It also includes questions on the changes of squirrel abundance, details of tree damage, squirrel control measures, and the number of squirrels killed. Using the data collected by the Forestry Commission, we wish to construct a model to predict the trends in the distribution of both species of squirrels in Great Britain.

Several researchers have studied the British squirrel populations, notably Reynolds [53] and Usher et al. [68]. The annual Forestry Commission data has been summarized in the form of distribution maps reflecting change over a two-year period.

Usher et al., [68] used an overlay technique to extract data from the distribution maps. Each 10-km square on the overlay map that contained Forestry Commission land is classified into one of four states:

R: only red squirrels recorded in that year.
G: only gray squirrels recorded in that year.
B: both species of squirrels recorded in that year.
O: neither species of squirrels recorded in that year.

In order to satisfy the Markov assumption, squares that were present only in two consecutive years were counted. Counting the pairs of years, squares are allocated to any one of 16 classes, e.g., **R → R, R → G, G → G, B → O**, etc.

A summary of these transition counts for each pair of years from 1973–74 to 1987–88 is given in table 3.3 and is reprinted by permission of Blackwell Science Inc.

A frequency interpretation is required to employ the Markov chain analysis. If the entries in each column are totaled, the corresponding matrix entry is found by division. For example, column **R** has a total $2{,}529 + 61 + 282 + 3 = 2{,}875$, so that the entry in the **R, R** position is $2{,}529/2{,}875 \approx 0.8797$. Care must be taken when calculating these frequencies. Inappropriate rounding will violate the requirement that the columns sum to

	R	G	B	O
R	2,529	35	257	5
G	61	733	20	91
B	282	25	4,311	335
O	3	123	310	5,930

TABLE 3.3 Red and Gray Squirrel Distribution Map Data for Great Britain.

3 Stages, States, and Classes

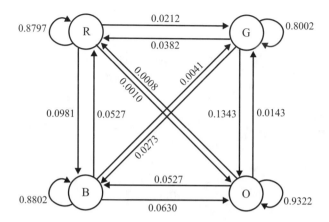

FIGURE 3.15 State diagram for the Markov Squirrels.

one. The transition matrix (rows and columns are in **R, G, B, O** order) is

$$T = \begin{pmatrix} 0.8797 & 0.0382 & 0.0527 & 0.0008 \\ 0.0212 & 0.8002 & 0.0041 & 0.0143 \\ 0.0981 & 0.0273 & 0.8802 & 0.0527 \\ 0.0010 & 0.1343 & 0.0630 & 0.9322 \end{pmatrix}.$$

The state diagram of this transition matrix T is given in Figure 3.15. We interpret these transition frequencies as conditional probabilities. For example, there is an 87.97% chance that squares that are currently in state **R** (red squirrels only) will remain in state **R**; similarly, there is a 2.73% chance of squares that are currently occupied by both squirrel species, state **G**, will become occupied by neither species, state **B**, after the next time step. Since the data taken from the annual Forestry Commission survey is summarized as pairs of years, each time step represents a two-year period. The matrix form of the transition probabilities is convenient for calculations. Using matrix multiplication, we compute the two-time-step transition matrix as $T^2 = T \times T$, which is given by

$$T^2 = \begin{pmatrix} 0.7799 & 0.0657 & 0.0930 & 0.0048 \\ 0.0360 & 0.6432 & 0.0089 & 0.0250 \\ 0.1733 & 0.0567 & 0.7834 & 0.0960 \\ 0.0108 & 0.2344 & 0.1147 & 0.8742 \end{pmatrix}.$$

The entries of this transition matrix are again interpreted as conditional probabilities. For instance, there is a 17.33% chance that squares currently occupied by only red squirrels, state **R**, will be occupied by both species, state **B**, in two time steps (four years).

Using the transition matrix T, it is possible to gain insight into the long-term behavior of the two species of squirrels. We compute the steady-state matrix L for the two squirrel populations. The question of interest in the study of the squirrel populations is what happens to the distribution of the squirrel populations over a long period of time.

For our squirrel model, the steady-state matrix is approximately

$$L \approx T^{200} = \begin{pmatrix} 0.1705 & 0.1705 & 0.1705 & 0.1705 \\ 0.0560 & 0.0560 & 0.0560 & 0.0560 \\ 0.3421 & 0.3421 & 0.3421 & 0.3421 \\ 0.4314 & 0.4314 & 0.4314 & 0.4314 \end{pmatrix}. \tag{71}$$

Thus the steady-state distribution is

$$\overline{X} = \begin{pmatrix} 0.1705 \\ 0.0560 \\ 0.3421 \\ 0.4314 \end{pmatrix}.$$

This result is interpreted as the long-term behavior of the squirrel populations in Great Britain as follows: 17.05% of the squares will be in state **R**, containing only red squirrels. There will be 5.6% of the squares in state **G** containing only gray squirrels. There will be populations of both squirrels in 34.21% of the squares (state **B**), with the majority of the squares, 43.14%, being occupied by neither species of squirrels (state **O**).

If the assumptions made in this model are correct, the red squirrel is not currently in danger. In fact, it will have sole possession of more regions than the gray squirrel will have. In the long term, the gray squirrels do not drive the reds to extinction. Actually this analysis says nothing about population sizes, only about the number of regions controlled by each type of squirrel. While it seems plausible that if the red squirrel territory (number of regions) is declining, then the population is declining; the opposite may be true. A problem in the projects section asks you to perform this analysis for the two squirrel species in Scotland, where the red squirrel is still widely distributed.

3.10 Harvesting Scot Pines: An Absorbing Markov Chain Model

Scot pine trees (*Pinus sylvestris*) are grouped into classes according to girth. M. B. Usher, [69], studied the dynamics of a Scot pine forest in Corrour, Scotland. We wish to develop a model to study movement of Scot pine trees from one girth size to another and determine the average number of years to harvest.

We group the trees into six classes or states, labeled $i = 0, 1, \ldots, 5$. State 0 corresponds to the thinnest trees, and state 5 represents the thickest trees. Tree girth measurements are taken every six years. The girth classes are chosen so that, at each measurement, the trees either remain in their current class or gain sufficient girth to move into the next class, but never gain enough girth to skip a class. We also assume that all trees eventually reach full girth and, once in that state, are harvested. Thus state $i = 5$ is an absorbing state.

The assumption that a tree either remains in its current girth class or moves to the next larger girth class gives the following transition matrix for this Scot pine forest model.

3 Stages, States, and Classes

Girth Class	Number of trees remaining at start of period	Number of trees remaining in class during period
0	4,461	3,214
1	2,926	2,029
2	1,086	813
3	222	171
4	27	17

TABLE 3.4 Scot Pine (*Pinus sylvestris*) Census Data from Corrour, Scotland.

$$T = \begin{pmatrix} p_0 & 0 & 0 & 0 & 0 & 0 \\ 1-p_0 & p_1 & 0 & 0 & 0 & 0 \\ 0 & 1-p_1 & p_2 & 0 & 0 & 0 \\ 0 & 0 & 1-p_2 & p_3 & 0 & 0 \\ 0 & 0 & 0 & 1-p_3 & p_4 & 0 \\ 0 & 0 & 0 & 0 & 1-p_4 & 1 \end{pmatrix}. \quad (72)$$

This matrix is lower triangular. The last state, $i = 5$, being absorbing is reflected here by the entry in position 6, 6 being one. We need to determine the values of p_0, p_1, p_2, p_3, and p_4. Table 3.4 provides the data for a single census period and is taken from the paper by Usher [69] and is reprinted by permission of Blackwell Science Inc.

The value of p_0 is the proportion of trees that remain in state 0, which is $p_0 = 3{,}214/4{,}461 \approx 0.7205$, so that $1 - p_0 \approx 0.2795$. We similarly calculate $p_1 = 2{,}029/2{,}926 \approx 0.6934$, $p_2 \approx 0.7486$, $p_3 \approx 0.7703$, and $p_4 \approx 0.6296$. The transition matrix is then given as

$$T = \begin{pmatrix} 0.7205 & 0 & 0 & 0 & 0 & 0 \\ 0.2795 & 0.6934 & 0 & 0 & 0 & 0 \\ 0 & 0.3066 & 0.7486 & 0 & 0 & 0 \\ 0 & 0 & 0.2514 & 0.7703 & 0 & 0 \\ 0 & 0 & 0 & 0.2297 & 0.6296 & 0 \\ 0 & 0 & 0 & 0 & 0.3704 & 1 \end{pmatrix}. \quad (73)$$

The steady-state matrix is computed numerically with T^n using a large value of n.

$$L \approx T^{200} = \begin{pmatrix} 0 & 0 & 0 & 0 & 0 & 0 \\ 0 & 0 & 0 & 0 & 0 & 0 \\ 0 & 0 & 0 & 0 & 0 & 0 \\ 0 & 0 & 0 & 0 & 0 & 0 \\ 0 & 0 & 0 & 0 & 0 & 0 \\ 1 & 1 & 1 & 1 & 1 & 1 \end{pmatrix}. \quad (74)$$

This steady-state matrix should not come as a surprise. The matrix (74) shows that for each of the tree girth classes $i = 1, \ldots, 4$, members eventually reach full girth size, state $i = 5$, and are harvested. This is an obvious observation from the assumption that all trees survive to full girth size.

Using the block decomposition shown in (68), we have $A_{5\times 5}$ is the submatrix given by

$$A_{5\times 5} = \begin{pmatrix} 0.7205 & 0 & 0 & 0 & 0 \\ 0.2795 & 0.6934 & 0 & 0 & 0 \\ 0 & 0.3066 & 0.7486 & 0 & 0 \\ 0 & 0 & 0.2514 & 0.7703 & 0 \\ 0 & 0 & 0 & 0.2297 & 0.6296 \end{pmatrix}.$$

The matrix $B_{1\times 5}$ is the row

$$B_{1\times 5} = \begin{pmatrix} 0 & 0 & 0 & 0 & 0.3704 \end{pmatrix},$$

and $O_{5\times 1}$ is the column vector of all zeros,

$$O_{5\times 1} = \begin{pmatrix} 0 \\ 0 \\ 0 \\ 0 \\ 0 \end{pmatrix}.$$

In this model, the matrix $I_{1\times 1}$ is just the 1×1 identity matrix.

We may verify that in the Scot pine tree model,

$$B_{1\times 5}(I_{5\times 5} - A_{5\times 5})^{-1} = \begin{pmatrix} 1 & 1 & 1 & 1 & 1 \end{pmatrix}.$$

For the Scot pine tree model, the fundamental matrix is

$$F = (I - A)^{-1} = \begin{pmatrix} 3.5778 & 0 & 0 & 0 & 0 \\ 3.2149 & 3.2616 & 0 & 0 & 0 \\ 3.9208 & 3.9777 & 3.9777 & 0 & 0 \\ 4.2912 & 4.3535 & 4.3535 & 4.3535 & 0 \\ 2.6611 & 2.6998 & 2.6998 & 2.6998 & 2.6998 \end{pmatrix}. \tag{75}$$

Recalling the block form of the transition matrix (68), the position of the submatrix A indicates that i and j have values $0,\ldots,4$, so that $f_{00} = 3.5778$ is the average number of time steps that a tree spends in the first girth state before moving up to the next girth state. Since each time step in this model represents a six-year period, we expect a tree to take an average of $3.58 \times 6 = 21.5$ years to move from the smallest girth size state 0 to state 1, the next largest state size. The sum of the entries for an individual column gives the average number of time steps to absorption. The Scot pine tree model predicts that on average, a tree of smallest girth size, state 0, will take about $(3.58 + 3.21 + 3.92 + 4.29 + 2.66) \times 6 \approx 106$ years to reach full girth (state 5).

Note that the lower-triangularity and the similarity of the entries in the fundamental matrix (75) reflect the assumptions of the Scot pine tree model.

We should expect that the average number of time steps required for a tree to move from one girth size to another would be independent of which state the tree began in, but once in a particular state, the average number of time steps depends upon that particular state. For example, we see in (75) that it takes about four time steps (24 years) for a tree to move from state 2 to state 3, regardless of which state the tree began in.

3.11 Life Tables

Life tables are a useful bookkeeping device frequently used in population modeling. The ideas are simple; the hardest aspect is understanding and remembering the terminology. Rather than defining abstract concepts and then giving an example, we give an example and use it to introduce the necessary definitions.

Table 3.5 shows a piece of a life table for Canadian females between the ages of 0 and 14. (Data is taken from Brown [7] and is reprinted by permission of ACTEX Publications, Inc.) This is actually a piece of a larger life table that goes to age 102. The first column, labeled with the symbol x, gives the age. A life table takes a hypothetical group of individuals born during one time step (here one year) called a *cohort,* and tracks them over their lifetimes. The number of individuals initially in a cohort (the birth cohort) is denoted by ℓ_0 and is called the *radix* of the population. The number left at the beginning of the xth year is denoted by ℓ_x. In this table, $\ell_0 = 100,000$, and $\ell_{10} = 99,081$ indicating that 99,081 of the original cohort survived to the beginning of the 10th year. Usually a radix is a multiple of 10. Sometimes the number one is chosen as a radix and the successive ℓ_x's can be interpreted as survival percentages, but usually the radix is chosen so that the desired number of significant digits is kept using whole numbers. In experimental data, the true number initially in a cohort is rarely a power of 10, so life tables based on experimental data may have an extra column containing the true population size (often given the symbol N_x). In this case, $\ell_x = \frac{N_x}{N_0} \times 10^n$, where n is the desired power of 10. In this context, the ℓ_x column tracks a standardized cohort rather than a hypothetical cohort.

Each year a certain number of individuals in a cohort die. These are given in the column headed d_x. Here $d_0 = 678$, indicating that 678 individuals in the cohort died

Age x	ℓ_x	d_x	p_x	q_x	L_x	T_x	$\overset{o}{e}_x$
0	100,000	678	0.99322	0.00678	99,415	7,972,923	79.73
1	99,322	62	0.99938	0.00062	99,286	7,873,508	79.27
2	99,260	41	0.99959	0.00041	99,235	7,774,222	78.32
3	99,219	30	0.99970	0.00030	99,204	7,674,987	77.35
4	99,189	25	0.99974	0.00026	99,175	7,575,783	76.38
5	99,164	22	0.99978	0.00022	99,153	7,476,608	75.40
6	99,142	18	0.99982	0.00018	99,133	7,377,455	74.41
7	99,124	16	0.99984	0.00016	99,116	7,278,323	73.43
8	99,108	14	0.99986	0.00014	99,101	7,179,207	72.44
9	99,094	13	0.99986	0.00014	99,087	7,080,106	71.45
10	99,081	15	0.99986	0.00014	99,073	6,981,018	70.46
11	99,066	15	0.99985	0.00015	99,059	6,881,945	69.47
12	99,051	17	0.99982	0.00018	99,043	6,782,886	68.48
13	99,034	21	0.99979	0.00021	99,023	6,683,844	67.49
14	99,013	27	0.99973	0.00027	98,999	6,584,820	66.50

TABLE 3.5 1985–87 Canadian Female Life Table.

during their first year of life. It follows that

$$d_x = \ell_{x+1} - \ell_x.$$

The next column gives p_x which is the probability that an individual who has lived x years (i.e., up to the beginning of the year x) will survive that year. Note that p_x is a conditional probability. It does not give the probability of someone's in the original cohort surviving the xth; rather it gives the probability of surviving the xth year given that one has lived up to the beginning of the xth year. It is evident that $p_0 = \frac{\ell_1}{\ell_0}$, and it follows from the definition of p_x that for any x

$$p_x = \frac{\ell_{x+1}}{\ell_x}.$$

Suppose we wanted to know the probability that someone five years old will survive to age nine. This is denoted by the obtuse notation $_4p_5$ indicating that the individual has already survived five years and that this is the probability of surviving four more years. Thinking about the table and what it means, we see that $_4p_5 = \frac{\ell_9}{\ell_4}$. In general, $_tp_x = \frac{\ell_{x+t}}{\ell_x}$. This concept is extremely useful for these conditional probabilities are the survival probabilities in a Leslie model or a staged model. Suppose we want to build a human population model, but do not want to deal with the 103 age classes. We could select some more manageable number of classes. Suppose we (somewhat capriciously) choose age classes $[0, 5)$, $[5, 10)$, $[10, 15)$, $[15, 30)$, $[30, 45)$, $[45, 60)$, $[60, 80)$, and $[80, 103)$. To construct a state diagram, we need the probability that someone who has lived no years will survive to age 10, the probability that someone who has lived 10 years will survive to 15 and so on. From the previous discussion, $_5p_0 = \frac{\ell_5}{\ell_0}$, $_5p_5 = \frac{\ell_{10}}{\ell_5}$, and so on. These transition probabilities or survival rates are the weights over the arrows in our state diagram.

The next column headed q_x is the probability of death during year x, given that one has lived to the beginning of year x. There are several ways to represent this. The notation p and q is reminiscent of probability theory and suggests that $q_x = 1 - p_x$, which is true. Directly from the definitions, we have

$$q_x = \frac{d_x}{\ell_x}.$$

The symbol L_x denotes a more complicated concept. It represents the *number of life-years lived* by the ℓ_x people who attain the age of x over the year of age from x to $x+1$. Another way to think of L_x is as the area under the ℓ_x curve between x and $x+1$. The units "life-years" represent a type of area. Recall that we use an integral to find the area under a curve. Thus

$$L_x = \int_0^1 \ell_{x+t} dt.$$

A common way to approximate L_x is to assume that the d_x deaths are uniformly distributed over the age interval x to $x+1$. With this assumption, ℓ_x is a straight line, and number of life-years lived is the area under a trapezoid, which is the average value of ℓ_x times one year. Thus $L_x \approx \frac{1}{2}(l_x + l_{x+1})$. If the uniformly-distributed-deaths assumption is not reasonable, the integral definition must be used.

The next column, T_x is the number of life-years left to live for a cohort or the *total future lifetime*. Thus
$$T_x = L_x + L_{x+1} + L_{x+2} \cdots,$$
or
$$T_x = \sum_{t=x}^{\infty} L_x.$$

The last column $\overset{o}{e}_x$ is the *average future lifetime* which is defined as
$$\overset{o}{e}_x = \frac{T_x}{\ell_x}.$$

Thus $\overset{o}{e}_{10} = 70.46$ means that a individual who lives to age 10 will live an average of 70.46 more years or have an average lifetime of 80.46. A newborn, however, has an average lifetime of only 79.73. This makes sense since deaths early in life lower the average lifetime. This trend continues. In the full table, we find that individuals who live to age 80 live an average of 8.93 more years, and thus have an average lifetime of 88.93.

The ideas and data of this section are taken from the book *Introduction to the Mathematics of Demography (Second edition)* by R. Brown [7]. Another book addressing these issues is *Introduction to the Mathematics of Population* by N. Keyfitz [31]. Both of these are highly recommended to the interested reader.

3.12 For Further Reading

A good introductory level linear algebra book is:

Howard Anton and Chris Rorres, *Elementary Linear Algebra* (Seventh edition), John Wiley & Sons, Inc., 1994. There is an "Applications" version of this book which covers the use of matrices in modeling diverse phenomena.

The best book that we are aware of on the subject of matrix models is:

Hal Caswell, *Matrix Population Models*. Sinauer Associates, Inc., 1989.

For those interested in modeling human populations specifically we recommend the following two books:

Robert L. Brown, *Introduction to the Mathematics of Demography* (Second Edition), ACTEX Publications, Inc., Winsted, Connecticut, 1993;

Nathan Keyfitz, *Introduction to the Mathematics of Population with Revisions*, Addison-Wesley, Reading, Mass., 1977.

3.13 Exercises

1. Suppose an animal lives three years. The first year it is immature and does not reproduce. The second year it is an adolescent and reproduces at a rate of 0.8 female offspring per female individual. The last year it is an adult and produces 3.5 female offspring per female individual. Further suppose that 80% of the first-year females survive to become second-year females, and 90% of second-year females survive to become third-year females. All third-year females die. We are interested in modeling only the female portion of this population.

  a. Draw a state diagram for this scenario.
  b. Construct the Leslie matrix.
  c. Compute the eigenvalues for this matrix. From these determine if the population will eventually grow or decline. What is the rate of this growth (or decline)?
  d. Suppose that a population of 100 first-year females are released into a study area along with a sufficient number of males for reproductive needs. Track the female population over 10 years.
  e. Compute the eigenvector associated with the dominant eigenvalue. Normalize both this eigenvector and the population distribution after 10 years. Compare these two vectors.

2. Showing that a model is valid can be difficult. Often it is much easier to show that a model is invalid. Consider the following model that has been proposed [4] to describe the growth of redwoods (*Sequoia sempervirens*). Redwoods frequently live 1,000 years or more. The model used classifies redwoods into three age stages: 0 to 200 years (young), 200 to 800 years (mature), and older than 800 (old). A current census in a particular stand finds that currently there are 1,696 young redwoods, 485 mature redwoods, and 82 old redwoods. The transition between stages over a 50-year period is given by the matrix T

$$\begin{pmatrix} 12 & 26 & 6 \\ 0.30 & 0.92 & 0 \\ 0 & 0.18 & 0.67 \end{pmatrix}.$$

Trace this stand of trees through five time steps (five 50-year periods for a total of 250 years) and explain how we know that this model cannot possibly be valid.

3. Two mathematical objects associated with a square matrix are its trace and determinant. The *trace* is the sum of the diagonal elements. The *determinant* is harder to define. The determinant of a two-by-two matrix

$$A = \begin{pmatrix} a & b \\ c & d \end{pmatrix}$$

is $ad - cb$. Often straight vertical lines are used to denote a determinant. Thus

$$\det(A) = \begin{vmatrix} a & b \\ c & d \end{vmatrix}.$$

For larger square matrices, we use the expansion-by-minors definition. To expand on the first row, we take the entry in the $(1,1)$ position and multiply it by the determinant of the matrix that remains when we cross out the first row and first column (called a minor) and subtract the entry in the $(1,2)$ position times the determinant of the matrix obtained by crossing out the first row and second column, plus the $(1,3)$ entry times the determinant of the matrix obtained by crossing out the first row and third column, and

3 Stages, States, and Classes

so on, alternating between adding and subtracting the next term. For example, if

$$A = \begin{pmatrix} a & b & c \\ d & e & f \\ g & h & i \end{pmatrix},$$

then

$$\det(A) = a \begin{vmatrix} e & f \\ h & i \end{vmatrix} - b \begin{vmatrix} d & f \\ g & i \end{vmatrix} + c \begin{vmatrix} d & e \\ g & h \end{vmatrix}.$$

Observe that this is a recursive definition since the determinant of a 4×4 matrix involves computing four determinants of 3×3 matrices, each of which requires the computation of three 2×2 matrices. Here we defined expansion on the first row. We can expand on any row or column and direct the reader to any introductory matrix algebra book for details or for other methods of computing a determinant.

Find the trace and determinant of each matrix. Determine whether each matrix is singular (has no inverse).

$$\begin{pmatrix} 1 & 3 \\ 0 & 2 \end{pmatrix} \qquad \begin{pmatrix} 1 & 2 & 3 \\ 2 & 3 & 1 \\ 1 & 2 & 1 \end{pmatrix} \qquad \begin{pmatrix} 1 & 2 & 3 \\ 2 & 3 & 1 \\ 3 & 5 & 4 \end{pmatrix}$$

4. This exercise steps the reader through the process of finding eigenvalues and eigenvectors by hand and then demonstrates their use. Let

$$A = \begin{pmatrix} 3 & 2 \\ -1 & 0 \end{pmatrix}.$$

We want to find λ and V so that $AV = \lambda V$. Equivalently $(A - \lambda I)V = \mathbf{0}$. The only non-trivial solutions occur when $(A - \lambda I)$ is singular. This happens when $\det(A - \lambda I) = 0$. Evaluate this determinant to obtain the characteristic equation. Next solve the characteristic equation to get the characteristic roots, or eigenvalues. Put one of the eigenvalues (call it λ_1) into the equation $(A - \lambda I)V = \mathbf{0}$, where $V = (v_1, v_2)$. The result should be two simultaneous equations in the unknowns v_1 and v_2. Try to solve this system. If your calculations are correct, the best you can do is to express v_2 in terms of v_1 (or vice versa). Write V in terms of only v_1. Factor v_1 out of V. The result is *an* (not *the*) eigenvector associated with the eigenvalue λ_1 you are using. We call it V_1. Notice that any multiple of this vector is still an eigenvector. Repeat using the other eigenvalue and find an eigenvector associated with it. Call the eigenvalue λ_2 and its eigenvector V_2.

Next suppose $X(0) = (1, 3)^T$. Write this as a linear combination of V_1 and V_2. This is done by writing $(1, 3)^T = aV_1 + bV_2$. This is a system of two linear equations in a and b. Solve for a and b. As long as an $n \times n$ matrix has n linearly independent eigenvectors (if you know what that means), we can write any vector as a linear combination of the eigenvectors.

Numerically verify the following equations:

$$AX(0) = a\lambda_1 V_1 + b\lambda_2 V_2.$$

$$A^2 X(0) = a\lambda_1^2 V_1 + b\lambda_2^2 V_2.$$

What should the right-hand side of

$$A^n X(0) =?$$

be? Prove it by induction.

5. Show that if U and V are eigenvectors of the matrix A associated with the eigenvalue λ, then so is any linear combination $cU + dV$.

6. A copy machine is always in one of two states, either working or broken (not working). If it is working, there is a 70% chance that it will be working tomorrow. If it is broken there is a 50% chance it will still be broken tomorrow. Assume that one day is a natural time step.
   a. Draw a state diagram for this scenario.
   b. Formulate the transition matrix.
   c. Assuming that the machine is working today, what is the probability that it will be working tomorrow? The next day? After one week? After one month?
   d. What is the long-term probability that the copy machine will be working on any given day?

7. A slightly more refined model of a copy machine has three states: working, broken and fixable, broken and unfixable. If it is working, there is a 69.9% chance it will be working tomorrow and a 0.1% chance it will be broken and unfixable tomorrow. If it is broken and fixable today, there is a 49.8% chance it will be working tomorrow and a 0.2% chance it will be unfixable tomorrow. Unfixable is, of course, unfixable, so the probability that an unfixable machine is unfixable tomorrow is 1 and the probability of its being anything other than unfixable is 0. Again assume a day as the time step.
   a. Draw a state diagram for this scenario. Categorize the states as absorbing and nonabsorbing.
   b. Formulate the transition matrix. Label the states so that the identity matrix is in the lower right-hand block.
   c. Compute the fundamental matrix and interpret the results. How long will this machine last? How much of that time will it be working, and how much of that time will it be under repair (broken and fixable)?

8. Using the definitions and notation from the life tables, show that if

$$p_x = \frac{\ell_{x+1}}{\ell_x} \quad \text{and} \quad q_x = \frac{d_x}{\ell_x},$$

then

$$q_x = 1 - p_x.$$

9. Pardon the morbid nature of this exercise. Using Table 3.5, find the following probabilities.
   a. The probability that a newborn survives to age 5.
   b. The probability that a 4-year old survives to age 5.
   c. The probability that a newborn dies between the ages of 5 and 6.
   d. The probability that a newborn dies between the ages of 5 and 10.

3 Stages, States, and Classes

e. The probability that a 5-year old survives to age 10.

f. The probability that a 5-year old dies between the ages of 10 and 13.

10. Using the first two columns of Table 3.5, reconstruct the columns labeled d_x, p_x, q_x, and L_x. Assume a uniform distribution of deaths to construct L_x which gives results close to (but usually not equal to) the values in Table 3.5. Note that you cannot complete the last two columns without the entire table (to age 102).

3.14 Projects

Project 3.1. Age Structured Bobcats

The bobcat was investigated in Projects 1.2 and 2.1. Here we continue this series of projects by constructing an age-structured model for the bobcat. Age class data from Cox et al. [13] is presented in Table 3.6.

1. Construct a state diagram for the bobcat using the best-case data in Table 3.6 as the weights over the arrows. Use this to construct the Leslie matrix for the bobcat best-case scenario.

2. Find the dominant eigenvalue of the Leslie matrix from part **1**. At what rate will the population eventually be growing? Find the corresponding eigenvector. Find the eventual age distribution (percentage in each age class) of bobcats if these parameters remain valid for a long time.

3. Starting with a population vector of $(0, 100, 50, 50, 25, 10, 0, 0, 0, \ldots 0)$ trace this bobcat population through a period of 10 years. How does the relative distribution (percentage in each age class) after 10 years compare to the distribution suggested by part **2**?

4. Repeat parts **1–3** using the worst case data.

5. Cox et al. also note that catastrophes that lower the reproduction rate by 30% occur every 25 years on average. Use this information and the best case data to create a

Age Class	Survival Worst Case	Survival Best Case	Reproduction Worst Case	Reproduction Best Case
1	0.32	0.34	0.60	0.63
2	0.68	0.71	0.60	0.63
3	0.68	0.71	1.15	1.20
4	0.68	0.71	1.15	1.20
5	0.68	0.71	1.15	1.20
⋮	⋮	⋮	⋮	⋮
16	0.68	0.71	1.15	0.27

TABLE 3.6 Demographic Data for the Bobcat. Data Represents Females Born to Females.

	BD	G	O
BD	0.3	0.2	0.1
G	0.6	0.5	0.1
O	0.1	0.3	0.8

TABLE 3.7 Transition Probabilities for Ecological Succession.

stochastic age-structured model. Run 100 trials of this simulation over a ten-year period. Display the results as both multiple line graphs and side-by-side box plots.

Project 3.2. Ecological Succession

In this project we consider modeling the ecological succession of barren dunes to an oak community using a Markov chain. In this simple model, it is assumed that there are three states for a specific ecological region. These states are bare dunes, which will be labeled **BD**; a grass community, which is labeled **G**; and an oak community, labeled **O**.

Transition probabilities are shown in Table 3.7. (Recall that this is read "from" columns "to" rows.)

1. Using this information, construct the state diagram representing the ecological succession from bare dunes to oak community.

2. Construct the transition matrix T.

3. Suppose initially 10% of the regions are bare dunes, 10% are grass communities, and 80% are oak communities. Determine the distribution of states after 2 time steps and after 10 time steps. Interpret the meaning of these results.

4. Determine the limiting matrix L and the steady-state distribution for this ecological succession. Interpret the meaning of these results.

Project 3.3. Changing Human Population Growth Rate

This project addresses the question of whether the human population growth rate can be reduced without limiting the number of children women have (China's method) or shortening life spans (unacceptable to most of the civilized world).

In the late 1800s it was noticed that England's human growth rate was lower than expected when looking at comparable countries. One explanation was that the Church of England required young couples to wait until they could afford a home before they got married. The idea was that children would fare better if their parents had the resources to take care of them. The effect was that couples began having children at a later age.

Specifically this project tests whether waiting to have children has any noticeable effect on the population growth rate. It should also convey that even crude models can be used to answer important questions.

1. Break the female population into three age classes: $[0, 20)$, $[20, 40)$, and $[40, 60)$. We will assume that the average female has 2.3 children, half of whom are female. First assume that the $[0, 20)$ women give birth to 0.5 girls, the $[20, 40)$ women give birth to 0.5

3 Stages, States, and Classes

girls, and the $[40, 60)$ women give birth to 0.15 girls. (Note that $0.5+0.5+0.15 = 2.3/2$). To avoid confusing birth and survival issues, assume that 100% of the $[0, 20)$ women survive to the $[20, 40)$ age class and 100% of the $[20, 40)$ women survive until the $[40, 60)$ class, and all women over 60 leave the studied population. Starting with 100 females in each age class (300 total), track the female population through 400 years (20 time steps).

2. Repeat the same analysis, but with births occurring later. Assume now that the birth rates are 0, 0.75, and 0.4 for the three age classes respectively. Notice that each woman still gives birth to 2.3/2 girls. Keep the survival rates the same.

3. Plot the results of these two runs on the same coordinate axes. What conclusions can you draw?

4. To each of the two models above, compute the percentage growth rate for each case (i.e., $(x[n + 1] − x[n])/x[n] \times 100\%$). Is the hypothesis that delaying births reduces growth rate supported by this model?

Open-ended part: Certain aspects of this model are crude. Does the basic message change if it is made more realistic? Try more realistic survival rates or initial populations. Try more age classes. Try keeping women in the system past the age of 60, but no longer having children. Try devising a reasonable model where the opposite conclusion can be drawn?

Project 3.4. Markovian Squirrels in Scotland

In Section 3.9, a Markov model for red and gray squirrels in England is presented. This project revisits this model with data collected in Scotland.

The red squirrel, *Sciurus vulgaris L.*, is still widely distributed in Scotland. However, the gray squirrel, *Sciurus carolinensis Gmelin*, is beginning to spread to Scotland. The following is the transition matrix for the squirrel populations in Scotland.

$$T = \begin{pmatrix} 0.874 & 0.095 & 0.077 & 0.059 \\ 0.015 & 0.723 & 0.007 & 0.193 \\ 0.109 & 0.095 & 0.905 & 0.143 \\ 0.002 & 0.087 & 0.009 & 0.605 \end{pmatrix}.$$

1. Construct the state diagram for this transition matrix.

2. Determine the long-term behavior of the squirrel populations in Scotland by finding the steady-state distribution for this transition matrix. Interpret the results.

Project 3.5. Hermit Crabs

The hermit crab (*Pagarus longicarpus*) does not have a hard protective shell to protect its body. It uses discarded shells to carry around as portable shelters. These empty shells are rare commodities in tide pools and are available only when their occupants die. In an experiment at Long Island Sound tide pools, an empty shell was dropped into the water to initiate a chain of vacancies. This experiment was repeated a large number of times as vacancies flowed from larger to generally smaller shells. The states in this experiment are the various shell sizes. Table 3.8 summarizes the number of moves between states.

There are seven states in this model. States 6 and 7 are absorbing states, which is reflected in the above table by the absence of moves from these states. State 6 represents

	1	2	3	4	5	6	7	Total
1	2	7	9	2	0	0	1	21
2	0	3	19	17	1	0	2	42
3	0	2	20	11	10	4	23	70
4	0	0	10	26	26	6	24	92
5	0	0	0	5	22	2	30	59

TABLE 3.8 Number of Moves between Shells in a Crab Vacancy Chain.

the situation in which a crab without a shell occupies an empty shell. If a shell is not occupied during a 45-minute observation period, we assume that the shell is abandoned and label this as absorbing state 7.

1. Construct the state diagram.
2. Construct the transition matrix for this absorbing Markov chain.
3. Determine the steady-state matrix and the absorption probabilities.
4. Calculate the expected length of the crab vacancy chains.

This project is adapted from work originally in the papers by Chase et al. [11], [72], and is discussed further in Beltrami's modeling text [3].

Project 3.6. California II: Using Matrices

We again model the population of California over time. This project revisits Project 1.8 from Chapter 1, puts it in a matrix framework, and addresses some different issues. It is not necessary to have done the earlier project, but we will refer back to Tables 1.3 and 1.4.

1. Rates. If you did not do Project 1.8, read the problem description and do Part **1** which instructs you to complete Table 1.4.

2. Matrix Model. Set up a matrix model of the form $X(t+1) = (I + B - D + M)X(t)$ where $X(t)$ is the population vector with two components $x_{\text{California}}$ and $x_{\text{Rest of U.S.}}$. The matrices B, D, and M are the birth, death, and net migration matrices respectively (all should be diagonal) and I is the identity matrix (why is it needed?). We will simplify computations by using a "growth" matrix $G = I + B - D + M$. Using the 1955 data as the initial information, verify that the model accurately gives the 1960 data.

3. Predictions. Run a simulation using the matrix model from 1960 to present (recall the five-year time step). Search the literature to find official estimates for the population of these regions for each prediction. Plot the model results with the observed results and discuss how well the model serves to predict these population sizes.

4. Long-Term Analysis. Compute the dominant eigenvalue and its associated eigenvector for the "growth" matrix G. Normalize this vector (divide by a constant so that the components add to one). Explain its meaning. What is the long-term behavior of this model? You should be answering two questions: At what rate are the populations growing? and What are the long-term proportions of people in and out of California?

3 Stages, States, and Classes

	To California	To U.S.	Total 1955 Population
California	12,174	814	12,988
Rest of U.S.	1,938	150,144	152,082

TABLE 3.9 1955–1960 Migrations in Thousands.

5. Higher Resolution Migration Model. Until now we have been working with *net migration*. In Table 1.3, we see that a net of 1,124,000 people moved into California. By *net* we mean that 1,124,000 more people moved in than moved out, not that 1,124,00 people moved in and no one moved out. Table 3.9 shows the actual transitions between regions (or "states" in the sense used in this chapter).

Ignoring births and deaths construct a matrix model for the transitions between regions (should have the form $X(n+1) = TX(n)$). Comment on why this is a model for a nonabsorbing Markov chain, and use the methods of this chapter to determine the stable proportions for people in California versus people outside of California. How do these proportions compare with the results found in part **4**? Find the eigenvalues (growth rates) for this matrix. Do they make sense under the stated assumptions?

6. Putting It All Together. Construct a model with the birth and death information from part **1** and the migration information from part **5**. This model should have the form $X(t+1) = (B - D + T)X(t)$. (What happened to the identity matrix I?) Again simplify the model using a "growth matrix" $G = B - D + T$. Using the 1955 populations as initial conditions, what does this model predict for 1960? What are the long-term proportions predicted by this model, and how do they compare with the models of parts **4** and **5**?

This project is based on Chapter 2 of the book by Rogers [55]. California data is from the California State Department of Finance, Financial and Population Research Section [8] and is reprinted by permission of the University of California Press.

Project 3.7. Equilibrium Temperature Distributions

Modeling the flow of heat through objects of various types and shapes is a very important problem, and many modelers (engineers) devote their careers to this one pursuit. We have generally avoided this problem because non-engineering students frequently have little intuitive knowledge about how heat flows and generally do not have the background to solve the partial differential equations involved. Finding the steady state or limiting temperature distribution, however, requires only the reasonable fact that the temperature at any point is the average of the points around it (more precisely, the temperature at a point is the average of the temperatures at all points on a circle centered about that point). While proving this fact requires knowledge beyond the scope of this course, what it states seems reasonable enough to take on faith. This idea is explored in this project and the next.

Here is the basic problem. Take a thin plate of some shape (thin means the effects of thickness can be ignored) and determine a temperature distribution around the edge of the plate which will remain fixed. Heat will flow through the plate until the temperature has stabilized. This equilibrium temperature distribution is called the *steady state* or *limiting temperature distribution*. As mentioned above, the only fact we need is that the temperature at any interior point is the average of the temperatures of the points on a circle centered about that point. We will solve this problem numerically, which immediately raises the problem of how to find the average at the infinite number of points on the interior. What is done in practice is to find the average only at a finite number of regularly spaced points called *nodes*.

Consider the following example. In Figure 3.16, we show a 4 cm by 4 cm square plate with the bottom edge held at 0 degrees and all other edges held at 5 degrees. For this example, we chose to place the nodes on a 5 by 5 grid. Thus there will be nine interior nodes, which are unknown, and 16 boundary nodes whose temperatures remain fixed. The temperature at each node is the average of the temperatures at the four nearest nodes (which is the closest thing to a circle when dealing with only the nodes on a rectangular grid). Clearly using more nodes will yield better results. Professional work usually involves repeating with more and more nodes until the solution stabilizes to within some accepted error bound. We chose a 5 by 5 grid, just to get the terminology and notation across. We will solve this problem and let the reader deal with larger grids and irregularly shaped plates.

We have nine unknowns which we will label as $T_1, \cdots, T_9$. Averaging at each node gives nine equations in nine unknowns.

$$T_1 = \tfrac{1}{4}(0 + 5 + T_2 + T_4),$$
$$T_2 = \tfrac{1}{4}(0 + T_1 + T_3 + T_5),$$
$$T_3 = \tfrac{1}{4}(0 + T_2 + 5 + T_6),$$
$$T_4 = \tfrac{1}{4}(T_1 + 5 + T_5 + T_7),$$
$$T_5 = \tfrac{1}{4}(T_2 + T_4 + T_6 + T_8),$$

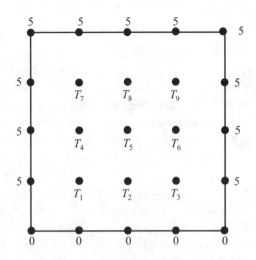

FIGURE 3.16 Node Points in a Square Plate.

3 Stages, States, and Classes

$$T_6 = \tfrac{1}{4}(T_3 + T_5 + 5 + T_9),$$
$$T_7 = \tfrac{1}{4}(T_4 + 5 + T_8 + 5),$$
$$T_8 = \tfrac{1}{4}(T_5 + T_7 + T_9 + 5),$$
$$T_9 = \tfrac{1}{4}(T_6 + T_8 + 5 + 5).$$

Rearranging gives

$$4T_1 - T_2 - T_4 = 5,$$
$$-T_1 + 4T_2 - T_3 - T_5 = 0,$$
$$-T_2 + 4T_3 - T_6 = 5,$$
$$-T_1 + 4T_4 - T_5 - T_7 = 5,$$
$$-T_2 - T_4 + 4T_5 - T_6 - T_8 = 0,$$
$$-T_3 - T_5 + 4T_6 - T_9 = 5,$$
$$-T_4 + 4T_7 - T_8 = 10,$$
$$-T_5 - T_7 + 4T_8 - T_9 = 5,$$
$$-T_6 - T_8 + 4T_9 = 10.$$

This can be written in matrix form as $AT = B$ where A is

$$\begin{pmatrix} 4 & -1 & 0 & -1 & 0 & 0 & 0 & 0 & 0 \\ -1 & 4 & -1 & 0 & -1 & 0 & 0 & 0 & 0 \\ 0 & -1 & 4 & -1 & 0 & -1 & 0 & 0 & 0 \\ -1 & 0 & 0 & 4 & -1 & 0 & -1 & 0 & 0 \\ 0 & -1 & 0 & -1 & 4 & -1 & 0 & -1 & 0 \\ 0 & 0 & -1 & 0 & -1 & 4 & 0 & 0 & -1 \\ 0 & 0 & 0 & -1 & 0 & 0 & 4 & -1 & 0 \\ 0 & 0 & 0 & 0 & -1 & 0 & -1 & 4 & -1 \\ 0 & 0 & 0 & 0 & 0 & -1 & 0 & -1 & 4 \end{pmatrix},$$

T is $(T_1, \ldots, T_9)^T$, and $B = (5, 0, 5, 5, 0, 5, 10, 5, 10)^T$.

We can solve this using matrix inversion to get $T = A^{-1}B$. Doing this yields $T = (2.988, 2.776, 4.106, 4.174, 4.011, 4.472, 4.699, 4.621, 4.773)^T$.

It is hard to visualize what is happening with this result, expressed as a list of numbers. We encourage the user to experiment with graphical methods of displaying results. Color contour plots are particularly effective for temperature distribution problems, but 3-dimensional graphs can also be used.

1. Re-do the analysis done above for the largest matrix that your computer can reasonably handle (see Project 4.11). Display the results making effective use of graphics.

2. Find the equilibrium temperature distribution for a 2 cm by 5 cm rectangular plate. Assume the sides of length 2 cm are held at 0 degrees and the sides of length 5 cm are held at 10 degrees. Again display the result using effective graphics.

3. Find the equilibrium temperature distribution for a 5 cm thin rod which has one end held at 3 degrees and the other end held at 10 degrees. Note that this is a one-dimensional problem so the analysis must be modified appropriately (temperature at each node is the average of the nearest two nodes). This is a natural use of the Thomas algorithm (see Project 4.12). Display results using effective graphics.

Project 3.8. The Ehrenfest Chain

The Ehrenfest chain is a Markov model for the exchange of heat or gas molecules between two bodies. The model consists of two boxes, labeled A and B, containing d balls labeled $1, 2, \ldots, d$. An integer is selected at random from $1, 2, \ldots, d$, and the ball labeled by this integer is removed from its box and placed in the other box. This process is repeated indefinitely, with the assumption of independence between trials. There are $d+1$ states in this model representing the number of balls in box A, namely some number between zero and d.

The transition matrix T for the Ehrenfest chain has entries

$$p_{i,j} = \begin{cases} \frac{i}{d} & j = i-1 \\ 1 - \frac{i}{d} & j = i+1 \\ 0 & \text{elsewhere,} \end{cases}$$

so for $d = 3$,

$$T = \begin{pmatrix} 0 & \frac{1}{3} & 0 & 0 \\ 1 & 0 & \frac{2}{3} & 0 \\ 0 & \frac{2}{3} & 0 & 1 \\ 0 & 0 & \frac{1}{3} & 0 \end{pmatrix}.$$

The transition matrix T is not regular.

1. Construct the state diagram for the Ehrenfest chain with $d = 3$.

2. Show that for large values of k, $T^{2k} \to E$ and $T^{2k+1} \to O$, where

$$E = \begin{pmatrix} \frac{1}{4} & 0 & \frac{1}{4} & 0 \\ 0 & \frac{3}{4} & 0 & \frac{3}{4} \\ \frac{3}{4} & 0 & \frac{3}{4} & 0 \\ 0 & \frac{1}{4} & 0 & \frac{1}{4} \end{pmatrix} \quad \text{and} \quad O = \begin{pmatrix} 0 & \frac{1}{4} & 0 & \frac{1}{4} \\ \frac{3}{4} & 0 & \frac{3}{4} & 0 \\ 0 & \frac{3}{4} & 0 & \frac{3}{4} \\ \frac{1}{4} & 0 & \frac{1}{4} & 0 \end{pmatrix}.$$

This shows that the Ehrenfest chain is not regular.

3. Numerically verify that for the Ehrenfest chain the steady-state distribution will be $\overline{X} = \frac{1}{2}(EX_0 + OX_0)$ for any vector X. For example, if one initially has one ball in box A, (i.e., $X_0 = (1, 0, 0, 0)^T$), then the steady state distribution is given by $(\frac{1}{8}, \frac{3}{8}, \frac{3}{8}, \frac{1}{8})^T$.)

Remarks on this project: A probabilistic interpretation of this approach to the steady-state distribution of the Ehrenfest chain is given by noting that regardless of the initial number of molecules in the first box, after a long time, the probability of finding k molecules in the box is the same as if the molecules had been distributed at random, each molecule having probability $\frac{1}{2}$ of being in box A. Asymptotically there will be two distinct distributions for the molecules, each distribution occurring on successive trials. The steady-state distribution is then viewed as the average of these two distributions. It is also interesting to note that the steady-state distribution will be binomial with $p = \frac{1}{2}$.

3 Stages, States, and Classes

Project 3.9. Impala Management

The purpose of this project is to produce a report examining the effects of several different types of game management on the population dynamics of an impala herd. In particular, it should report on herd population if impala are managed by 1) predation, 2) trophy hunting, or 3) game ranch management. For each of these three cases, it will consider the effects of low levels of hunting or predation (6% of the population) and high levels of hunting or predation (16% of the population). In all, it should compare the effects of six management strategies on the population over time. Aside from the data and definitions below, you are on your own to construct the appropriate model and run it for an appropriate length of time.

Definitions. *Predation* means male and female impala are killed equally and 45% of the kill is juvenile, 20% of the kill is sub-adult and 35% of the kill is adult. *Trophy hunting* means 70% of the impala killed are male and of all the animals killed 2.5% are juvenile, 2.5% are sub-adult, and 95% are adult. Basically the old males with horns are the desired game. *Game ranching* means 70% of the impala killed are males, but the age proportions are 5% juvenile, 75% sub-adult, and 20% adult. Here the desired product is meat. More males are killed because the females give birth. Impala live 11 years. A *juvenile* is less than one year old, a *sub-adult* is at least one year old and less than five years old, and an *adult* is five years or older.

For example, if we were game ranching at a high level and the population was at 100, we would kill 16 animals. Of those 16, 11 (actually 11.2) would be male, and 5 (actually 4.8) would be female. Of the males, 1 (actually 0.56) would be juvenile, 8 (actually 8.4) would be sub-adult, and 2 (actually 2.24) would be adults. Of the females, 0 (actually 0.25) would be juvenile, 4 (actually 3.6) would be sub-adult, and 1 (actually 0.96) would be adult. To simplify matters, work with partial animals (e.g., 3.25 impala), and take sub-adults (and adults) equally from the various age classes. If hunting at a certain level wipes out an age class, just kill all the animals within it and don't worry about the fact that fewer than 6% or 16% are removed that year. Be careful not to create negative animals!!

Impala facts:

Fifty percent of the impala born are female, and fifty percent are male.

Only adult males breed, and one male can breed with three females. If the population of breeding males drops below a third of the breeding females, then the female birth rates must be adjusted downward to compensate. Assume that downward compensation follows a linear relation with birth rates of 0 if no males are present.

Assume to start with that there are 220 animals equally distributed among age classes and sexes. Let your model run for 15 years without management to stabilize the natural growth rate. This will ensure that the system is close to some equilibrium state before it is perturbed by management, and this will eliminate the effects of the choice of the initial population distribution. It will also tell how an unmanaged herd will grow, which will give a baseline to compare with the management results.

It might be advisable to start with as simple a model as possible and then add complications.

Age Class	Male survival rates	Female survival rates	Birth rates (females born to females)
1	0.60	0.60	0
2	0.80	0.90	0.35
3	0.95	0.95	0.45
4	1.00	0.97	0.45
5	1.00	0.97	0.45
6	1.00	0.95	0.45
7	1.00	0.95	0.45
8	0.75	0.95	0.45
9	0.34	0.70	0.45
10	0	0.80	0.45
11			0.45

TABLE 3.10 Birth and Survival Rates for Impala Age Classes.

Note: The initial population size and distribution were made up. All other parameters including the numbers associated with different types of hunting and predation were taken from the article by Ginsberg and Milner-Gulland entitled *Sex-Biased Harvesting and Population Dynamics in Ungulates: Implications for Conservation and Sustainable Use* [21] and is reprinted by permission of Blackwell Science Inc.

Project 3.10. Barren Ground Caribou

Modelers and biologists from the Canadian Wildlife Service held a workshop to develop a model of barren-ground caribou population dynamics which resulted in the paper [71]. This project constructs a simplified version of this model. Assume you are part of a biological modeling team hired by the Canadian Wildlife Service to address the following questions. 1) What is expected from an unmanaged caribou population under normal circumstances? 2) What is the effect of harsh springs and winters? 3) What is the effect of the current hunting strategy under normal circumstances and with weather effects? 4) What is the effect of a revised hunting plan under both normal and weather effects? The parts below isolate and give details on these issues. Your report should address all these points, but you are in no way constrained by the organization presented here.

1. Baseline model. Using the demographic data in Table 3.11, construct a model for the caribou. Assume calves (fawns) are equally likely to be male as female. Further assume that the male population is always sufficient for siring purposes. Although caribou occasionally live more than 10 years, there is one class for caribou of age 10 and over. You should experiment and use your judgment to determine the number of years to run your model.

2. Weather Effects. Bad winters slightly affect the adult survival rate. With high snowfalls, the caribou are slowed and more susceptible to wolf attacks. Winters are rated on an index ranging from 0 to 1 with 0 being mild and 1 being the most severe. Figure 3.17 shows the change in survival rate for one adult class. Assume the others

3 Stages, States, and Classes

Age	Number of male	Number of female	Mortality rate	Percentage of females calving
1	4,500	4,500	0.60	0.00
2	4,000	4,000	0.03	1.8
3	3,500	3,800	0.03	48.0
4	3,100	3,600	0.03	82.0
5	2,700	3,400	0.03	92.0
6	2,400	3,200	0.03	92.0
7	2,100	3,000	0.02	92.0
8	1,900	2,900	0.02	90.0
9	1,700	2,700	0.02	90.0
10+	2,700	5,000	0.50	90.0

TABLE 3.11 Starting Population Size and Normal Rate Values for Caribou Simulation. Reprinted from Ecological Modelling, *Computer Simulation of Barren-Ground Caribou Dynamics* by C.J. Walters, R. Hilborn and R. Peterman, 1975, with permission from Elsevier Science.

are shifted so that the mild winters match the survival rates shown in Table 3.11. (In other words the third age class ranges from 0.97 to 0.94, with the transition points at 0.25 and 0.5.) Bad springs severely affect the survival of fawns, and can result in the death of the entire cohort of offspring (the transition point is at 0.85). The fawn survival rate as a function of a spring index is shown in Figure 3.18. Assume the winter and spring indices are random variables following a uniform distribution. Further assume that the severities of winters and springs are independent of each other (i.e., use separate random numbers for winter and spring). Add this stochasticity to your model and run an appropriate number of simulations.

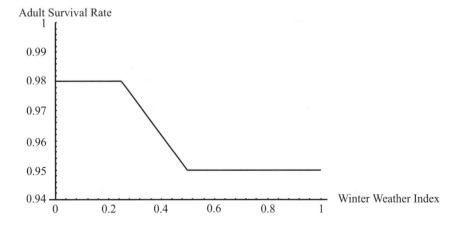

FIGURE 3.17 Effects of Winters on Adult Survival. Index Is Inches of Snow Divided by 100.

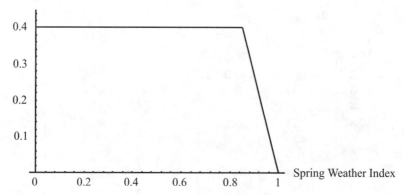

FIGURE 3.18 Fawn Survival Rate is 40% Unless Spring is Extremely Severe.

3. Hunting. There are two hunting seasons for caribou: summer and winter. Hunting rates for the various sexes, age classes, and seasons are shown in Table 3.12. Re-do the models from parts **1** and **2** to include hunting.

4. Hunting Redux. A new management plan proposes doubling the male harvest and halving the female harvest. Consider the effects of this both with and without weather effects.

Age	Summer Male	Summer Female	Winter Male	Winter Female
1	0.005	0.005	0.005	0.005
2	0.0285	0.016	0.06655	0.012
3	0.0285	0.016	0.06655	0.012
4	0.0285	0.016	0.06655	0.012
5	0.0285	0.016	0.06655	0.012
6	0.0285	0.016	0.06655	0.012
7	0.0285	0.016	0.06655	0.012
8	0.0285	0.016	0.06655	0.012
9	0.0285	0.016	0.06655	0.012
10+	0.0285	0.016	0.06655	0.012

TABLE 3.12 Hunting Rates for Various Ages, Sexes, and Seasons. Reprinted from Ecological Modelling, *Computer Simulation of Barren-Ground Caribou Dynamics* by C.J. Walters, R. Hilborn and R. Peterman, 1975, with permission from Elsevier Science.

4

Empirical Modeling

All useful models must incorporate information from the real world to some degree. This information frequently takes the form of numerical data. In Chapter 2, data was used to establish parameter values and parameter distributions. In this chapter, data will be used to establish the way in which one variable depends on another (or on several other variables). The first part of this chapter deals with fitting functions to data. It begins with fitting linear functions, which are the simplest and most commonly used functions. In this context, the ideas of covariance, correlation, linear regression, and the least-squares method are discussed. Following this, the fitting of nonlinear functions is investigated. Nonlinear functions fall into one of three cases: intrinsically linear, intrinsically nonlinear but linearizable, and intrinsically nonlinear but not linearizable. The word "linear" in linear regression refers to the function's being a linear combination of basis functions. With the basis functions $\{1, x\}$, linear combinations have the form $y = \beta_0 + \beta_1 x$ which is, coincidentally, a line. With other basis functions (say $\{1, x^2\}$), the result is a curve other than a line. So *intrinsically linear* functions are those which can be written in the form $y = \beta_0 f_0 + \beta_1 f_1 + \cdots + \beta_n f_n$, where the β_i's are constants and the f_i's are the basis functions. Many important nonlinear functions, such as the exponential function $y = Ae^{rx}$ and the power function $y = Ax^r$ where A and r are parameters to be fit, are not intrinsically linear. They can, however, be converted to intrinsically linear functions by suitable transformations, such as taking logs of one or both variables. We call these functions *intrinsically nonlinear but linearizable*. Finally, some functions such as the logistic curve

$$y = \frac{Ae^{rx}}{1 + e^{rx}}$$

are both nonlinear and cannot be transformed into a linear function. In these cases, we can sometimes apply the least-squares criterion directly, though frequently there are numerical difficulties. In Section 4.8.1, the least-squares method is used directly to fit logistic curves. Another approach to nonlinear data is interpolation. Rather than trying to fit a single curve to data, lines, quadratics, or cubics are fit to pieces of the data.

The result is a series of piecewise defined polynomials. This method is a little "dirtier" than least-squares fits in the sense that the curve explains the data, but there is no "nice" theoretically pleasing formula as a result.

For many purposes, the equation we fit to the data is the desired model. For others the equation is just a piece of the model. It might, for example, represent a density dependent growth rate that then gets used in a difference or differential equation. We consider these applications to differential equations in the next chapter.

There are a number of statistical issues involved in linear regression including the amount of variance explained by a regression curve and the significance of the model. These issues are discussed in the final part of this chapter where the elements of a regression or ANOVA table are discussed. Once readers understand these issues, they are prepared to understand the fundamentals of multiple regression and curvilinear regression.

4.1 Covariance and Correlation— A Discussion of Linear Dependence

Intuitively, we think two random variables are dependent if one either increases or decreases as the other changes. The degree of dependence changes from one set of variables to another. Consider the scatter plots of two data sets shown in Figures 4.1 and 4.2.

If all the points fall along a straight line, as indicated in Figure 4.1, then x and y are highly dependent. In contrast, the data shown in Figure 4.2 would indicate little or no dependence between x and y. In this section, we consider two ways of measuring the degree of linear dependence between two variables: covariance and the coefficient of correlation.

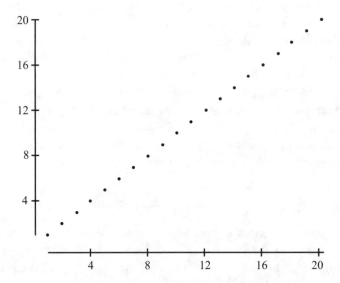

FIGURE 4.1 Dependent Observations.

4 Empirical Modeling

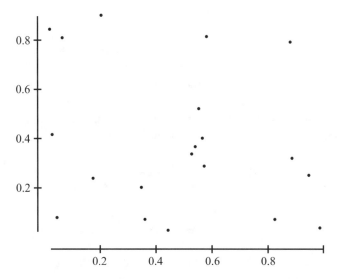

FIGURE 4.2 Independent Observations.

Suppose we know the values of $\bar{x}$ and $\bar{y}$, the means of the sample's x and y coordinates respectively. If we plotted $(\bar{x}, \bar{y})$ on the axes of the graph of Figure 4.1, we could then locate one of the plotted points (x_i, y_i), and consider the deviations $(x_i - \bar{x})$ and $(y_i - \bar{y})$. Notice that both deviations assume the same sign for a particular point and hence their product $(x_i - \bar{x})(y_i - \bar{y})$ is positive. This is true for *all* points on the graph of Figure 4.1. Points to the right of $\bar{x}$ yield pairs of positive deviations; points to the left produce pairs of negative deviations; and the average of the product of the deviations $(x - \bar{x})(y - \bar{y})$ is large and positive. If the linear relation indicated in Figure 4.1 sloped downward, all corresponding pairs of deviations would be of opposite sign, and the average of $(x_i - \bar{x})(y_i - \bar{y})$ would be a large negative number.

The situation is different for the graph of Figure 4.2, where little dependence exists between x and y. Corresponding deviations, $(x_i - \bar{x})$ and $(y_i - \bar{y})$, assume the same algebraic sign for some points and opposite signs for others. Thus the product $(x_i - \bar{x})(y_i - \bar{y})$ is positive for some points, negative for others, and averages to some value near zero.

From the previous discussion, it should be clear that the average of the product $(x_i - \bar{x})(y_i - \bar{y})$ provides a measure of the linear dependence of x and y. This quantity is called the *sample covariance* of x and y. The sample covariance is

$$\hat{\text{Cov}}(x, y) = \frac{\sum (x_i - \bar{x})(y_i - \bar{y})}{(n - 1)},$$

where n is the sample size. The hat indicates that this is the sample covariance, not the population covariance, and we divide by $n - 1$ instead of n to make this an unbiased estimator of the population covariance.

The larger the absolute value of the sample covariance, the greater the linear dependence between x and y. Positive values of the covariance indicate that y increases as x increases;

negative values indicate y decreases as x increases. A zero value of the covariance would indicate no linear dependence between x and y.

Unfortunately, it is difficult to employ the covariance as an absolute measure of dependence because its value depends upon the scale of measurement and, consequently, it is hard to determine whether a particular covariance is large at first glance. For example, suppose the covariance is 7. Is 7 a large number or not? This problem is corrected by standardizing the covariance values. The sample *correlation coefficient* r is the standardized version of the sample covariance and is defined as

$$r = \frac{\hat{\text{Cov}}(x,y)}{s_x s_y}$$

or

$$r = \frac{\sum (x_i - \bar{x})(y_i - \bar{y})}{(n-1) s_x s_y} \tag{76}$$

where s_x and s_y are the standard deviations of x and y.

It can be shown (Exercise 1) that the correlation coefficient r satisfies the inequality $-1 \leq r \leq 1$. Thus r values of -1 or $+1$ imply perfect correlation which means that all points fall on a straight line. An r value of zero implies a zero covariance and there is no dependence between the variables. The sign of the correlation coefficient depends upon the sign of the covariance. Positive values of the correlation coefficient indicate that y increases as x increases; negative values indicate y decreases as x increases.

One very important point is that saying x and y are highly correlated means only that x and y vary together. There is no implication that changes in x *cause* changes in y. We repeat: **CORRELATION DOES NOT IMPLY CAUSATION**. For example, you are probably familiar with the folklore fact that most accidents happen close to home. If you collected data, you would probably find a high negative correlation between distance from home and number of accidents. This does not imply that being close to home causes accidents. Most likely, people do the greatest proportion of their driving near their homes. They may be just as likely to be in an accident on their street as they would be on a street 2,000 miles away, but they are on their street much more often. Generally, experiments are required to show causation.

Up until this point, we have been talking about covariance and correlation of samples which are what we work with in practice. In theory, we assume that these are samples taken from a population with random variables X and Y. The above discussion holds when dealing with populations after making the standard conversions: μ_X and μ_Y replace $\bar{x}$ and $\bar{y}$, σ_X and σ_Y replace s_x and s_y, the covariance is $\text{Cov}(X,Y)$, the correlation is ρ, and, instead of averages, we compute expectations. Thus

$$\text{Cov}(X,Y) = E[(X - \mu_X)(Y - \mu_Y)],$$

where μ_X and μ_Y are the means of the random variables X and Y.

$$\rho = \frac{\text{Cov}(X,Y)}{\sigma_X \sigma_Y}$$

where σ_X and σ_Y are the standard deviations of X and Y.

To be technically correct, the terms covariance and correlation coefficient refer to $\text{Cov}(X, Y)$ and ρ, and the terms sample covariance and sample correlation coefficient refer to $\hat{\text{Cov}}(x, y)$ and r. Frequently people are imprecise and use the terms covariance and correlation for both sample and population and let the context make clear which is meant. We follow this convention throughout this chapter.

4.2 Fitting a Line to Data Using the Least-Squares Criterion

In the previous section, we discussed two measures of linear dependence, the covariance and the correlation coefficient. When the correlation coefficient is large, it indicates a dependence of one variable on the other. Several sections of this chapter are devoted to finding an appropriate curve to use to explain this dependence. The simplest non-trivial relationship, and by far the most frequently used, is the straight line. We begin by attempting to find a linear equation of the form

$$y = \beta_0 + \beta_1 x$$

to explain the variation in the data. For this discussion, consider the plot of x versus y shown in Figure 4.3.

There are a number of ways to determine the best line through these points. If we drew a line through these points by eye, this line is the best according to some hard to define set of criteria. In some sense, our line minimizes the distances between the points and the line. We can make this process rigorous by specifying that the (perpendicular) distance between the points and the line be minimized, or the vertical distance between each point and the line be minimized, or the horizontal distance be minimized. We can also choose between minimizing the sum of the true distances (which involves derivatives of absolute

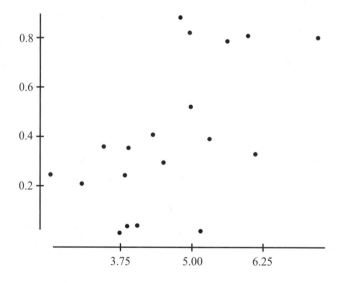

FIGURE 4.3 Observations of x and y.

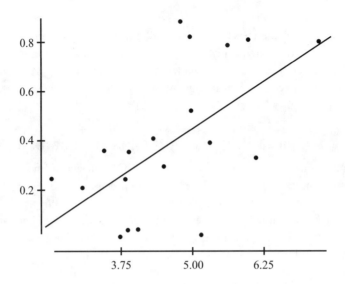

FIGURE 4.4 Fitting a Line.

values) or minimizing the sum of squares (which is suggested from the distance formula and has nice derivatives).

Of the various criteria for a best fit line, the following is used most frequently. The *least-squares criterion* for fitting a line through a set of data points minimizes the sum of the squares of the vertical deviations from the fitted line. The resulting equation is called the *regression equation*, and its graph is called the *regression line*. See Figure 4.4. We will use this criterion exclusively. It has the advantage that we usually want to fit a y value, given an x value; so this criterion minimizes the distance between observed y values and predicted y values for each x value (or rather minimizes the sum of these distances). The use of squares over absolute values to measure distances eases the computational work considerably and can be thought of as an analog to the distance formula.

The line which represents the true relationship between x and y is denoted by

$$\hat{y} = \beta_0 + \beta_1 x. \qquad (77)$$

In real situations we never know the true relationship, but are able only to estimate it from data. Thus we need to estimate β_0 and β_1, and we denote the estimated parameters by $\hat{\beta}_0$ and $\hat{\beta}_1$. The ideas of the true relationship versus the estimated relationship are discussed in much greater detail in Section 4.10 below.

The symbol $\hat{y}_i$ denotes the predicted value of the ith y value (i.e., when $x = x_i$). The deviation of the observed value of y_i from the $\hat{y}_i$ line, $y_i - \hat{y}_i$, is called the *residual* or *error*. The sum of the squares of the residuals is denoted as SS_{Res}, the SS standing for "sum of squares", and Res abbreviating "residual". The least-squares procedure finds the values of $\hat{\beta}_0$ and $\hat{\beta}_1$ which minimize the sum of the squares of the residuals.

4 Empirical Modeling

The function to be minimized is

$$SS_{\text{Res}} = \sum_{i=1}^{n}(y_i - \hat{y}_i)^2 = \sum_{i=1}^{n}[y_i - (\hat{\beta}_0 + \hat{\beta}_1 x_i)]^2. \tag{78}$$

If the function SS_{Res} has a minimum, it occurs when the partial derivatives of SS_{Res} with respect to $\hat{\beta}_0$ and $\hat{\beta}_1$ are both zero. This is given by the *least-square equations*:

$$\frac{\partial SS_{\text{Res}}}{\partial \hat{\beta}_0} = 0 \text{ and } \frac{\partial SS_{\text{Res}}}{\partial \hat{\beta}_1} = 0. \tag{79}$$

Solving these equations gives us

$$\hat{\beta}_1 = \frac{n\sum_{i=1}^{n} x_i y_i - \sum_{i=1}^{n} x_i \sum_{i=1}^{n} y_i}{n\sum_{i=1}^{n} x_i^2 - (\sum_{i=1}^{n} x_i)^2}, \tag{80}$$

and

$$\hat{\beta}_0 = \overline{y} - \hat{\beta}_1 \overline{x}, \tag{81}$$

where $\overline{x}$ and $\overline{y}$ are the arithmetic average of the data points x_i and y_i.

Thus we have estimates for the parameters β_0 and β_1 used in the equation (77). The equation we use as a model is the regression equation

$$\hat{y} = \hat{\beta}_0 + \hat{\beta}_1 x. \tag{82}$$

In practice, we use software packages to find the $\hat{\beta}_i$. The current discussion is intended to give an understanding of what a package is doing. Exercise 3 requires the use of these equations by hand with a small data set, just to developing understanding.

Occasionally we may drop the hats from our equations for either notational convenience or notational convention, but it is always understood that the regression equation is a prediction of the relation and that the coefficients are estimates found using the data.

4.3 R^2: A Measure of Fit

An important question is how well a model fits observed data. This section discusses one such measure of fit known as R^2. (Some texts distinguish between r^2 in single independent variable models and R^2 in multivariate models. Since there is one definition, we use one symbol, R^2, to avoid confusion. Our choice of R^2 corresponds to the output from most statistics packages.) The R^2 statistic measures the amount of variation in the data that is explained by a model. Although most of this chapter looks at R^2 in connection to results of fitting algorithms, it can be used to measure the amount of variance explained by any model (which has a single dependent variable which varies with one or more independent variables) including, for example, the curve produced by a spreadsheet in Chapter 1.

For the following discussion observed data points are denoted as (x, y) or (x_i, y_i). The function produced by the model is $\hat{y} = f(x)$, and the value predicted by the model at x_i is $\hat{y}_i$. Thus the predicted or modeled points are $(x, \hat{y})$ or $(x_i, \hat{y}_i)$. While we occasionally drop this *hat* notation for simplicity in later sections, it is necessary to be precise in this section.

The simplest useful model we can fit to data is a constant function. With this model, the dependent variable y does not change as the independent variable x changes. While different constants could be chosen (for example, any measure of center), the mean $\bar{y}$ is the most commonly chosen constant. If all the data actually has the same y coordinate, then the data has no variation and a constant function explains the data completely. If, however, the y values are not constant, then, clearly, there is some variation in the data about the mean. One way to measure this variation is called the *total sum of squares,* or SS_{Tot}, which is defined

$$SS_{\text{Tot}} = \sum(y_i - \bar{y})^2.$$

Notice that $SS_{\text{Tot}}/(n-1)$ was called the sample variance in Chapter 2, reinforcing the idea that SS_{Tot} is a measure of the variance in the data.

If we are building a model to explain the variation seen in the data, we need to use a model more complicated than a constant function; we call it $\hat{y} = f(x)$. If the data lie on this function exactly, then it explains all the variation. Usually, however, there is some noise to the data causing it to lie about a model function. Sometimes this noise is due to randomness. Sometimes there is curve which is a better model. Using a line, for instance, to model perfectly parabolic data, results in data points not lying on the model curve even in the absence of any random effects. In any case, there are two types of variations in the data: variation explained by the model and variation not explained by the model. The deviations between the predicted and actual y values $(y_i - \hat{y}_i)$ are called *residuals.* The variation not explained by the model is called the *residual sum of squares* or SS_{Res}. This is formally defined as

$$SS_{\text{Res}} = \sum(y_i - \hat{y}_i)^2.$$

One common form of data fitting is called regression. Because of this, the variation explained by the model (regression function) is called the *regression sum of squares* or

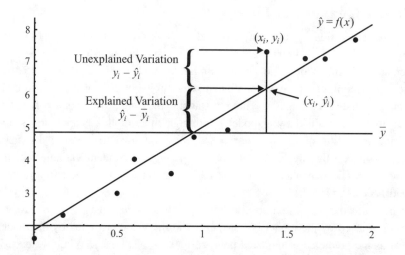

FIGURE 4.5 Total Variation: Explained and Unexplained.

4 Empirical Modeling

SS_{Reg}. This is defined by

$$SS_{\text{Reg}} = \sum (\hat{y}_i - \bar{y})^2.$$

Again this is the deviation from the mean explained by the model.

We have included an exercise (see Exercise 2) to show algebraically that

$$SS_{\text{Tot}} = SS_{\text{Reg}} + SS_{\text{Res}}.$$

Intuitively this means that the total variation in the data is the sum of the variation explained by the model as well as the variation not explained by the model. Thus the ratio

$$\frac{SS_{\text{Reg}}}{SS_{\text{Tot}}}$$

is the fraction (or percentage if multiplied by 100%) of the variation in the data explained by the model. This ratio is called R^2. Thus

$$R^2 = \frac{SS_{\text{Reg}}}{SS_{\text{Tot}}}.$$

Some books call R^2 the *coefficient of determination*. If the data is almost completely random, then almost none of the variance in the data is explained by the model. In this case, $SS_{\text{Reg}} \approx 0$ and hence $R^2 \approx 0$. On the other hand if the data has almost no noise and lies very nearly on the model or regression curve, then $SS_{\text{Res}} \approx 0$, so $SS_{\text{Tot}} \approx SS_{\text{Reg}}$, and hence $R^2 \approx 1$.

The observant reader may notice that there are some similarities between this discussion and the discussion of correlation. Indeed the notation R^2 comes from the fact that if we fit a line to data which minimizes the unexplained variance, the statistic R^2 is exactly the correlation r squared. For this reason, in simple linear regression, R^2 is denoted by r^2.

4.4 Finding R^2 for Curves Other Than Lines

Most introductions to the R^2 statistic follow the discussion of the preceding section which presents the model curve as a line and the observations as points about the line. As we will see shortly, we occasionally want to use model curves which are not lines. As an example, look ahead to Figures 4.11, 4.12, 4.14, and 4.16. In these figures, we have data with an evident nonlinear trend, and, using techniques discussed in the next several sections, we fit a power function and an exponential function to the data. For the purposes of the current discussion let us focus on the power function

$$y = e^{-25.7094} x^{8.35392}$$

(which we will derive later and take as a given for now). A graph of this function and the data is shown in Figure 4.16.

The question is how to compute R^2, the amount of variance in the data accounted for by the model, which in this case is this power function model. There are two ways of doing this. The most straightforward method is to observe that the definition of R^2 ($= SS_{\text{Reg}}/SS_{\text{Tot}}$) does not require that the model be a line and to apply the

4.4 Finding R^2 for Curves Other Than Lines

definition directly. This is done in some of the Chapter 4 examples in the *Mathematica* Appendix. While this is perfectly acceptable, it is usually not a built-in feature of easy-to-use statistical packages and must be computed with a user-written routine, as we have done in the *Mathematica* Appendix. (*Mathematica* can compute R^2 automatically for a special class of curves. See Section 4.13 on curvilinear regression for details.) While this is not especially difficult, it does add an extra layer of complexity to the process. Fortunately for models with a single independent variable, we can use the routines built into statistical packages for working with lines, provided we view the problem from a different angle.

The normal setting for regression is to have the independent variable on the x-axis and both the predicted and observed dependent variable values on the y-axis. For curves, we alter this to have the predicted data values on the x-axis and the observed values on the y-axis. Thus we ignore the independent variable and look at ordered pairs of the form $(predicted, observed)$. Take a moment to convince yourself that if the model predicted results with 100% accuracy, then the points should lie on the line $y = x$. It is also clear that we can fit a regression line to these points, and R^2 of this regression line explains the amount of variance in the observed values explained by the predicted values (i.e., explained by the model). It should be plausible, though perhaps not obvious, that this R^2 value is the same as would be found using $SS_{\text{Reg}}/SS_{\text{Tot}}$.

In Figure 4.6 we see the data from Figure 4.16 presented with predicted values on the x-axis and observed values on the y-axis. We note that the points fall very nearly on a line. If they fell exactly on a line, we would expect $R^2 = 100\%$, but as they are nearly, but not quite linear, we expect a R^2 value close to, but less than, 100%. Using the built-in regression features of the statistical package *Data Desk,* we find $R^2 = 99.3\%$. This value is also obtained directly from the definition of R^2 in the *Mathematica* Appendix.

We conclude this section by noting that both of the methods described here work for any deterministic model, whether obtained by minimizing least-squares or another method such as drawing a line through data by eye or using a curve suggested by theory. All that is needed is a predicted value for every observed value and vice versa. The interpretation

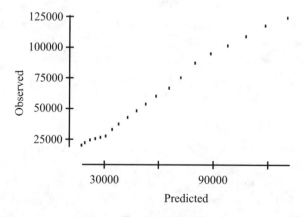

FIGURE 4.6 Cost of Advertising: Predicted Versus Observed.

4 Empirical Modeling

is still the same; R^2 gives the amount of variance in the data that is explained by the model.

4.5 Example: Opening *The X-Files* Again

In Example 2.2.3 of Chapter 2, we considered the number of viewers for the television show *The X-files*. Now we are in a position to develop an empirical model to predict the number of weekly viewers. The data from Example 2.2.3 is reproduced in Table 4.1 and shows the number of viewers in millions for each of the first two seasons.

The dependent variable is the number of viewers, and the independent variable is time, which could either be measured by episode number or by the number of the week that an episode aired. In regression analysis, we want both variables to have scales which are appropriately represented by the real numbers. In particular, the distance between 1 and 2 should be the same as the distance between 3 and 4. With this requirement, we use the week the episode aired instead of the episode number. Thus the first show, aired September 10, 1993, is labeled week 1; the second show which aired September 17, 1993, is labeled week 2; etc. The sixth episode has a first coordinate of 7 because it aired during the seventh week. There are two 23's since two new episodes were shown during the 23rd week. This data, in the form of ordered pairs, is presented in Table 4.2.

This data is plotted in Figure 4.7. This graph indicates a linear trend and suggests that a linear model would be appropriate. The gap in the graph between week 36 and week 53 is due to summer reruns between the two seasons.

The correlation coefficient of this data, computed using equation (76), is $r = 0.8605$, which indicates a reasonable degree of correlation of the data. Further, since $r^2 = R^2$ we have $R^2 = 0.7405$ so this linear model accounts for about 74% of the variance in the data.

Season 1	Season 2	Season 1	Season 2
7.4	9.8	6.2	8.8
6.9	9.3	6.8	10.2
6.8	8.7	7.2	10.8
5.9	8.2	6.8	9.8
6.2	8.5	5.8	10.7
5.6	9.2	7.1	9.6
5.6	9	7.2	10.2
6.2	9.1	7.5	9.8
6.1	8.6	8.1	7.9
5.1	9.9	7.7	8.5
6.4	8.5	7.4	8.1
6.4	9.7	8.3	9
			9.6

TABLE 4.1 Number of Viewers of *The X-files* (in millions).

(1,7.4)	(2,6.9)	(3,6.8)	(4,5.9)	(5,6.2)	(7,5.6)	(8,5.6)
(9,6.2)	(10,6.1)	(11,5.1)	(14,6.4)	(15,6.4)	(18,6.2)	(20,6.8)
(22,7.2)	(23,6.8)	(23,5.8)	(24,7.1)	(28,7.2)	(30,7.5)	(32,8.1)
(33,7.7)	(35,7.4)	(36,8.3)	(53,9.8)	(54,9.3)	(55,8.7)	(56,8.2)
(57,8.5)	(58,9.2)	(60,9)	(61,9.1)	(62,8.6)	(65,9.9)	(66,8.5)
(69,9.7)	(70,8.8)	(72,10.2)	(73,10.8)	(74,9.8)	(75,10.7)	(76,9.6)
(78,10.2)	(80,9.8)	(82,7.9)	(84,8.5)	(85,8.1)	(86,9)	(87,9.6).

TABLE 4.2 *The X-files* Data in Ordered Pairs (Week Number, Viewer (millions)).

We saw in Example 2.2.3 that the second season had a much higher average number of viewers than the first season. How well can we predict the second season from the first? (Here we are pretending that the second season has not happened yet, and we are using the first season model to make predictions. Then we will check our model with the second season data.) Performing regression analysis on the first season data

(1,7.4)	(2,6.9)	(3,6.8)	(4,5.9)	(5,6.2)	(7,5.6)	(8,5.6)
(9,6.2)	(10,6.1)	(11,5.1)	(14,6.4)	(15,6.4)	(18,6.2)	(20,6.8)
(22,7.2)	(23,6.8)	(23,5.8)	(24,7.1)	(28,7.2)	(30,7.5)	(32,8.1)
(33,7.7)	(35,7.4)	(36,8.3)				

gives the regression equation $\hat{y} = 5.8937 + 0.0466x$. This equation is plotted in Figure 4.8 along with the data points for only the first season. Visually, this line looks like a good fit to the data, but the data has some noise. The statistic R^2 is found to be 0.4103, which says that this model explains 41% of the variance seen in the data.

How well does this equation predict viewing during the second season? A plot of the regression equation together with the data for both seasons is shown in Figure 4.9. Visually, the prediction appears good. Using the linear model, we predict the number of

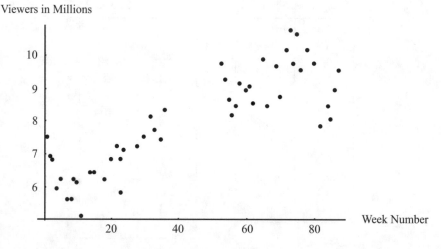

FIGURE 4.7 Scatter Plot of The First Two Seasons of *The X-files*.

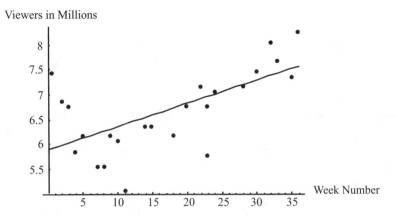

FIGURE 4.8 Regression Line and the First Season Data.

viewers at any time during the second season. For instance, the model predicts that the number of viewers for the first episode of the second season is

$$5.8937 + 0.0466 \times 53 = 8.36 \text{ million}.$$

The actual number of viewers was 9.8 million.

Computing R^2 for the model of the first season and the data of the second season gives a value of $R^2 = 0.2532$. In other words, this model actually explains only 25% of the variation in the data. Interestingly, if we compute the regression line for the second season, it has $R^2 = 0.0098$. Thus the first season model explains much more of the second season variance than the second season model. This seeming contradiction raises a very important point. Regression optimizes according to the least-squares criterion; thus minimizing unexplained variance. This is not the same as maximizing R^2 which

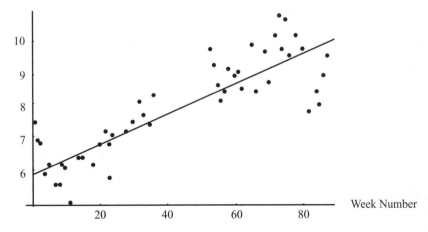

FIGURE 4.9 Regression Line and the First Two Seasons of Data.

is the ratio of explained variance to the total variance. In this example, $SS_{\text{Res}} = 19.91$ for the first season model with second season data, and $SS_{\text{Res}} = 14.95$ for the second season model with the second season data. Indeed, the regression model for the second season has a smaller sum of squares as required. However, the total and regression sums of squares have also changed resulting in a lowered R^2. These calculations are in the *Mathematica* Appendix.

Next we consider how well the first season model fits all the data. This is shown in Figure 4.9. The statistic $R^2 = 0.7466$; this implies that this model explains 74% of the variation seen in the data—which is not only good, but better than the amount of variation the first season model explains in the first season data alone.

Finally, a regression line for both seasons is

$$\hat{y} = 5.9968 + 0.0457x.$$

Again it is interesting that $R^2 = 0.7404$, which is slightly worse than the first season model alone. Concluding, the model for the first season alone does a respectable job of explaining variance observed over both seasons.

Which model should we use? That, of course, depends on the purpose. Our purpose is to demonstrate the R^2 statistic, which is the reason we fit the data every which way. If we had only the first season data, then clearly we would fit only it. If we had both seasons (as we do), there are two common approaches. One is to fit a model to all the data. This was the first analysis we performed. The problem is that we need to collect more data to verify our predictions. Another approach is to split the data (as we have into first and second seasons) and fit a model to the first half and use the second half to test the first.

On a final note, the regression line for the second season data alone has slope $\hat{\beta}_1 = 0.0071$ while the regression line for the first season data alone has slope $\hat{\beta}_1 = 0.0466$, which might indicate that the growth is tapering off. On these grounds, a logistic function is worth examining. This is done in Project 4.4.

4.6 Curvilinear Models

We can fit any function to data, but we want a function that behaves in a manner similar to the data. For many purposes a line is adequate, but often the data clearly exhibits nonlinearities such as concavity or periodic behavior. The purpose of this section is to explore ways of extending the least-squares criterion to nonlinear data.

4.6.1 A Catalog of Functions

This section reviews basic functions and function construction from precalculus. These will be used as building blocks in our discussion of empirical model construction.

A Ladder Of Powers. Deciding which function to fit to data is an important step in the modeling process. For data which is growing steadily, natural candidates are linear functions ($y = mx + b$), power functions ($y = x^p$), logarithmic functions ($y = \log_b x$), and exponential functions ($y = e^x$ or $y = b^x$). These functions are familiar to the reader

4 Empirical Modeling

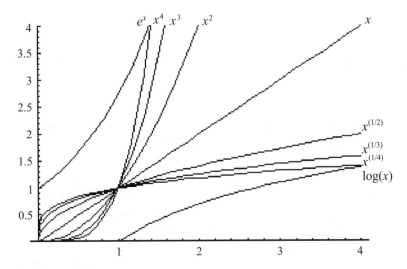

FIGURE 4.10 Ladder of Functions.

from precalculus and calculus. For reference, their graphs are shown in Figure 4.10. The most important thing to remember is the order of dominance. For x large enough, the exponential functions grow faster than any power function (or finite sum of power functions), power functions with degree greater than one grow faster than lines, lines grow faster than power functions with degree between zero and one, and all the aforementioned grow faster than logarithm functions. Keeping this in mind, and recalling the basic shapes (especially concavity) helps us choose the function to fit to a curve.

Periodic Functions. Periodic functions repeat themselves. More precisely, a function $f(x)$ is *periodic* if there is a positive constant k so that $f(x+k) = f(x)$ for all x. The smallest value of k for which $f(x+k) = f(x)$ is called the *period* of the function $f(x)$.

The sine and cosine functions are the familiar examples of periodic functions (with period 2π) and are frequently used in modeling periodic data. In Fourier analysis, it is shown that any periodic function is expressible as a (possibly infinite) sum of sine and cosine functions of different periods. For our purposes, we put a single sine function and a single cosine function together to model simple periodic functions.

The general form of the sine function is

$$y = A \sin b(x - d),$$

where the constant $|A|$ is known as the *amplitude*, $2\pi/b$ is the period, and d is the *phase shift* or the amount the sine curve is translated to the right, if $d > 0$. As seen below, fitting the phase shift can be very difficult so we frequently make use of the fact (see Exercise 6) that $A \sin b(x - d)$ can be expressed as $C_1 \sin bx + C_2 \cos bx$, provided $A = \sqrt{C_1^2 + C_2^2}$ and $\tan(-bd) = \frac{C_1}{C_2}$. The constants C_1 and C_2 can be easily found using least-squares regression, so we always use the form $C_1 \sin bx + C_2 \cos bx$.

New Curves From Old. One of the most important topics from precalculus is understanding what happens to the graph of a function under simple modifications. These ideas can be very helpful in creating a function to model data, and we briefly review these ideas.

If we know the graph of $f(x)$ and $a > 0$, then $f(x - a)$ has the same shape only shifted by a units to the right. The graph of $f(x + a)$ is the same as $f(x)$ only shifted to the left. The graph of $f(x) + a$ and $f(x) - a$ shift $f(x)$ up and down respectively. The graph of $-f(x)$ is the graph of $f(x)$ flipped about the x-axis, while the graph of $f(-x)$ is the graph of $f(x)$ flipped about the y-axis.

Multiplication is a little more complicated. The graph of $af(x)$ is the graph of $f(x)$ stretched in the vertical direction ($a > 1$ is literally a stretch, while $0 < a < 1$ is a compression). The graph of $f(ax)$ is a stretch in the horizontal direction, only this time $a > 1$ is a compression, while $0 < a < 1$ is a true stretch.

These ideas were seen in the discussion on the sine curve above. The general form of a sine curve is $y = A \sin b(x - d)$ (we assume all parameters are positive). This function has the graph of the usual sine curve $y = \sin x$ only stretched by A units in the y-direction (the amplitude), the period is $2\pi/b$ which is longer if $b < 1$. Finally, after stretching the period, the curve is shifted right d units, which is the *phase shift*.

The ideas of multiplication by a constant extend to the multiplication of two functions. Namely, $g(x)f(x)$ has a graph similar to $f(x)$, only stretched and compressed according to $g(x)$. In general this can be nasty, but there are some important special cases. The function $g(x)\sin(x)$ has a graph consisting of oscillations between $g(x)$ and $-g(x)$. The function $g(x)$ is called an *envelope*, and the function oscillates within the envelope. To hand draw, say $x^2 \sin(x)$, we draw the parabolas x^2 and $-x^2$ with a dotted line (the envelope) and then draw the curve between them in an oscillatory manner observing the phase and period. A common function is $e^{ax} \sin(bx)$ which represents either an exponentially growing or decaying oscillation as might be observed with a spring oscillating on a surface with friction.

Use Of Scatter Plots. How do we know which curve to try to fit to data? There are two important methods. One is to use theory which suggests how the data should behave. For example, a young child's height may be growing exponentially, but we know this model is valid for a limited time because people stop growing after a point, so a logistic type model would be more appropriate for long-term purposes. In the absence of such theory, or if the theory appears not to hold, or even if it does, visually examining the data is a useful method. A scatter plot of the data often reveals the type of model we should try to fit. The concavity of the scatter plot suggests which form should be fit. Scatter plots that are concave up suggest fitting with functions e^x or x^p, with $p > 1$. Scatter plots that are concave down suggest fitting logarithmic or power (x^p, with $0 < p < 1$) functions. Intercepts may suggest shifting one of these basic components. Asymptotes might suggest using either some type of logistic curve or some type of rational function such as $y = \dfrac{x}{Ax - B}$. Periodic behavior suggests sines and cosines. If none of these seems appropriate, we need to try something else, such as combinations of these elements or interpolation.

4.6.2 Intrinsically Linear Models

The previous sections developed fitting a line according to the least-squares criterion. This method provides a powerful modeling tool for functions other than lines. Begin by observing that while the equation $\hat{y} = \beta_0 + \beta_1 x$ is a linear function in the sense of being a line, it is also a linear combination of 1 and x. Viewed this way it is a linear function in the two variables β_0 and β_1. This linearity in the parameters β_0 and β_1 is the key to obtaining the least-squares equations (80) and (81), and any model that is linear in the parameters β_0 and β_1, but not necessarily in the independent variable x can use the method of least squares. For instance,

$$y = \beta_0 + \beta_1 \sin x$$

is linear in β_0 and β_1 so the least-squares method can be applied. Similarly, if more than one independent variable, say $x_1, x_2, \ldots, x_k$, are of interest, then we may model

$$y = \beta_0 + \beta_1 x_1 + \beta_2 x_2 + \cdots + \beta_k x_k$$

which is called a *multiple* least-squares model. The least-squares method can be applied to any model that is a linear combination of the β_i's, namely

$$y = \beta_0 f_0(x) + \beta_1 f_1(x) + \beta_2 f_2(x) + \cdots + \beta_k f_k(x). \tag{83}$$

Thus we can apply the least-squares technique to any function which is a linear combination of functions of x, which we call *basis functions*. Technically basis functions satisfy a condition called independence, but for our purposes, think of them as building blocks used to construct a model. The least-squares equations for estimating the parameters $\beta_0, \beta_1, \cdots, \beta_k$ in equation (83) are obtained in a manner analogous to equations (81) and (80); however, more complicated machinery is involved. For our purposes, software packages will be used to estimate the β_i parameters. To effectively apply the method of least squares in building nonlinear models, we must have a good sense of which functions combine to exhibit a desired behavior. A warning that will be repeated several times is that we always obtain better and better fits by adding more and more basis functions, but the resulting equation often does a worse job of predicting than a simple function would have done. A huge equation also conveys little theoretical information. The goal is to find the best possible simple equation. One way to do this is to understand elementary functions and what happens when we do simple operations with them, which is why Section 4.6.1 is important.

4.6.3 Intrinsically Nonlinear Models Which Can Be Linearized

There are many real-world situations for which the most appropriate deterministic model is not linear, but which can be transformed into a linear relation. As we have seen, many populations of plants or animals tend to grow exponentially. If $Y(t)$ denotes the size of a population at time t, we might employ a model

$$Y(t) = Ae^{rt}.$$

In the Cost of Advertising example below, a model is fit to a polynomial or power function of the form

$$Y(t) = At^p.$$

In both these cases the models are not linear in the parameters A, r, or p.

Exponential And Power Law Models. Both exponential and power law equations can be transformed into linear equations. In the case of $y = Ae^{rx}$, if we take logarithms of both sides, the result is $\ln y = \ln A + rx$ which is a linear equation in the variables x and $\ln y$. Thus if an exponential model is to be fit to a set of paired data, $\{(x_1, y_1), (x_2, y_2), \ldots, (x_n, y_n)\}$, we begin by taking the logarithm of the y-coordinates. This yields

$$\{(x_1, \ln y_1), (x_2, \ln y_2), \ldots, (x_n, \ln y_n)\}. \tag{84}$$

Next we find the regression line through the points in (84), which yields the equations

$$\ln y = \hat{\beta}_0 + \hat{\beta}_1 x.$$

Solving this equation for y gives

$$y = \tilde{\beta}_0 e^{\hat{\beta}_1 x},$$

where $\tilde{\beta}_0 = e^{\hat{\beta}_0}$. This is the form we desire.

Next suppose we want to fit a power law equation, (an equation of the form $y = Ax^p$). This time we take logarithms of both sides, yielding $\ln y = \ln A + p \ln x$ which is a linear equation in the variables $(\ln x, \ln y)$.

Thus if a power model is desired, fitting a regression line through

$$\{(\ln x_1, \ln y_1), (\ln x_2, \ln y_2), \ldots, (\ln x_n, \ln y_n)\}$$

gives the equation

$$\ln y = \hat{\beta}_0 + \hat{\beta}_1 \ln x.$$

Solving for y yields

$$y = \tilde{\beta}_0 x^{\hat{\beta}_1},$$

where $\tilde{\beta}_0 = e^{\hat{\beta}_0}$.

We emphasize exponential and power law models because they are commonly used, but there are other equations which can be linearized by an appropriate transformation. Here are several others:

Logarithmic Models. If the desired equation has the form $y = A + B \ln x$, then observe that this is linear in the variable $(\ln x, y)$. Thus, fitting a regression line through

$$\{(\ln x_1, y_1), (\ln x_2, y_2), \ldots, (\ln x_n, y_n)\}$$

gives the equation

$$y = \hat{\beta}_0 + \hat{\beta}_1 \ln x.$$

4 Empirical Modeling

Models With A Vertical And Horizontal Asymptote. The equation

$$y = \frac{x}{Ax - B}$$

has a vertical asymptote at $\frac{B}{A}$ and a horizontal asymptote at $\frac{1}{A}$. Whether the function is increasing or decreasing depends on the sign of B. The substitution $v = \frac{1}{y}$, $u = \frac{1}{x}$ transforms this equation into $v = A - Bu$. Thus fitting a regression line through

$$\{(\frac{1}{x_1}, \frac{1}{y_1}), (\frac{1}{x_2}, \frac{1}{y_2}), \ldots, (\frac{1}{x_n}, \frac{1}{y_n})\}$$

gives the equation

$$\frac{1}{y} = \hat{A} - \hat{B}\frac{1}{x}.$$

This can be converted back into the original form.

Almost Logistic. The equation

$$y = \frac{e^{A+Bx}}{1 + e^{A+Bx}}$$

is essentially a logistic curve with carrying capacity of 1. It has applications in probability and statistics (called a logistic distribution). This equation can be transformed into $v = A + Bx$ by the substitution

$$v = \ln \frac{y}{1-y}.$$

The true logistic function, however, has the form

$$y = \frac{Ae^{Bx}}{1 + e^{Bx}}$$

and cannot be transformed into a linear equation. One use of this transformation, if you are willing to estimate the carrying capacity (A in this equation) by some other method (eyeballing the data, theoretical ecological limits, etc.), is to find the intrinsic growth rate (B in this case). Fitting a true logistic curve to data is discussed in Section 4.8.1.

4.7 Example: The Cost of Advertising

Every day we are bombarded with advertisements. We are always being told which brand of cat litter our cats prefer or which politician we prefer. The frequency and intensity of these advertisements seem to be on the increase. One measure of this increase is the amount of money spent on advertising. The annual total expenditure in millions of dollars of advertising in the United State from 1970 to 1989 is given in Table 4.3. The data for 1970–1986 is from the *Statistical Abstract of the United States: 1988*[47] and is reprinted by permission of Crain Communications Inc. The data for 1987–1989 is due to Jonathan Jernigan.

A graph of this data is presented in Figure 4.11. For convenience, the year is represented by the last two digits. We construct three models of this data: a linear, a power model, and

Year	Expenditure (in millions)
1970	19,550
1971	20,700
1972	23,210
1973	24,980
1974	26,620
1975	27,900
1976	33,300
1977	37,440
1978	43,330
1979	48,780

Year	Expenditure (in millions)
1980	53,550
1981	60,430
1982	66,580
1983	75,850
1984	87,820
1985	94,750
1986	102,140
1987	109,650
1988	118,050
1989	123,930

TABLE 4.3 U.S. Advertising Expenditures. Reprinted with permission of Crain Communications Inc.

an exponential model. All calculations are performed with *Mathematica*, and a sample of these calculations can be found in the *Mathematica* Appendix.

We begin with a linear model which we find to be

$$y = -40{,}1900 + 5{,}809.16x.$$

This equation is plotted together with the data in Figure 4.12.

For this model, $R^2 = 0.95307$, which is very respectable. From the graph, however, it is evident that the unexplained variance has a trend, as opposed to being due to noise. One observation is the visually evident fact that the data is concave up. This suggests that we consider fitting a nonlinear model; a power law model or an exponential model are likely candidates. If we assume that the United States population is growing exponentially and

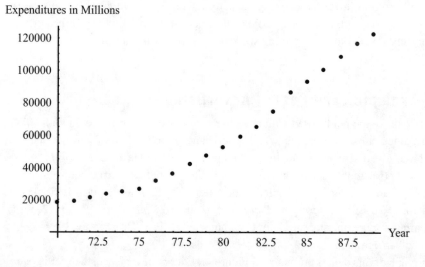

FIGURE 4.11 U.S. Advertising Expenditures.

4 Empirical Modeling

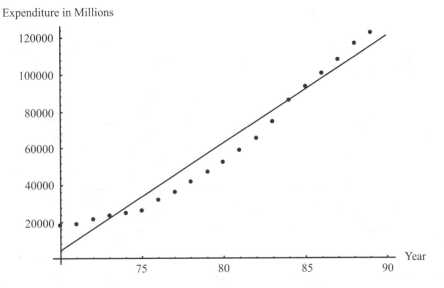

FIGURE 4.12 Linear model of Expenditures Versus Year.

that many other things associated with the population are growing exponentially as well, on theoretical grounds we can reasonably suspect that the annual expenditure is growing exponentially. We begin with an exponential model. Taking the natural logarithm of the expenditures (see Table 4.4) and plotting this data gives Figure 4.13.

Figure 4.13 appears to be more or less linear. Fitting a regression line to this transformed data, we obtain

$$\ln y = 2.43694 + 0.105489x$$

Year	Log of Expenditures
70	9.88073
71	9.93789
72	10.0523
73	10.1258
74	10.1894
75	10.2364
76	10.4133
77	10.5305
78	10.6766
79	10.7951

Year	Log of Expenditures
80	10.8884
81	11.0092
82	11.1062
83	11.2365
84	11.383
85	11.459
86	11.5341
87	11.605
88	11.6789
89	11.7275

TABLE 4.4 Logarithm of U.S. Advertising Expenditures.

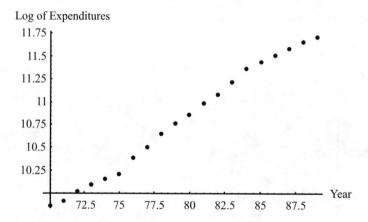

FIGURE 4.13 Logarithm of Expenditures Versus Year.

which, upon solving for y, gives the exponential model

$$y = \exp(2.43694 + 0.105489x)$$
$$= 11.43799 e^{0.105489x}. \qquad (85)$$

This curve is plotted with the original data in Figure 4.14. The model reflects the upward trend in the data, but also preserves the concavity of the data. This model has $R^2 = 0.988$ which implies that 98.8% of the variance in the data is explained by this model, which is very good.

The concavity of the scatter plot Figure 4.11 suggests that we might also consider a power model for this data. To fit a power model, the natural logarithm of both the year and expenditure must be calculated. This data is shown in Table 4.5. The plot, Figure 4.15, appears to be almost linear.

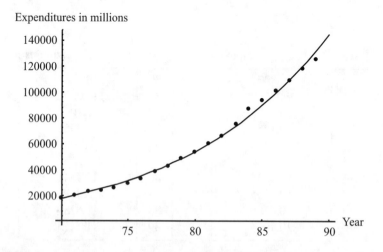

FIGURE 4.14 Exponential Model Of Expenditures Versus Year.

4 Empirical Modeling

Fitting a regression line to this transformed data, we obtain

$$\ln y = -25.7094 + 8.35392 \ln x,$$

which upon solving for y gives the power model

$$y = \exp(-25.7094 + 8.35392 \ln x)$$
$$= e^{-25.7094} x^{8.35392}. \tag{86}$$

Log of Year	Log of Expenditure	Log of Year	Log of Expenditure
4.2485	9.88073	4.38203	10.8884
4.26268	9.93789	4.39445	11.0092
4.27667	10.0523	4.40672	11.1062
4.29046	10.1258	4.41884	11.2365
4.30407	10.1894	4.43082	11.383
4.31749	10.2364	4.44265	11.459
4.33073	10.4133	4.45435	11.5341
4.34381	10.5305	4.46591	11.605
4.35671	10.6766	4.47734	11.6789
4.36945	10.7951	4.48864	11.7275

TABLE 4.5 Logarithm of U.S. Advertising Expenditures.

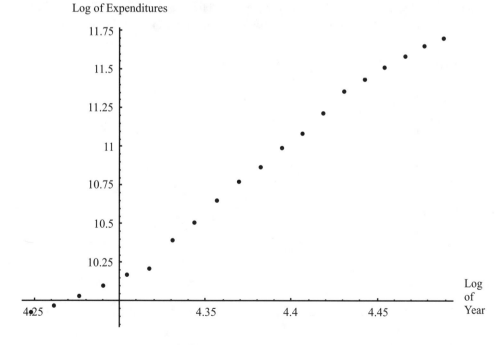

FIGURE 4.15 Logarithm of Expenditures Versus Logarithm of Year.

This power model is plotted in Figure 4.16 along with the data. Graphically, this model appears to represent the data better than the exponential model shown in Figure 4.14. We can quantitatively compare these models. The SS_{Res} equation (78), provides a useful measure when comparing two models for the same data. The model with the smaller SS_{Res} is the better choice. The SS_{Res} for the exponential model (85) is

$$SS_{\text{Res}} = \sum_{i=1}^{20}(y_i - 11.43799 e^{0.105489 x_i})^2$$
$$= (19550 - 18700.4)^2 + (20700 - 20594.92)^2 + \ldots + (123930 + 116984.85)^2$$
$$= 3.07135 \times 10^8$$

where each term of the sum is the square of the difference between the actual and predicted values. The SS_{Res} here seems very large. We must keep in mind that when calculating the error in some approximation, the error is relevant only when reported relative to the measurement. In this context, we wish to compare this error to the SS_{Res} for the power model (86). The SS_{Res} for the power model is computed in a similar manner and is

$$SS_{\text{Res}} = 1.76738 \times 10^8.$$

The SS_{Res} of the power model is smaller then the SS_{Res} of the exponential model. It is often more helpful to compare R^2 values. For the exponential model $R^2 = 0.988$, while for the power model $R^2 = 0.993$. Thus the power model explains 99.3% of the variation in the data, compared to 98.8% for the exponential model. Thus, while the exponential model is good, the power model is even better—excellent, in fact. From Figure 4.12 it is apparent that the linear model does not reflect the curvature in the data; we further note

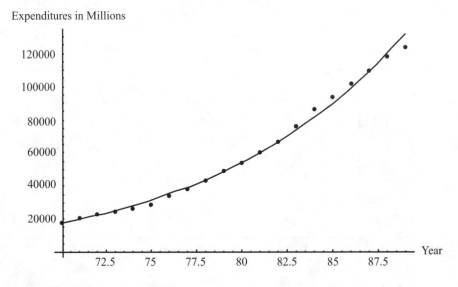

FIGURE 4.16 Power Model of Expenditures Versus Year.

4 Empirical Modeling

that the SS_{Res} for the linear model is $SS_{\text{Res}} = 1.105 \times 10^9$, a factor of 10 larger than the SS_{Res}'s for the exponential and power models.

4.8 Example: Lynx Fur Returns

This model combines many of the ideas discussed above.

Extensive records are kept on the sale of animal pelts or furs. These are often used by biologists as a measure of the size of the population. The assumption is that the higher the population level, the more animals will be trapped. A large collection of fur return and other animal time series data is found in the book *Wildlife's Ten-year Cycle* [30] by Keith. One of the animals covered in this source is the Canadian lynx. Lynx fur returns for eight provinces or territories of Canada from 1919 to 1957 are given in Table 4.6.

A scatter plot of this data is shown in Figure 4.17. Examination of this graph reveals that the data exhibit periodic behavior. (Technically, the definition of periodic is not satisfied, but some periodic building block is evident. We will continue to abuse the word periodic in this way.) The fur returns oscillate with a period of about 10 years. This periodic behavior suggests that there is a cyclical nature to fur returns. However, the data is not purely a periodic function, since there is a decreasing trend in the maximum or minimum returns over each 10-year period. For instance, in 1925 the return was 33,027, while in 1935 the return was 22,448.

Building an empirical model for the lynx data requires several steps. For a baseline model, a line is fit to the data. This function models the linear trend in the data. The year is represented by the last two digits as we have done in several previous examples. Fitting a regression line to this data gives

$$y = 24,212.7 - 306.03x. \tag{87}$$

This equation is plotted together with the data in Figure 4.18.

Year	Return	Year	Return	Year	Return	Year	Return
1919	8,378	1930	7,957	1940	6,583	1950	9,592
1920	6,456	1931	8,410	1941	6,979	1951	6,653
1921	11,617	1932	11,916	1942	7,544	1952	12,636
1922	17,202	1933	16,781	1943	10,164	1953	10,876
1923	26,381	1934	22,012	1944	12,259	1954	13,876
1924	29,529	1935	22,448	1945	9,306	1955	9,660
1925	33,027	1936	17,534	1946	8,129	1956	8,397
1926	28,619	1937	10,523	1947	6,548	1957	8,958
1927	21,363	1938	8,079	1948	4,083		
1928	11,582	1939	7,411	1949	3,714		
1929	7,572						

TABLE 4.6 Lynx Fur Returns from 1919 to 1957. Reprinted by permission of The University of Wisconsin Press.

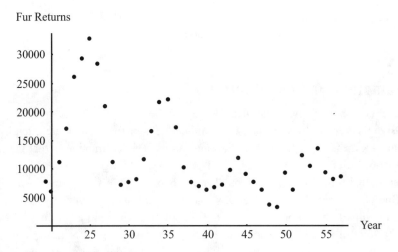

FIGURE 4.17 Scatter Plot of Lynx Fur Return Data.

The linear model represents the decreasing trend in the data. The next model of the lynx data incorporates the periodic behavior. From the catalog of basic functions, $\sin bx$ and $\cos bx$ are the periodic building blocks. The process of numerically fitting the period falls into the category of nonlinear and nonlinearizable. Since the period of the data has been estimated to be 10 years, we will use this value. The basis functions that we use for the second modeling attempt are $\sin 2\pi x/10$ and $\cos 2\pi x/10$ along with the linear building blocks 1 and x. The implementation of this is best accomplished using a software package. For example, the **Fit** command in *Mathematica* applied to the data with the basis functions $\{1, x, \sin(2\pi x/10), \cos(2\pi x/10)\}$ gives the equation

$$y = 26986.1 - 381.682x - 6994.41\cos(\pi x/5) + 1880.02\sin(\pi x/5)$$

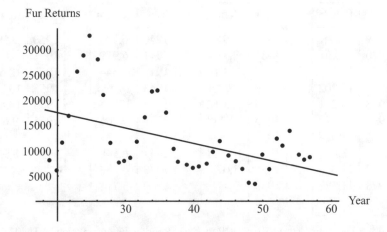

FIGURE 4.18 Linear Model of Lynx Fur Returns.

4 Empirical Modeling

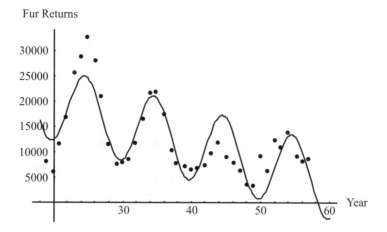

FIGURE 4.19 A Periodic Model of Lynx Fur Returns.

as the second empirical lynx model. A plot of this equation along with the data is given in Figure 4.19.

This model reflects the cyclical nature and the decreasing trend, but does a poor job representing the amplitude of each cycle. The third model incorporates these amplitudes. The data in Figure 4.17 does not have constant amplitude; rather, the amplitude of each cycle is decreasing. The next model incorporates nonconstant amplitude by being a linear combination of the basis functions $\{1, x, f(x)\sin(2\pi x/10), f(x)\cos(2\pi/10)\}$.

To model the amplitude or envelope function $f(x)$, the data must be manipulated to find the extrema of $f(x)$. The first step is to remove the linear trend from the data. This "detrending" of the data is accomplished by subtracting the value of the linear model (87) at each time step from the observed data value. For instance, the detrended value for the year 1919 is given by

$$8,378 - (24,212.7 - 306.03 \times 19) = -10,020.13.$$

The remaining detrended values are calculated similarly and are given in Table 4.7. A plot of this detrended data is given in Figure 4.20 and shows the decreasing amplitude.

To model the envelope of the data, we only need to consider the extrema of the data. Since the period of the data has been estimated as 10 years, the minima and maxima occur every five years. Since $\sin x$ and $\cos x$ oscillate between 1 and -1 in magnitude, to model the envelope, we take the absolute value (i.e., find the magnitude) of these data points. The envelope data is shown in Table 4.8. The envelope data is plotted in Figure 4.21. A question about an appropriate choice for the form of $f(x)$ now arises. The simplest form for $f(x)$ is a linear function. Since $f(x)$ represents the amplitude of the lynx data, its values should be non-negative, that is $f(x) \geq 0$ for all x. Further, since the amplitude of the lynx data is decreasing, a decreasing linear $f(x)$ will not always be non-negative. So, a linear envelope is valid for only a limited time into the future.

A theoretically more pleasing choice for a non-negative decreasing $f(x)$ is an exponential function. To fit an exponential, the data must be transformed by taking the

Year	Return	Year	Return	Year	Return	Year	Return
19	-10,020.1	30	-7,074.71	40	-5,388.37	50	680.97
20	-11,636	31	-6,315.67	41	-4,686.33	51	-1,952.
21	-6,169.01	32	-2,503.64	42	-3,815.3	52	4,337.04
22	-277.977	33	2,667.4	43	-889.267	53	2,883.07
23	9,207.06	34	8,204.43	44	1,511.77	54	6,189.1
24	12,661.1	35	8,946.46	45	-1,135.2	55	2,279.14
25	16,465.1	36	4,338.5	46	-2,006.17	56	1,322.17
26	12,363.2	37	-2,366.47	47	-3,281.13	57	2,189.21
27	5,413.19	38	-4,504.44	48	-5,440.1		
28	-4,061.77	39	-4,866.4.	49	-5,503.06		
29	-7,765.74						

TABLE 4.7 Detrended Lynx Fur Returns.

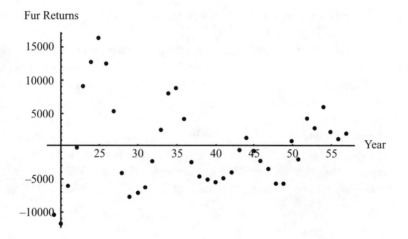

FIGURE 4.20 Detrended Lynx Data.

natural logarithm of the observed values of the dependent variable y. The logarithm of the envelope data is given in Table 4.9. Fitting a regression line to this transformed data, we obtain

$$\ln f(x) = 11.3409 - 0.07937x$$

which, upon solving for $f(x)$, gives the exponential model

$$f(x) = \exp(11.3409 - 0.07937x)$$
$$= 84,195.77 e^{-0.07937x}. \tag{88}$$

A plot of this along with the envelope data is shown in Figure 4.22. It is now possible

4 Empirical Modeling

Year	Envelope
20	11,636.0
25	16,465.1
30	7,074.71
35	8,946.46
40	5,388.37
45	1,135.2
50	680.97
55	2,279.14

TABLE 4.8 Magnitude of the Extrema of the Detrended Lynx Data.

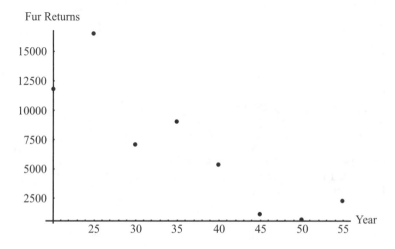

FIGURE 4.21 Envelope for Lynx Data.

Year	Envelope
20	9.36186
25	9.709
30	8.86428
35	9.09901
40	8.592
45	7.03456
50	6.52352
55	7.73155

TABLE 4.9 Logarithm of the Extrema of the Detrended Lynx Data.

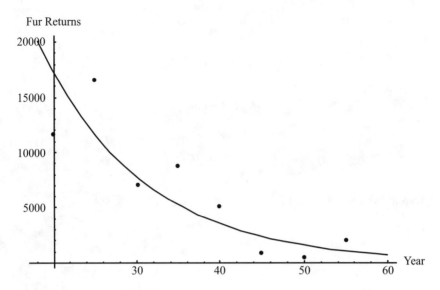

FIGURE 4.22 Exponential Envelope.

to build the lynx model using the least-squares method with

$$\{1, x, f(x)\sin(2\pi x/10), f(x)\cos(2\pi x/10)\}$$

as the basis, where $f(x)$ is given by equation (88). The implementation of this is again best accomplished using a software package. Using the **Fit** command in *Mathematica* with this basis gives the equation

$$y = 28,951.525 - 414.862\,x - 0.994326\,e^{11.3409 - 0.0793708\,x}\cos(\frac{\pi x}{5}) +$$
$$0.139518\,e^{11.3409 - 0.0793708\,x}\sin(\frac{\pi x}{5})$$

as the empirical lynx model with an exponential envelope. A plot of this equation along with the data is given in Figure 4.23. The model and the data visually agree. For this model $R^2 = 0.845098$, so about 85% of the variability in the data is accounted for by this model.

4.8.1 Intrinsically Nonlinear Models Which Cannot Be Linearized: The Logistic Equation

Some equations are neither a linear combination of basis functions nor can they be transformed into a linear combination of basis functions. An important example is the logistic equation which has the form

$$y(t) = \frac{K}{1 + C\exp(-rt)}. \tag{89}$$

Several typical logistic curves are shown in Figure 5.4. We talked about the logistic difference equation in Chapter 1 and will talk about the logistic differential equation in

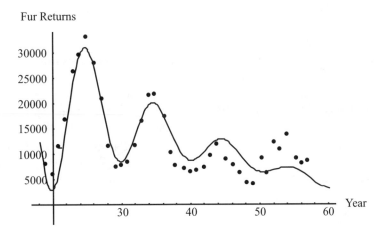

FIGURE 4.23   A Model of Lynx Returns.

Chapter 5. The logistic equation (89) is the solution of the logistic differential equation. It is frequently used to model phenomena in which exponential growth is exhibited initially, but then the growth cannot be sustained and slows until ultimately there is no growth at all. Examples include the height of a child over time (infants grow rapidly; adults do not grow taller) and the growth of a population with limited resources such as bacteria in a petri dish (grows rapidly at first; stops growing when the carrying capacity is reached).

Equations which can not be linearized frequently require complicated numerical procedures to fit their parameters to data. Here we discuss several methods of fitting a logistic curve to data. Similar methods might also apply to other models that you encounter. If they do not, you are encouraged to explore the vast literature on the subject.

The first method is relatively easy and can be applied to any model where the parameters have physical meanings. The method is simply to estimate the parameters. The resulting model will probably not be the best in terms of the least-squares criterion, but will probably be acceptable. A graph or plot of the fitted curve with the data can be used as an intuitive measure of how well the curve works. In a logistic curve, for example, there are two undetermined parameters other than the initial value: the intrinsic growth rate r and the carrying capacity (or saturation level) K. If the data is clean enough, we estimate the carrying capacity from the graph. If there is enough data at the beginning of the curve, we fit an exponential model to the portion of the data that looks like an exponential function, to get r. This r will be a little low since the growth rate is not constant in a logistic model (if it were constant, it would be an exponential model), but this value may be adjusted a little (which we check by comparing the plot with the data).

The second method is to directly use the least-squares criterion by computing an error function and applying a minimum-finding algorithm (which are part of packages such as *Mathematica*). While this sounds straightforward, finding the absolute minimum of a function of more than one variable is much more complicated than the one variable case (see the absolute extrema section in any good calculus book for the one variable case).

For a two-parameter model (like the logistic's r and t_0 described below), a least-squares procedure attempts to find the absolute minimum of a surface (the least of the squared error function), where the surface may have many relative minima. A minimum-finding algorithm is generally given a starting guess and the search proceeds from there. Different guesses may result in different answers. Consider, for example, putting a drop of water on a sheet of wavy, dimpled glass. Depending on where you put the drop, it will run into a different dimple. Only drops which begin in a certain region end up in the absolute lowest dimple (if there is one). Thus the key to this method is to find an appropriate place to start the search (to place the water drop) so that the minimum found is the absolute minimum. There are other complications, like nonconvergence, which we do not mention here except to say that if a particular initial guess results in something weird (errors, nonconvergence), try another initial guess.

For the case of fitting the logistic, there is a visual technique for choosing a good initial guess. This discussion follows a paper by Cavallini [10]. Cavallini's paper was written with *Mathematica* in mind, and this example is presented in the *Mathematica* Appendix. Converting to other computer algebra systems should be straightforward, if the ideas are understood.

Depending on your background, the next several paragraphs may be hard to follow. If this is the case, at the very least you should be able to understand the method well enough to make changes to the *Mathematica* code to use the method with other data.

We start with the logistic equation in the form

$$y(t) = \frac{K}{1 + C\exp(-rt)}, \tag{90}$$

where C is the arbitrary constant of integration which can be determined using an initial condition (this should be familiar from integral calculus and will be revisited in Chapter 5). It is left as an exercise (Exercise 4) to show that (90) has an inflection point when $\exp(rt) = C$. We put $C = \exp(rt_0)$ where t_0 is the inflection point and rewrite (90) as

$$y(t) = \frac{K}{1 + \exp(-r(t-t_0))}. \tag{91}$$

The purpose of this step is to transform the problem from finding C, r, and K to finding t_0, r, and K. The method below allows for K to be computed from t_0 and r, thus reducing the problem to finding the two parameters r and t_0.

If the data is given as a set of n ordered pairs (t_i, y_i), then the error we want to minimize is

$$e = \sum_{i=1}^{n}(y(t_i) - y_i)^2. \tag{92}$$

For computational convenience we define a new function $h(t) = y(t)/K$; thus

$$h(t) = \frac{1}{1 + \exp(-r(t-t_0))}.$$

The rest of this analysis is cleanest using vector notation. Let

$$H = (h(t_1), \ldots, h(t_n)),$$

4 Empirical Modeling

and
$$Y = (y_1, \ldots, y_n).$$

It is left to the reader (Exercise 4) to verify that the error equation (92) can be rewritten as

$$e = K^2 <H,H> - 2K<H,Y> + <Y,Y>, \tag{93}$$

where $<,>$ is the dot or scalar product. From the ideas of least-squares presented in Section 4.2, in order for the error to be minimized, we require that $\partial e/\partial K = 0$. This gives

$$K = \frac{<H,Y>}{<H,H>}. \tag{94}$$

Substituting equation (94) into equation (93) produces

$$e = <Y,Y> - \frac{<H,Y>^2}{<H,H>}. \tag{95}$$

We use a minimum-finding algorithm to minimize this equation with respect to parameters r and t_0.

If you dozed off, come back now. The important issue is that equation (95) must be minimized. While equation (95) may look hairy, it is easily represented in a computer algebra system as seen in the example in the *Mathematica* Appendix. The next step is to find initial guesses for r and t_0. With nonlinear problems such as the logistic, choosing good guesses is essential. We use data visualization to make these initial guesses. Creating a 3D plot of equation (95) gives an overall sense of the shape of the error surface, but is awkward for determining coordinate values of the minimum. A contour plot is useful for more precise guesses of the minimum. Starting with these guesses, we use *Mathematica*'s built-in **FindMinimum** function to find the minimum of equation (95). Next we use the values of r and t_0 that we just found to find K using equation (94). Now K, r, and t_0 are known. Equation (91) can now be written with all parameter values specified. Finally, we plot this equation along with the original data.

Let us demonstrate the method with an example. Table 4.10 shows data comparing the weight of female black bears (*Ursus americanus*) in Arkansas with their age class. The results are taken from many bears and only the average is reported. Data is from a paper by Smith and Clark [62]. These data are plotted in Figure 4.24. As expected, as bears get older, they get heavier, up to a point. This suggests that a logistic curve might be a good model.

We begin with a 3D plot of equation (95) as a function of the parameters r and t_0 (see Figure 4.25). Since t_0 is the inflection point, it must be within the range of t data values. Finding a good range for r requires some experimentation. Figure 4.25 gives a sense of where a minimum might be and where it definitely is not. The minimum can be made more obvious by choosing different scales. In Figure 4.26, the function shown in Figure 4.25 is graphed with the scale on the error variable restricted to $[0, 1{,}000]$. While odd looking, it gives a clear region within which the minimum lies (all the flat areas have value greater than 1,000). Using this scaled plot, we set the ranges for the contour plot shown in Figure 4.27. Sometimes several plots of different types must be used before the

Age Class (years)	Weight (kg)
1	35
2	55
3	68
4	70
5	71
6	75
7	79
8	82
9	81
10	80
11	78
12	99
13	99
15	82

TABLE 4.10 Female Black Bear Weight Versus Age Class. Reprinted by Permission of the American Society of Mammalogists.

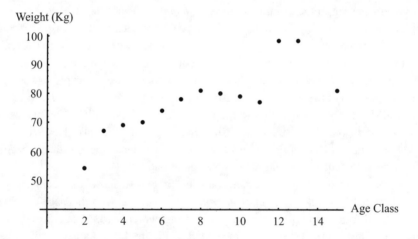

FIGURE 4.24 Plot of Black Bear Data.

behavior of the error is understood. From the contour plot, we make an initial guess of $r = .4$ and $t_0 = 1$. *Mathematica*'s **FindMinimum** command is then used

$$\text{FindMinimum}[\text{error}, \{r, 0.4\}, \{t0, 1\}]$$

which gives the following output

$$\{585.75, \{r \to 0.440661, t0 \to 1.15245\}\}.$$

4 Empirical Modeling

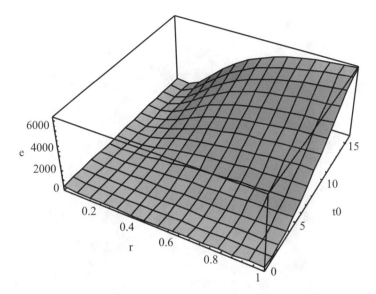

FIGURE 4.25 3D Plot of Error Function.

This output tells us that the minimum error is 585.75 at $r = 0.440661$ and $t_0 = 1.15245$. (Notice that our guess was very close.) Putting these values into equation (94) we find that $K = 87.238$. Finally, putting all parameters into equation (91) and plotting this equation with the data gives Figure 4.28.

We have highlighted the key ideas from Cavallini's paper [10], and the reader interested in more details is directed there. In principle his method can be applied to other functions,

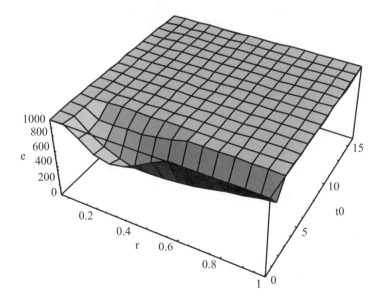

FIGURE 4.26 3D Plot of Error Function Between $[0, 1000]$.

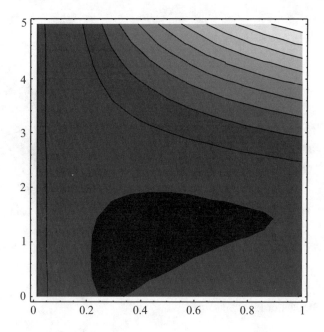

FIGURE 4.27   Contour Plot of Error Function.

but getting the correct form of the error equation could prove difficult. Further, the error surface may be much more complicated. Given the logistic curve's importance, however, this method is a useful tool to have in our bag of modeling tricks.

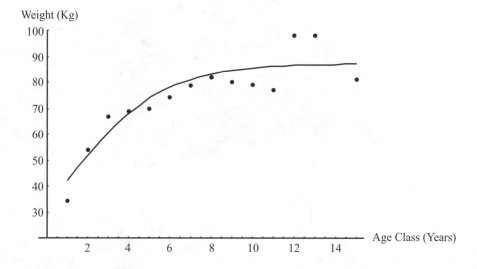

FIGURE 4.28   Data with Fitted Logistic Curve.

4.9 Interpolation

Fitting curves using the least-squares criterion is appropriate for many purposes, including predicting the value of a function in the future (outside the range of data) and determining a rule or law to explain the behavior of the dependent variable. Indeed, in the hard sciences, a curve fit to data frequently becomes a law. Sometimes, however, the purpose is only to understand the function values between observed data points and there is no need for a law or theoretical understanding. Consider the population data in Figure 4.29. This data will be discussed more fully in Example 4.9.3, and the data values are presented there in Table 4.11. No simple fit would be appropriate for this data; though perhaps one might break the data into two parts. If no combination of elementary functions is needed and one has interest only in the population between 1670 and 1966, then regression analysis is unnecessary. Interpolation is the tool of choice.

The word "interpolation" has the prefix "inter" which indicates that it explains behavior between data points. The interpolating function passes through the data points and is used to predict what the function values would have been if observations had been made between the recorded data. Using an interpolation curve to predict function values that extend beyond the data is called *extrapolation* and is not recommended in most circumstances since, unlike regression, the interpolating function may have extreme variations beyond the scope of the data. In fact, some computer packages implementing interpolation routines will not compute values outside of the range of data. Thus interpolation predicts function values between data points, but not beyond the data. It does not give a "nice" equation that gives understanding of the underlying dynamics. In general, interpolation is not suitable for data with a large amount of random fluctuation or noise, though there are ways around this; one of these is discussed in Section 4.9.4.

4.9.1 Simple Interpolation

The simplest type of interpolation is to fit a line between two points. We are familiar with the method of doing this from precalculus. Linear interpolation is often used to predict where a function will be between two data points. In the days before scientific

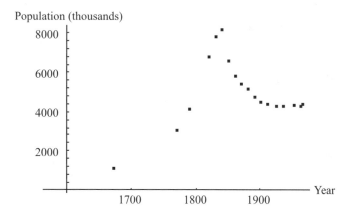

FIGURE 4.29 Interpolation Data.

calculators, trig and log tables were used find values of the trigonometric and logarithm functions. Frequently one wanted a value between values given in the table, and the line between the two nearest data points was computed. This line was then used to approximate the desired function between these two values. While this method is no longer used to compute trigonometric functions, it can be used for other tables, especially if the table consists of observed data. For example, the data used to generate Figure 4.29 includes the points $(1861, 5{,}799)$ and $(1871, 5{,}412)$. Suppose we would like a reasonable estimate for the population in 1865. Using the equation for a line passing through two points,

$$y - y_1 = \frac{y_2 - y_1}{x_2 - x_1}(x - x_1),$$

we obtain

$$y = 77{,}819.7 - 38.7\,x.$$

When $x = 1865$, $y = 5{,}644.2$. Thus we estimate that the population was $5{,}644.2$ thousand or $5{,}644{,}200$ in 1865.

Looking again at the data generating Figure 4.29, we find $(1851, 6552)$, $(1861, 5799)$, and $(1871, 5412)$. Thus between 1851 and 1861 the drop in population was 753; whereas, from 1861 to 1871, the drop was only 387. A plot of these three points is shown in Figure 4.30. This figure shows that while there is no line passing through these three points, there is a parabola. The curvature of the parabola gives more information about 1865 than we have from the two nearest data points alone. Figure 4.31 shows the four nearest data points and the cubic (third degree) interpolation equation through them. Notice that due to the concavity, we expect a quadratic or cubic prediction for the population in 1865 to be lower than the linear prediction. Indeed, the estimate for the population in 1865 using the quadratic interpolation function is $5{,}600.28$ thousand, and using the cubic is $5{,}612.37$ thousand. The cubic estimate is higher than the quadratic

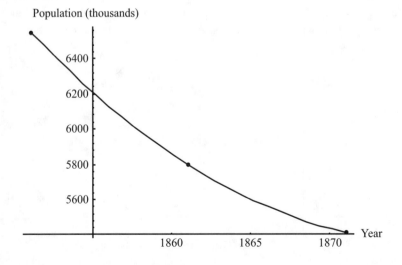

FIGURE 4.30 Quadratic Interpolation Function.

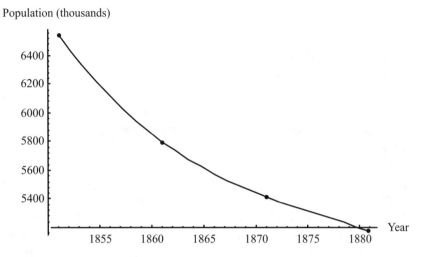

FIGURE 4.31 Cubic Interpolation Function.

estimate because the population does not fall as much from 1871 to 1881 as it does from 1851 to 1861. Many software packages compute interpolating functions, and for the purpose of this book we recommend their use. The interested reader can find the equations used in nth-degree interpolation in any standard numerical analysis book.

In principle a polynomial of degree $n - 1$ can be fit to n data points. While this can be done, the results for n larger than 3 or 4 frequently are meaningless.

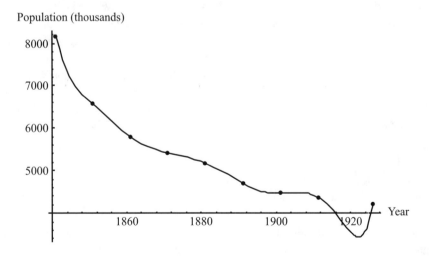

FIGURE 4.32 Eighth-Degree Interpolation Function.

4.9.2 Spline Interpolation

After experimenting with interpolating functions on a data set of more than a few points, one quickly realizes that merely connecting the dots gives a better sense of what is happening than a single polynomial of very large degree. Actually, connecting the dots is interpolation. It is using piecewise linear interpolation functions to connect each successive pair of dots. This is a powerful technique, which gives remarkable graphs, and which uses mathematics no more complicated than finding the equation of a line between two points.

The use of many piecewise linear functions to form a continuous function is called a *linear spline*, and we call this method *linear spline interpolation.* In general a spline is a function, built piecewise out of curve segments, with some smoothness conditions. The smoothness condition for linear splines is continuity. As discussed earlier, there are advantages to quadratic and cubic interpolations; namely, the curvature of the graph is taken into account, usually resulting in better estimates. Again we can piece together quadratic equations between data points, and the result is a piecewise quadratic interpolation function; we can use cubic pieces to form a piecewise cubic interpolation function. When doing quadratic interpolation, we frequently relax the requirement that the parabola pass through three data points. Indeed, the three points may be different as we move through the interval. There are infinitely many parabolas passing through two points. We choose the one whose first derivative at the left endpoint matches the first derivative at the right endpoint of the spline used in the interval immediately to the left. The result is a smooth interpolation curve, with no corners or cusps. In particular, the interpolation curve is differentiable. Using cubics we can get both first and second derivatives of the splines to agree at the interpolation points. The result is even greater smoothness (the interpolation function is both differentiable and second differentiable). A *quadratic spline* is a piecewise quadratic interpolating function which is differentiable; a *cubic spline* is a piecewise cubic interpolating function which is twice differentiable. In general an nth-order spline is a piecewise nth-order interpolating function which is $n-1$ times differentiable. Recall that differentiability is geometrically realized as smoothness, so higher-degree splines exhibit higher degrees of smoothness. While in theory we can create splines of higher degree, they often exhibit aberrant behavior similar to that observed in simple interpolation. As a general practice, we should always plot an interpolation function with the data and visually observe whether there is any strange behavior. The theory of splines is further explored in Exercise 7. For more details, the calculus book (Volume I) by Ostebee and Zorn [49] has a chapter on splines which is presented at a level consistent with this book.

While spline interpolation can be performed by hand, it is a technique which is more appropriately done by machine and is, in general, simple to do. In *Mathematica,* for example, the command **Interpolation[data][x]** interpolates a cubic (the default) spline function through the data. The command **Interpolation[data, InterpolationOrder->4][x]** puts a quartic spline interpolating function to the data. Despite the simplicity of using the interpolating features of a program, the output can be a hairball. For 30 data points, there are 29 polynomials which are pieced together. *Mathematica* just acknowledges that a function has been defined which is an interpolation function, treats it like any other function, but does not show the equations to the user. (In fact, you can manipulate *Mathematica* to show what it stores for an interpolating function, but again the result is a hairball.)

These ideas are illustrated in Example 4.9.3.

4.9.3 Example: The Population of Ireland

Table 4.11 (data from Mitchell [43] and Edwards [16]) presents the population of Ireland from 1672 to 1966. This data was used in the interpolation discussion above. This is an interesting set of data for it reflects a relatively low growth rate for several centuries due to famine, disease, and war; the boom in Irish population with the introduction of the potato from the United States together with a long period of peace; and the population crash due to the Great Irish Potato Famine of 1847. After the famine of 1847 there was a large emigration from Ireland and this data shows the population dropping over the next 50 years to approximately half of its peak value. This is a rare human population curve in that the population declines from the middle of the 19th century through to the middle of the 20th century. This data is plotted in Figure 4.33.

We begin our interpolation analysis with a piecewise-linear interpolation to the data. This is shown in Figure 4.34. As mentioned earlier, this is nothing more than putting a line between each successive pair of data points. In many programs, this can be done

Year	Population (in thousands)
1672	1,100
1770	3,000
1791	4,100
1821	6,802
1831	7,767
1841	8,175
1851	6,552
1861	5,799
1871	5,412
1881	5,175
1891	4,705
1901	4,459
1911	4,390
1926	4,229
1936	4,248
1951	4,332
1961	4,243
1966	4,369

TABLE 4.11 Population of Ireland From 1672 to 1966: Northern Ireland and the Republic of Ireland are Combined from 1926 on. The 1936 Census is 1936 for the Republic of Ireland and 1937 for Northern Ireland. Used by Permission of the Central Statistics Office, Ireland.

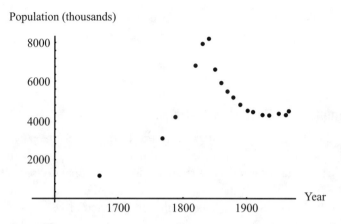

FIGURE 4.33 Population of Ireland.

merely by selecting a "connected graph" option. This model gives us a very good sense of the overall behavior of the population. Next we interpolate a cubic spline function to this data. This is shown in Figure 4.35. This curve connects the points in a smooth manner and maintains the pattern we observe in the data. Figure 4.36 shows a quartic (fourth degree) spline interpolation. Notice that while it is smoother than the third degree, it also show a population boom in the late 1600s and a population decline in the early 1700s which do not exist and are not suggested by the data. Thus we consider the third-degree interpolation function ideal for this data.

One use of interpolation data is to estimate annual growth rates. If $f(x)$ is the interpolation function for this data, the annual growth rate is $f(x + 1) - f(x)$. Here we are using the interpolation function to predict the population level at each of the years

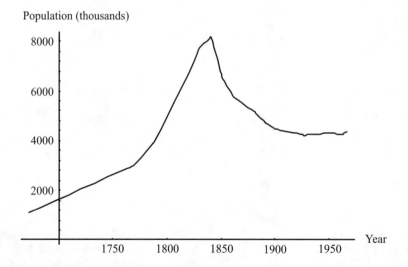

FIGURE 4.34 Population of Ireland: Linear Spline Interpolation.

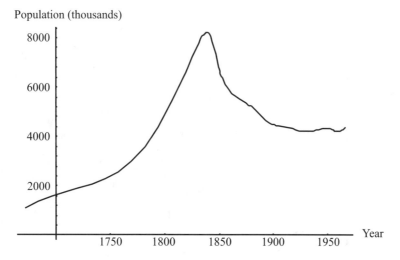

FIGURE 4.35 Population of Ireland: Cubic Spline Interpolation.

between a census. The growth rate used in Chapter 1 is the percent growth rate, which is

$$r(x) = \frac{f(x+1) - f(x)}{f(x)}.$$

Technically we should multiply $r(x)$ by 100% to justify calling it the percentage growth rate, but we will consider this to be understood. These rates are shown in Figures 4.37 and 4.38. Thus we could construct a discrete model for the population of Ireland between 1672 and 1966 using the just constructed $r(x)$ function in the difference equation $x(n+1) - x(n) = r(n)\,x(n)$. Further, in Chapter 5, we will explore continuous growth. We will be

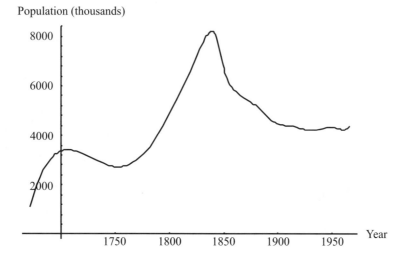

FIGURE 4.36 Population of Ireland: Quartic Spline Interpolation.

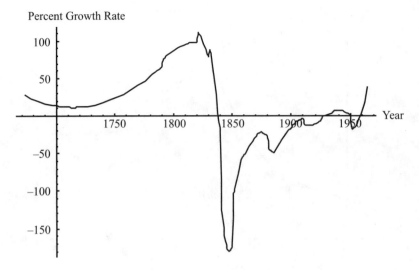

FIGURE 4.37 Growth Rate for the Population of Ireland Computed from a Cubic Spline.

interested in the instantaneous growth rates $f'(x)$ and the instantaneous percentage growth rate $f'(x)/f(x)$. If we perform quadratic or cubic spline interpolation, the interpolation curve is differentiable at all points and we obtain instantaneous rates by differentiating the interpolation function. With the Ireland data, the continuous growth rates are not visually different from the discrete growth rates, so we do not include their graphs here. Just to illustrate a point, Figure 4.39 shows the discrete growth rate computed from the linear interpolation. Again, higher degree spline interpolations result in smoother functions.

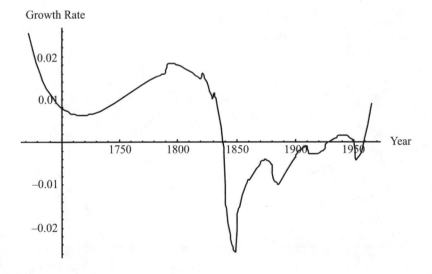

FIGURE 4.38 Percent Growth Rate for the Population of Ireland Computed from a Cubic Spline.

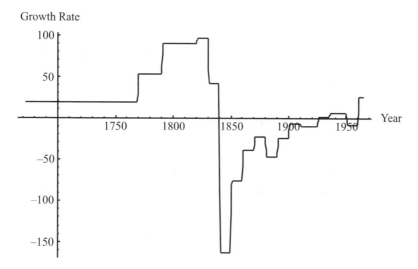

FIGURE 4.39 Growth Rate for the Population of Ireland Computed from a Linear Spline.

Let us turn our attention back to Figure 4.38 which shows the percentage growth rate. This graph illustrates historical events much more dramatically than Figure 4.35. The loss of population due to the famine of 1847 is startlingly obvious as a downward spike. The 1880s famine is now evident, whereas in Figure 4.35 it was not noticeable. It is also clear that its relative impact was much less than that of the 1847 famine. The population boom in the late 1700s and early 1800s is also evident. Note that the apparent decline in the late 1600s and early 1700s is, in fact, a decline in growth, not a decline in population. Notice also the population rose during the second world war when people hoping to escape the war in Britain fled to Ireland.

We conclude with examples of interpolation models' going bad. There is an estimate that the population in 8th century Ireland was 500,000. Figures 4.40, 4.41, and 4.42, show graphs of the linear, quadratic, and cubic spline interpolating functions for the Ireland data with this data point added. Due to the distance between this data point and the other points, both the quadratic and cubic interpolations lead to absurd population functions. Again it is always good practice to plot the interpolation function with the data to see if it makes sense.

4.9.4 Interpolating Noisy Data

Interpolation can also be used with noisy data, but in a more limited way. In particular, we interpolate to a perceived trend curve, not to the actual data. As mentioned earlier in the chapter, the human eye can be as good at curve fitting as any algorithm. In particular, it can see trends in noisy data or appropriately weight extreme data. There are a number of smoothing algorithms that are used, but if the user has knowledge of the data, smoothing by eye may give more desirable results than a mechanical algorithm. Further, some smoothing algorithms smooth essential details from the data. The method described here is easy to use and implement.

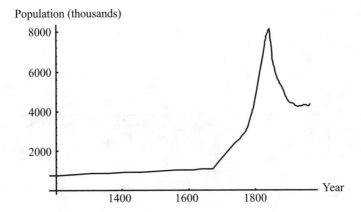

FIGURE 4.40 Linear Spline Interpolation on Extended Data Set.

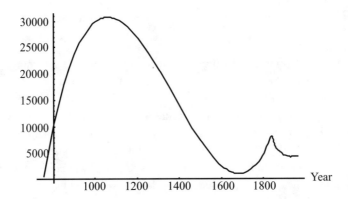

FIGURE 4.41 Quadratic Spline Interpolation on Extended Data Set.

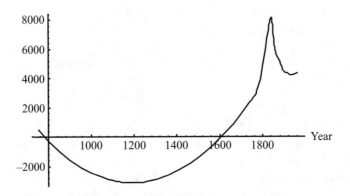

FIGURE 4.42 Cubic Spline Interpolation on Extended Data Set.

4 Empirical Modeling

1. Plot the data carefully. Precision is important, so it is recommended to do this on a large sheet of paper with as fine a scale on the axes as possible.
2. By hand, draw the best perceived trend curve to the data.
3. Pick a series of reference points on the hand-drawn curve and determine their coordinates. These points should be spaced throughout the curve with more points taken near rapid changes in the curve.
4. Fit an interpolation function to the coordinates of reference points. Use this as the trend curve.

The method is illustrated in Figures 4.43, 4.44, 4.45, and 4.46.

4.10 The Statistics of Simple Regression

Linear regression is usually presented from one of two distinct perspectives. In Section 4.2, the mathematical viewpoint was presented. It is primarily interested in fitting a

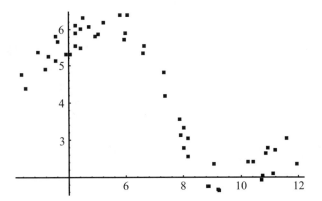

FIGURE 4.43 Noisy Data.

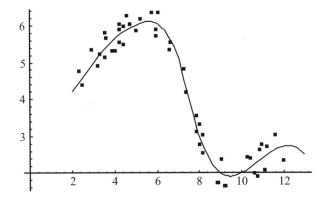

FIGURE 4.44 Drawing a Curve through the Data.

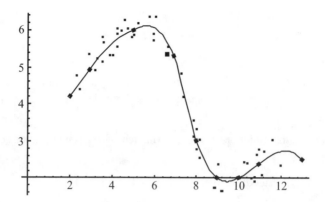

FIGURE 4.45 Picking Reference Points on the Curve.

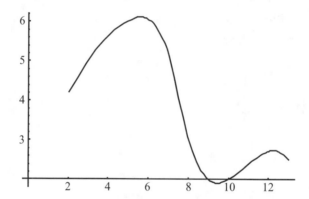

FIGURE 4.46 An Interpolation Function through the Reference Points.

curve to data using the least-squares criterion with little stress placed on the significance of the fit. The other viewpoint, common in applied statistics, is presented here. Its primary interests are the meaning, significance, and interpretation of the regression equation. The actual equation is often secondary.

Throughout this section we are considering primarily *simple regression* which means that the regression equation has the form $\hat{y} = \hat{\beta}_0 + \hat{\beta}_1 x$. This means two things: 1) there is only one independent variable, and 2) the equation is truly an equation for a line, not just a linear combination of basis functions which is what the "linear" means in "linear regression." These conditions are relaxed in the later sections on multiple regression and curvilinear regression.

One issue, mentioned earlier, is reiterated here, and will be emphasized again later on. In simple regression, there are two lines which we talk about. One is the true linear trend (linear trend of the population), and the equation for it is $y = \beta_0 + \beta_1 x$. The other is the best linear fit to the data (linear trend of the sample) using the least-squares criterion, which has equation $\hat{y} = \hat{\beta}_0 + \hat{\beta}_1 x$. Data is usually noisy, and the same process would give different data if it were possible to repeat it. Thus the line fit to the data (the hat

4 Empirical Modeling

equation) is an approximation based on a sample to the linear relation in the population (the non-hat equation). The reader should recall that the methods of the earlier sections gave the equation $\hat{y} = \hat{\beta}_0 + \hat{\beta}_1 x$.

Some of the experimental approach comes from Franklin [17].

4.10.1 Reading a Regression Table: I

Most statistical packages and many other programs that do statistics (such as *Mathematica*) create a type of output known as a regression or ANOVA (ANalysis Of VAriance) table. An example of such a regression table (created by *Data Desk*) is shown in Figure 4.51 below. Regression tables from different programs all have the same basic elements. This section contains detailed information on what these elements are and how to interpret them. The reader may want to keep Figure 4.51 in mind and refer back to it while reading through the detailed information. After this detailed information, Section 4.10.8 entitled *Reading a Regression Table: II*, will examine the elements of a regression table for several examples.

This material is simultaneously easy and hard. Once understood, it is easy to generate and interpret a regression table; it is hard to get the necessary understanding because many topics are involved that depend on one another. When we teach this material in class, we have a computer generate regression tables for all to see and go back and forth between describing the elements of a regression table and generating regression tables on sample data. Unfortunately this interactive approach is not conducive to book writing, but the reader may want to skim through to the end of *Reading a Regression Table: II*, then move back and forth between the example and the discussion of table elements. It is also helpful to read this with a statistical package available and explore on your own. Learning regression analysis is sort of like learning to ride a bike; once you learn to coordinate everything happening simultaneously, it is really very easy.

In this section, we provide formulas or definitions for the pieces of a regression table. For our purposes most of the formulas are not important, but we feel it is an important part of the learning process to be able to compute the various entries on your own. To a novice, the table is just a bunch of numbers, several of which are actually used. Even if we do not use all the entries, it adds to our comprehension to say that "this entry is obtained by dividing these two entries." In Exercise 8 you are asked to construct a regression table from the definitions for a small data set.

4.10.2 The Sums of Squares and R^2

This discussion parallels a similar discussion from the least-squares material earlier and is reviewed here for the reader's convenience. We recall the three sums of squares: SS_{Tot}, SS_{Res}, and SS_{Reg}. The first is the *total sum of squares* which measures the deviation between the data and the mean of the dependent variable.

$$SS_{\text{Tot}} = \sum (y_i - \bar{y})^2.$$

The second is the *residual sum of squares* which measures the deviation between the regression line and the data. Thus

$$SS_{\text{Res}} = \sum (y_i - \hat{y}_i)^2.$$

Finally, there is the *regression sum of squares* which measures the deviation between the regression line and the mean. In particular

$$SS_{\text{Reg}} = \sum (\hat{y}_i - \bar{y})^2.$$

Recall that the total sum of squares is the sum of the residual and regression sum of squares, thus

$$SS_{\text{Tot}} = SS_{\text{Res}} + SS_{\text{Reg}}.$$

Notice that if $\sum (y_i - \hat{y}_i)^2 = 0$ (i.e., $SS_{\text{Res}} = 0$), then the regression line explains all of the variance. On the other hand, if $\sum (\hat{y}_i - \bar{y})^2 = 0$ (i.e. $SS_{\text{Reg}} = 0$), then none of the variance in the data is explained by the regression line. Rephrased, if the data is very nearly linear, then $SS_{\text{Res}} \approx 0$, and if the data is random, then $SS_{\text{Reg}} \approx 0$.

In simple regression the ratio

$$\frac{SS_{\text{Reg}}}{SS_{\text{Tot}}}$$

is usually denoted by r^2; in multiple regression it is denoted by R^2; and most statistical packages report using the notation R^2. To avoid either confusion or awkward sentences, we have adopted the convention of always using R^2. The reader will understand that if the regression is actually simple regression, then R^2 should be interpreted as r^2.

The statistic R^2 is the percentage of the variance in the data accounted for by the regression equation. Naturally, $1 - R^2$ is the percentage of the variance not accounted for by the regression equation. If $SS_{\text{Reg}} \approx 0$, then $R^2 \approx 0$. If $SS_{\text{Res}} \approx 0$, then $R^2 \approx 1$. Thus the statistic R^2 is used to measure how good a model is, at least in the sense of explaining the variation in the data.

4.10.3 Regression Assumptions

Until now, the regression analysis discussed was valid for any data set. By this we mean that a line can be fit to any data set (of more than one point) according to the least-squares criterion. Of course it may not mean anything. Next we discuss the significance of these results. To test for this significance, we need to develop some statistical tests. In order for these tests to be valid, several assumptions must satisfied, at least approximately.

1. There is a true linear relationship underlying the data. The population (not the sample) satisfies the equation $y = \beta_0 + \beta_1 x + \epsilon$. Note this does not say that all the points are on a line, but rather lie about a line with some noise ϵ.
2. The true residuals (or noise) ϵ are mutually independent, which means that each one is not affected by the others.
3. The true residuals all have the same variance.
4. The true residuals follow a normal distribution with mean zero.

If these assumptions are met, then the following discussion involving the F- and t-ratios is valid.

4 Empirical Modeling

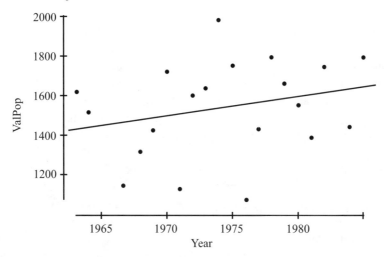

FIGURE 4.47 Elk of the Grand Tetons.

4.10.4 F-tests and the Significance of The Regression Line

In Figure 4.47, a regression line is shown for a series of elk data. In particular the data is the annual estimated population of elk in the Central Valley area of Grand Teton National Park. The estimated values are computed by taking actual census counts and adjusting them to correct for errors in counting. In this example, the regression line is sloping upward, which might lead one to conclude that the elk population is, overall, on the rise despite obvious ups and downs in the population. Next consider the data and regression line shown in Figure 4.48. Without knowing what the variables represent, one might again conclude that there is an upward trend to this data. This data set, however, was generated by a random number generator (Uniform[0, 2,000]). Knowing this, one's interpretation of

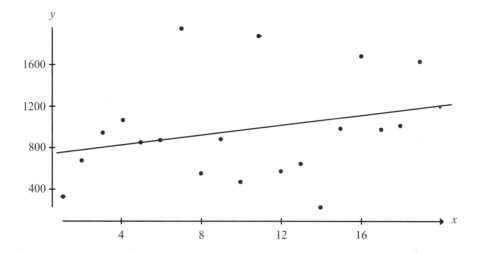

FIGURE 4.48 Unspecified Data.

the regression line changes completely. Now one can argue that the positive slope is due to random chance and that more data would change the trend towards zero slope. One might also argue that with any set of random points (from a symmetric distribution with mean zero), the probability of actually getting a regression line of zero slope is extremely rare and that essentially all of the time one would expect some non-zero slope, either positive or negative. Now let us go back and reconsider the elk data. Is this upward trend meaningful, or is it an artifact of the noise in the data? The F-ratio statistic helps us answer these questions.

With data, especially noisy data, we should not expect to get an absolute answer to the question of significance. If lines with slope of at least this magnitude are observed in random data 50% of the time, then we are doubtful of the significance of the conclusion. If slopes of this magnitude happen once every 20 times (i.e., 5% of the time), then we are more confident of the significance. Clearly if slopes of this magnitude are observed once every 100 or once every 1,000 times, we become even more confident that the number of elk is increasing. On the other hand, even if the slope randomly occurs once every 10,000 times, this might be that once, so we are never certain.

We use the idea of hypothesis testing to help us determine levels of significance. This discussion is similar to the discussion of the chi-square test in Section 2.4. In a general hypothesis test, a hypothesis is formulated. Using an appropriate statistical test, this hypothesis is either accepted or rejected at some level of significance α or confidence $((1-\alpha) \times 100\%)$. In the case of the regression line, we are trying to reject the hypothesis that the regression line is a constant. Formally the hypothesis (called the *null hypothesis*) is that $\beta_1 = 0$ (in multiple regression the null hypothesis is that $\beta_1 = \cdots = \beta_n = 0$).

The statistic we need is the *F-ratio* which is constructed out of the sums of squares and the degrees of freedom. The regression degrees of freedom df_{Reg} is the number of independent variables under consideration. In a simple regression it is always one, but it takes other values in multiple regression. The residual degrees of freedom df_{Res} is the number of valid cases (we do not count missing data) less the number of independent variables less one. Frequently we write $df_{\text{Reg}} = k$ and $df_{\text{Res}} = N - k - 1$. Then the F-ratio is defined to be

$$F = \frac{SS_{\text{Reg}}/df_{\text{Reg}}}{SS_{\text{Res}}/df_{\text{Res}}} = \frac{SS_{\text{Reg}}/k}{SS_{\text{Res}}/(N-k-1)}.$$

The expression $SS_{\text{Reg}}/df_{\text{Reg}}$ is called the *regression mean sum of squares* and is denoted by MS_{Reg}. Similarly the expression $SS_{\text{Res}}/df_{\text{Res}}$ is the *residual mean sum of squares* and is denoted by MS_{Res}. With this notation

$$F = \frac{MS_{\text{Reg}}}{MS_{\text{Res}}}.$$

To understand how the F-ratio statistic is used, consider Figure 4.49 which shows F-distributions for several different degrees of freedom. Starting with the appropriate curve, we mark the point which separates the area under the curve into two regions, one with 95% of the area and one with 5% of the area. This point is called a *cutoff* or *critical point* and is denoted by F_0. If the computed F-ratio value is larger than the cutoff, then

4 Empirical Modeling

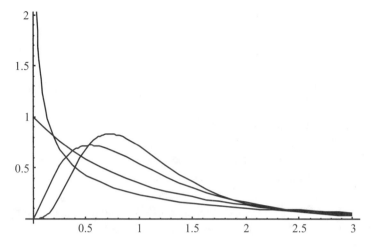

FIGURE 4.49 F-Distributions with $r_1 = 1, 2, 5$, and 10, and $r_2 = 20$.

there is a less than 5% chance of this slope's occurring randomly. Equivalently we are more than 95% confident that this slope did not occur by chance. Formally we say that we reject the null hypothesis at the $\alpha = 0.05$ level. If the F-ratio lies below the cutoff, we say that the slope is not significant at an $\alpha = 0.5$ level. This means that there is a greater than 5% chance of the slope's occurring randomly, and we consider this to be insufficient evidence for concluding that the regression line has a non-zero slope. The bottom line is that big F-ratios are good, small F-ratios are bad, and the cutoff F_0 is the cutoff point between good and bad.

If we have a numerical integration program handy, we can integrate the F-distribution from 0 to the calculated F-ratio value and compute the exact level of significance or *p-value* (which is 1 minus the value of the integral). Some packages compute this for you as part of the regression table.

Generally there are three ways to evaluate an F-ratio statistic for significance.

1. To look up the F-ratio in an F-table. Until recently this was essentially the only method available to the average person. An F-table is reproduced in Appendix B. Generally an F-table is set up for one or two levels of significance, commonly $\alpha = 0.05$ or $\alpha = 0.01$. For the table in Appendix B, the rows correspond to the residual degrees of freedom and the columns correspond to the regression degrees of freedom. For a simple regression, the first column and the row corresponding to the number of valid cases less two are used.

2. Many statistical packages give the level of significance corresponding to the F-ratio with the appropriate degrees of freedom. Commonly this is quoted as a p-value.

3. Computer algebra systems either do hypothesis testing or numerical integration. With numerical integration we can integrate to find the area under the F-distribution.

The reader is encouraged to figure out how to use each of these methods since each is instructive in some way. Looking at the F-table in the Appendix, we formulate the following rule of thumb.

A Rule of Thumb. For doing simple regression, if there are more than 15 cases, the regression line is significant at the $\alpha = 0.05$ level if the F-ratio value is greater than 4.5. If the F-ratio is less than 3.5, then the regression is not significant for any number of cases.

The idea of the F-test is that, if the F-ratio is large, then the regression is significant ($\beta_1 \neq 0$) and, if the value of the F-ratio is small, then the regression is not significant ($\beta_1 = 0$). The rule of thumb says that the cutoff between large and small is somewhere between 3.5 and 4.5. Thus if a data set has 25 cases and an F-ratio value of 20, we know that the regression is significant at a 0.05 level without needing to look up anything. Conversely, if the F-ratio value is 2.5, then we know that the regression is not significant at a 0.05 level regardless of the number of cases.

Returning to our elk example, we compute an F-ratio value of 1.20. Thus the upward trend is not significant at a 0.05 level. Remember, we can fit a line to any data, but there is no guarantee that the line has any importance.

4.10.5 The t-test for Testing the Significance of Regression Coefficients

In a simple regression, the regression line has the form $y = \beta_0 + \beta_1 x$. In multiple regression, the regression equation has the form $y = \beta_0 + \beta_1 x_1 + \cdots + \beta_n x_n$. Of interest is whether the β_i's are significantly different from zero. The t-test is used for this. The alert reader will recognize that in a simple regression, testing whether $\beta_1 = 0$ is the same as testing whether the regression line differs from a constant. Indeed in this setting the F-test and the t-test give the same conclusions. In multiple regression, however, the F-test tests the hypothesis that all the β_i's are zero (i.e., $\beta_1 = \cdots = \beta_n = 0$), whereas the t-test is used to test whether each $\beta_i = 0$. For simple regression there is no need to use the t-test, but in multiple regression it helps us decide which variables to leave in or remove from a model. We will use notation for multiple regression here, though if exploring the facets of simple regression, just remove the i subscripts. The t-ratio is

$$t = \frac{\hat{\beta}_i}{se_i}.$$

Here $\hat{\beta}_i$ is an estimate of the regression coefficient of the ith independent variable, and se_i is the standard error of $\hat{\beta}_i$. See Section 4.10.6 for the definition of standard error.

Using the t-ratio is almost the same as using the F-ratio, but with a few important exceptions. First we use the appropriate t-distribution. Here the degrees of freedom is the residual degrees of freedom $N - k - 1$, where N is the number of cases and k is the number of independent variables. Secondly, since the t-distribution is a two-tailed distribution (see Figure 4.50) like the normal, we need to use half the desired α value. The reason is that α, by definition, is the area in one tail and we want the area in both tails to equal the magnitude of α. Thus to test for confidence at the $\alpha = 0.05$ level, we must test at the $\alpha = 0.025$ level. If a statistical package is being used, this adjustment is probably already made, so we must be careful when interpreting computer output.

A t-table is reproduced in Appendix C. Take a few moments and compare the 0.025 column of the t-table with the first column of the F-table for $\alpha = 0.05$. You will find that one is the square of the other. Readers with some mathematical statistics background

4 Empirical Modeling

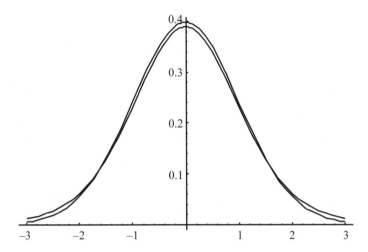

FIGURE 4.50 t Distributions with 10 and 100 Degrees of Freedom.

should recall that the square of the t-distribution is the F-distribution with $df_1 = 1$. This is the mathematical justification for our earlier assertion that for simple regression we can use either a t-test or an F-test to test for significance of the regression equation.

4.10.6 Standard Errors of the Regression Coefficients and Confidence Intervals

In simple regression, we assume that the data has a true linear underlying trend which can be expressed by the equation $y = \beta_0 + \beta_1 x$. What we are attempting to do with the linear regression process is to find β_0 and β_1. The only information we have is the data which is just a sample of the entire population (in the statistical sense of population). Consequently we can only estimate β_0 and β_1, which is what $\hat{\beta}_0$ and $\hat{\beta}_1$ are. The regression line $\hat{y} = \hat{\beta}_0 + \hat{\beta}_1 x$ is our best guess of $y = \beta_0 + \beta_1 x$.

Consider the following example. We get a computer to generate a sample according to the relation $y = 3x + 1 + \text{Normal}(0, 1)$. Thus the linear trend of the population is $y = 3x + 1$, and there is noise associated with each measurement which has a normal distribution with mean 0 and standard deviation 1. Our simulation generated the data in Table 4.12.

Now we fit a line to this data and get $y = 0.182637 + 3.11514x$. This regression analysis has $R^2 = 98.4\%$ and $F = 505$ with 8 degrees of freedom which is significant at better than $\alpha = 0.0001$. The regression coefficients, however, are not the true coefficients. Here $\beta_0 = 1$ and $\beta_1 = 3$, but $\hat{\beta}_0 = 0.1826374$ and $\hat{\beta}_1 = 3.11514$.

Next we repeat this experiment two more times. The data are listed in Table 4.13.

For one of these simulations the regression line is $y = 0.473978 + 3.03465x$, with $R^2 = 99.0\%$, $F = 795$, which is significant at a better than $\alpha = 0.0001$ level. For the other, the regression line is $y = 1.82789 + 2.77718x$ with $R^2 = 98.5\%$, $F = 536$, which is again significant at a better than $\alpha = 0.00001$ level.

These examples should make the following points clear. When performing regression analysis we are approximating the true linear trend using only the data at hand. The result

X	Y
1	2.6101912
2	6.0251801
3	9.9790073
4	12.279711
5	18.097476
6	18.516488
7	21.913528
8	25.350903
9	25.996769
10	32.389885

TABLE 4.12

X	Y2	Y3
1	4.0126983	5.5312794
2	5.6119796	5.2527634
3	10.185525	10.077944
4	12.778222	14.003516
5	15.358985	15.188173
6	17.447020	19.745863
7	22.801732	21.059856
8	24.413597	24.559109
9	29.252741	26.897840
10	29.783032	28.707645

TABLE 4.13 Results of Experiment.

is a line which differs from the true line because $\hat{\beta}_0$ and $\hat{\beta}_1$ are statistics which estimate the true parameters β_0 and β_1. In particular we have three data sets generated from the same distribution and $\hat{\beta}_1$ values ranging from 2.78 to 3.11. In practice, we usually have only one data set and never know the true line. There is a tendency to claim that the regression line fitted to the data is the underlying trend, but be aware that this is just an approximation. We fit a line to the sampled data, not to the underlying population. One way to emphasize this point and deal with the problem is to report confidence intervals for $\hat{\beta}_0$ and $\hat{\beta}_1$. Commonly, a 95% (or higher) confidence interval is reported with the regression coefficient.

The first step in computing a confidence interval for β_i is to compute the standard error associated with $\hat{\beta}_i$. Let us take a minute and discuss what a standard error is. If we were to repeat the experiment at the beginning of this section many times (see Exercise 5), we would have a (sample) distribution of $\hat{\beta}_i$ values which has a mean and standard deviation. The standard error of $\hat{\beta}_1$ is the standard deviation of the (population) distribution of estimates. This discussion holds for any parameter. The standard error of any estimate of a parameter is the standard deviation of the distribution of the estimates. We present the equations for computing the standard errors without derivation. These equations hold only for simple regression.

$$se_{\hat{\beta}_0} = \sqrt{\frac{SS_{\text{Res}}(\sum x_i^2)}{(n-2)(n\sum x_i^2 - (\sum x_i)^2)}} \quad \text{and} \quad se_{\hat{\beta}_1} = \sqrt{\frac{n\, SS_{\text{Res}}}{(n-2)(n\sum x_i^2 - (\sum x_i)^2)}}.$$

Next we find the cutoff or critical number t_0 from a t-table corresponding to the degrees of freedom and confidence level appropriate to our situation. In the case of eight degrees of freedom (as in the examples in this section) and a 95% confidence level, we use the $\alpha = 0.025$ (remember half of 0.050) column and the eighth row and find the cutoff number of 2.306.

The confidence intervals are then

$$CI_{\hat{\beta}_0} = \hat{\beta}_0 \pm t_{\text{cutoff}} * se_{\hat{\beta}_0} \quad \text{and} \quad CI_{\hat{\beta}_1} = \hat{\beta}_1 \pm t_{\text{cutoff}} * se_{\hat{\beta}_1}.$$

4 Empirical Modeling

For the first set of simulated data the 95% confidence intervals are $CI_{\hat{\beta}_0} = (-1.80, 2.17)$ and $CI_{\hat{\beta}_1} = (2.80, 3.43)$. For the second set we have $CI_{\hat{\beta}_0} = (-0.16, 1.10)$ and $CI_{\hat{\beta}_1} = (2.79, 3.28)$, and for the third set we have $CI_{\hat{\beta}_0} = (0.11, 3.54)$ and $CI_{\hat{\beta}_1} = (2.50, 3.05)$. In each case the computed confidence intervals cover the values $\beta_0 = 1$ and $\beta_1 = 3$, which is what we asserted.

4.10.7 Verifying the Assumptions

In Section 4.10.3, several requirements were made for the regression statistics to be meaningful. However there are several simple methods which we can use to get an intuitive sense of whether the most of the assumptions are met, at least approximately.

The first assumption is that the population has a true underlying linear trend. The human eye is a good judge of linearity. It is advisable to graph the data whenever possible to see if there are any nonlinear trends. Frequently nonlinear data can be transformed to linear data using methods discussed in Section 4.6.

The rest of the assumptions deal with the nature of the residuals $y_i - \hat{y}_i$. We deal only with the assumption that the distribution of the residuals is normal with mean zero. A minimal effort to examine this assumption is to construct a histogram of the residuals and see if it has roughly a normal shape centered about zero. Again showing something is normal is difficult, but obvious non-normalities easily stand out to the eye. Things to look for are non-zero center, non-bell curve shape, and skews to the left or right. If your software can construct normal probability plots, or q-q plots, make a normal probability plot of the residuals. If the data is perfectly normal, your plot should look like a line. Deviations from a linear graph are indications of non-normality.

4.10.8 Reading a Regression Table: II

In this section, we identify the elements of a regression table.

With statistical packages, we generate a regression table from our data. Figure 4.51 shows a regression table for the elk data generated using the statistical package *Data Desk*. The data is from Boyce [5] and is presented in Table 4.14 if you wish to follow along. Other packages would produce tables with roughly the same information.

The first line identifies the dependent variable which in this example is ValPop, the population of elk in the central valley. The independent variable(s) is(are) listed at the bottom with their statistics. In this case the year is the only independent variable. The second line tells us the number of valid cases. In this data set, there are 23 cases overall, but 3 have missing data, so there are $N = 20$ valid cases. The next line gives R^2, which in this case tells us that 6.3% of the variance in the data is accounted for by the regression line. R^2 adjusted has not been discussed and is used in multiple regression. When doing multiple regression, adding more independent variables always makes R^2 go up. The R^2 adjusted statistic will get larger as independent variables are added, to a point, and then will start to decline. It thus gives a criterion for the optimum set of independent variables added to a model.

The next line gives the standard error of the regression s and the degrees of freedom $N - k - 1$ where k is the number of independent variables (in this case 1). The standard

Dependent variable is: ValPop

No Selector

23 total cases of which 3 are missing
R squared = 6.3%   R squared (adjusted) = 1.1%
s = 245.0 with 20 - 2 = 18 degrees of freedom

Source	Sum of Squares	df	Mean Square	F-ratio
Regression	72256.6	1	72256.6	1.20
Residual	1080050	18	60002.8	

Variable	Coefficient	s.e. of Coeff	t-ratio	prob
Constant	-17425.8	17286	-1.01	0.3268
Year	9.60756	8.755	1.10	0.2869

FIGURE 4.51 Regression Table for Elk of the Grand Tetons.

Year	Valley Population
1963	1,627
1964	1,527
1965	*824
1966	*891
1967	1,140
1968	1,322
1969	1,431
1970	1,733
1971	1,131
1972	1,611
1973	1,644
1974	1,991
1975	1,762
1976	1,076
1977	1,442
1978	1,800
1979	1,667
1980	1,558
1981	1,396
1982	1,753
1983	*
1984	1,453
1985	1,804

TABLE 4.14 Summer Elk Trend Counts: Central Valley of Grand Teton National Park. The Symbol * Indicates Missing or Suspect Data. Used by Permission of Cambridge University Press.

error of regression gives the "give or take" in a prediction. If, for example, the regression line predicts that in 1972 there will be 1,520 elk, we can report this as 1,520 give or take 245 (which is s).

The next part of the table gives the SS_{Reg}, SS_{Res}, df_{Reg}, df_{Res}, MS_{Reg}, MS_{Res}, and F-ratio. Recall that $MS = SS/df$ and $F = MS_{\text{Reg}}/MS_{\text{Res}}$, so with the information in the first two columns of statistics, the rest of the table can be generated. Since we have more than 15 cases and $F = 1.20 < 3.5$, we use our rule of thumb to conclude that the regression line is not significant at a 95% confidence level. This means that we cannot assert that the elk population is growing; the upward trend we see could be due to chance.

The final section of the table gives the regression coefficients and their related statistics. Here $\beta_0 = -1{,}7425.8$ and $\beta_1 = 9.0756$. Their standard errors are reported in the next column which is used to compute confidence intervals. The numbers in the t-ratio column are the quotients of the preceding two columns ($t_{\hat{\beta}_0} = a/se_{\hat{\beta}_0}$ and $t_{\hat{\beta}_1} = b/se_{\hat{\beta}_1}$). The t-ratio values can be tested with a t-table, or we can use the p-values reported in the final column. Here we see that the hypothesis $\beta_1 = 0$ is rejected at a $\alpha = 0.2869$ level. Thus for most any purpose, we would say that β_1 is not statistically different from 0.

An instructive exercise is to generate data with varying degrees of randomness and observe the effects on the regression tables. Here we present four such sets of generated data. In all four, the underlying trend line is $y = 3x + 1$. In all four sets, the independent variable X has integer values from 1 to 10. The first data set is truly linear, and the Y values are determined from the equation $y = 1 + 3x$. The second, third, and fourth add a random term generated from normal distributions with means zero and standard deviations 1, 10, and 100 respectively. Thus the generating equation is $y = 1 + 3x + \text{Normal}(0, \sigma)$. Graphs of the data are presented in Figure 4.52 and regression tables are presented in Figure 4.53. The Y data sets go by the names $Y1$, $Y2$, $Y3$, and $Y4$ respectively.

First look at the graphs of the data as shown in Figure 4.52. The first two data sets are visually very linear. The third has an obvious linear trend, though the noise is quite apparent. The fourth has no trend discernible to the eye. Next, let us look at the statistics given by the regression tables shown in Figure 4.53. Looking first at R^2 we observe that the regression line explains 100% of the variance in a true line, which is to be expected because the line is the only source of variance in the data. With increases in noise, the R^2 becomes progressively smaller, dropping to 98.9%, 40.1%, and finally to 0.6%. The last data set has far more of its variance coming from the noise than from the linear trend.

Next compare the sums of squares. The absolute numbers do not mean much, but observe that as the randomness increases, the residual sum of squares increases. Recall that the total sum of squares is the sum of the residual and regression sum of squares. Notice that as the noise increases, a greater proportion of the total error is accounted for by the residual sum of squares. Conversely, when there is no noise, the total sum of squares is entirely accounted for by the regression sum of squares, which means all of the variance is accounted for by the regression line.

The degrees of freedom are the same for all four data sets, and the mean sums of squares tell us the same information as the sums of squares. The next object of interest is the F-ratio. In the case of perfectly linear data the F-ratio is infinite (resulting from

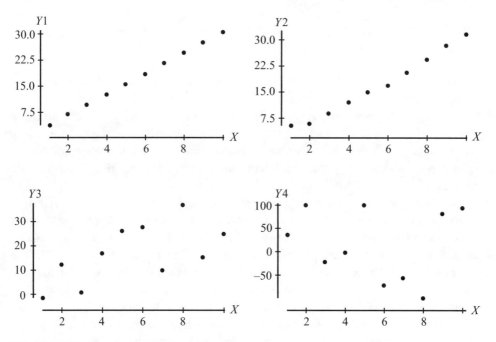

FIGURE 4.52 Line Plus Noise Regression Scatter Plots.

$MS_{\text{Res}} = 0$ in the denominator of the F-ratio). This affirms our assertion that the larger F is, the more significant the regression analysis, and with perfectly linear data the F-ratio should be as high as it can possibly be, which is, of course, infinite. With successive additions of noise, the F-ratio drops. We either look up the cutoff points in the F-table in the Appendix or, recalling that in simple regression the t-ratio gives the same significance test as the F-ratio, look down to the next section of the table and read the p-values for the β_1 coefficient. For the first two cases, significance is at a better than $\alpha = 0.0001$ level which is very good. The third case ($\sigma = 10$) is barely significant at a $\alpha = 0.05$ level (95% level), and when $\sigma = 100$, the regression is not significant at any acceptable level ($\alpha = 0.8289$ or a 17% level).

The final section of the tables deals with the regression equation. In the perfectly linear case we find that β_0 and β_1 are found correctly, with no standard error, and have infinite t-ratios which implies significance at every level. Again this is to be expected. As the noise increases, the estimates of the coefficients deviate from their true values, the standard errors increase, the t-ratios fall, and the significance (p-values) drop. Notice that in the case $\sigma = 100$ the trend line is actually found to be decreasing. Fortunately the statistics suggest that we should not consider this analysis to be meaningful.

4.11 Example: Elk of The Grand Tetons

One of the examples above involves population data for the elk herd of the Central Valley area of Grand Teton National Park. Here we comment further on this data set and look at

4 Empirical Modeling

Dependent variable is: Y1
R squared = 100.0%   R squared (adjusted) = 100.0%
s = 0 with 10 - 2 = 8 degrees of freedom

Source	Sum of Squares	df	Mean Square	F-ratio
Regression	742.500	1	742.500	∞
Residual	0	8	0	

Variable	Coefficient	s.e. of Coeff	t-ratio	prob
Constant	1.00000	0	∞	≤ 0.0001
X	3	0	∞	≤ 0.0001

Dependent variable is: Y2
R squared = 98.9%   R squared (adjusted) = 98.8%
s = 1.024 with 10 - 2 = 8 degrees of freedom

Source	Sum of Squares	df	Mean Square	F-ratio
Regression	773.710	1	773.710	738
Residual	8.38721	8	1.04840	

Variable	Coefficient	s.e. of Coeff	t-ratio	prob
Constant	0.397612	0.6995	0.568	0.5853
X	3.06240	0.1127	27.2	≤ 0.0001

Dependent variable is: Y3
R squared = 40.1%   R squared (adjusted) = 32.6%
s = 10.08 with 10 - 2 = 8 degrees of freedom

Source	Sum of Squares	df	Mean Square	F-ratio
Regression	544.095	1	544.095	5.36
Residual	812.656	8	101.582	

Variable	Coefficient	s.e. of Coeff	t-ratio	prob
Constant	3.40513	6.885	0.495	0.6342
X	2.56809	1.110	2.31	0.0494

Dependent variable is: Y4
R squared = 0.6%   R squared (adjusted) = -11.8%
s = 82.04 with 10 - 2 = 8 degrees of freedom

Source	Sum of Squares	df	Mean Square	F-ratio
Regression	335.472	1	335.472	0.050
Residual	53843.6	8	6730.45	

Variable	Coefficient	s.e. of Coeff	t-ratio	prob
Constant	29.2758	56.04	0.522	0.6156
X	-2.01651	9.032	-0.223	0.8289

FIGURE 4.53 Line Plus Noise Regression Tables.

another data set collected from 1912 to 1986 involving the entire elk herd, not just the elk of the Central Valley. If we look at Figure 4.51, we see that three cases are missing. One is truly missing data, the other two are data points which have been excluded. Leaving out data is dangerous business, often employed by people misusing statistics to argue a point. As second- or third-hand users of the data, we have no reason to exclude these points. In fact they lie well within the data and further lead to the conclusion that the regression equation is significant. The collectors of the data know more about the data than we do, and report that during the years in question insect control procedures spooked the elk and resulted in counts that were lower than they would have been otherwise. Thus these two data points were recorded under conditions that were different from the conditions present for all the other data points. Indeed the fact that these data points were too low is what makes the regression line that includes them significant.

The next data set is the elk population for the entire Grand Teton area from 1912 until 1986 and is from Boyce [5]. The data is reproduced in Table 4.15, and plotted in Figure 4.54 with the regression line. Our naive conclusion is that the elk population has an overall growth trend. A regression table for this data is shown in Figure 4.55. From this we notice that the F-ratio is 5.37, which is significant at a respectable 0.0241 level. Recall that this means we believe that the slope of the trend line is significantly different from zero or that the positive correlation is significant.

Looking again at the scatter plot (Figure 4.54), we observe that, even though there is a significant linear trend to the data, there is also a lot of noise. A reasonable question to ask is how much of the variation in the data does the regression line account for. The answer is given by R^2 which is 8.6%. Thus even though the regression line is significant and we are highly confident in the upward trend to the data, this trend accounts for very little of the fluctuation that we see. Other factors unknown to us, collectively called noise, account for about 91% of the variation that we see.

In order for us to make claims of significance, the data needs to satisfy the conditions of Section 4.10.3. Figure 4.56 shows a histogram of the residuals which are consistent with a sample drawn from a normal distribution of mean 0. This normality of the residuals is further examined in the normal probability plot shown in Figure 4.57, which is nearly linear. We do not address the issue of the residuals' being mutually independent.

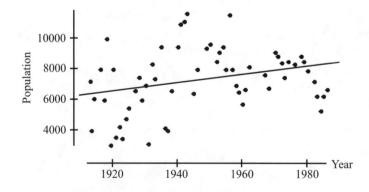

FIGURE 4.54 Elk of the Grand Tetons From 1912 to 1986.

4 Empirical Modeling

Year	Elk Population	Year	Elk Population	Year	Elk Population
1912	7,250	1937	4,000	1962	8,222
1913	4,000	1938	6,655	1963	*
1914	6,150	1939	*	1964	*
1915	*	1940	9,500	1965	*
1916	8,000	1941	11,000	1966	*
1917	6,000	1942	11,185	1967	7,665
1918	10,000	1943	11,700	1968	6,809
1919	3,000	1944	*	1969	*
1920	8,000	1945	6,443	1970	9,196
1921	3,500	1946	8,000	1971	8,877
1922	4,300	1947	*	1972	8,450
1923	3,400	1948	*	1973	7,480
1924	4,800	1949	9,423	1974	8,556
1925	5,500	1950	9,700	1975	*
1926	*	1951	*	1976	8,373
1927	6,625	1952	8,547	1977	*
1928	7,500	1953	9,200	1978	8,891
1929	6,000	1954	9,530	1979	8,552
1930	7,000	1955	8,000	1980	7,980
1931	3,110	1956	11,612	1981	*
1932	8,400	1957	8,000	1982	7,235
1933	7,460	1958	7,000	1983	6,269
1934	*	1959	6,543	1984	5,366
1935	9,500	1960	5,776	1985	6,266
1936	4,200	1961	6,705	1986	6,726

TABLE 4.15 Elk Population at the National Elk Refuge. The Symbol * Indicates Missing or Suspect Data. Used by Permission of Cambridge University Press.

Dependent variable is: Population
75 total cases of which 16 are missing
R squared = 8.6%   R squared (adjusted) = 7.0%
s = 2023 with 59 - 2 = 57 degrees of freedom

Source	Sum of Squares	df	Mean Square	F-ratio
Regression	21970215	1	21970215	5.37
Residual	233241264	57	4091952	

Variable	Coefficient	s.e. of Coeff	t-ratio	prob
Constant	-46130.8	23078	-2.00	0.0504
Year	27.4547	11.85	2.32	0.0241

FIGURE 4.55 Regression Table for Elk of the Grand Tetons From 1912 to 1986.

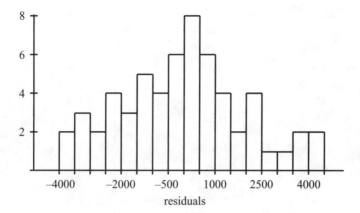

FIGURE 4.56 Residual Histogram for Elk of the Grand Tetons from 1912 to 1986.

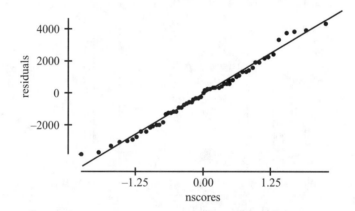

FIGURE 4.57 Normal Probability Plot Elk of the Grand Tetons from 1912 to 1986.

4.12 An Introduction to Multiple Regression

One of the goals for curve-fitting is to explain as much of the variance in the data as possible. In situations outside of carefully controlled experiments, it is frequently the case that two or more variables explain more of the variance than one variable alone. (In multiple regression, the independent variables are frequently called *predictors*.) Thus rather than fitting a line or curve to data, we want to fit a plane (or hyperplane) or surface to the data. There are several aspects to this process, some of which are much more complicated than simple regression, others of which are about the same. The derivation of the statistics and even their numerical computation requires the development of a matrix-based theory which we will not go into. On the other hand, the use of computer statistical packages for multiple regression is almost identical to their use in simple regression, and we follow this approach. Finally the interpretation of the regression is similar to the simple regression, but can also involve many complications. Here we discuss the interpretation which is analogous to simple regression and make mention

4 Empirical Modeling

of only some of the complications. These complications include which and how many variables to add to the model. In general, there may not be a single right way to answer these questions, which makes for heated discussions. Further, while there may be criteria that people are willing to agree on, finding the model that satisfies them may not be practical.

4.12.1 Example: A Testing Model

At the authors' university, students in college algebra classes are given a standardized placement exam over algebra and trigonometry. All entering students are also required to take the ACT exam. The purpose of the examinations is to try to predict student performance in college algebra. Two questions are: Which test is the better predictor of student performance? Is there any advantage in using both tests as predictors over just one? Also recorded is the student's year in school (freshman, sophomore, junior, senior) at the time of taking both the course and the pretest. It is also of interest whether this factor is important. If it is, then the same test scores might result in different departmental recommendations for freshmen and seniors.

The second question addresses one of the important issues in multiple regression. If one variable accounts for 50% of the variance and the other variable also accounts for 50%, then do they collectively account for 100% of the variance? The answer is generally no unless the two variable are completely uncorrelated. With examinations, we might expect good or poor performance on the ACT to be related to good or poor performance on the pretest. Thus there is some common amount of variance explained by both exams. This variance is explained only once in a two variable model. As a simple analogy, suppose the variance were the cards of a standard 52-card deck of cards. Suppose one variable were Hearts explaining 13 of the 52 cards, and the other variable were Aces explaining 4 of the 52 cards. Together Aces and Hearts do not explain 17 of the 52 cards, because the Ace of Hearts counts only once. Together these variables account for only 16 of the 52 cards. Next suppose that the variables are red cards and Hearts. Red cards account for half of the deck, and Hearts account for a quarter. Since all Hearts are reds, together Hearts and reds still only account for half of the cards. Finally suppose one variable is red cards and the other variable is black cards. Each accounts for half of the cards, and since there is no overlap, together they account for all of the cards. The situation is similar with regression. Any variance which is accounted for by two variables does not get counted twice.

The dependent variable GRADE has been encoded as follows: A is a 4, B is a 3, C is a 2, D is a 1, and failure to complete the class for any reason (F or withdrawal) is 0. Even though the grades take only discrete values we allow for averages to take any value over $[0, 4]$; hence we consider the dependent variable as being continuous. Similarly, YEAR has been encoded with 1 for freshman, 2 for sophomores, 3 for juniors, and 4 for seniors. Class rank is decided by number of hours completed, not number of years in attendance. The pretest variable PRETEST records the score on the pretest, a score between 0 and 25. Finally, ACT is the math ACT pretest score, which ranges from 0 to 32.

We will run seven regression analyses. Three are simple regressions: GRADE versus PRETEST, GRADE versus ACT, and GRADE versus YEAR. Three are multiple

Dependent variable is: GRADE
No Selector
R squared = 31.9%   R squared (adjusted) = 31.5%
s = 1.174 with 178 - 2 = 176 degrees of freedom

Source	Sum of Squares	df	Mean Square	F-ratio
Regression	113.778	1	113.778	82.5
Residual	242.745	176	1.37923	

Variable	Coefficient	s.e. of Coeff	t-ratio	prob
Constant	-3.90280	0.5598	-6.97	≤ 0.0001
ACT	0.263162	0.0290	9.08	≤ 0.0001

FIGURE 4.58 Regression Table for GRADE vs. ACT.

regressions with two predictors: GRADE versus PRETEST and ACT, GRADE versus PRETEST and YEAR, and GRADE versus YEAR and ACT. Finally there is one regression on all three predictors, namely GRADE versus ACT, PRETEST, and YEAR. Each of these regression runs is essentially the same as doing a simple regression run, only with one, two, or three variables specified; the complexity of multiple regression is the interpretation. The results of these runs are reported in Figures 4.59, 4.61, 4.63, and 4.64. We observe that the three simple regressions are all significant at a 0.05 level or better, and the ACT and PRETEST simple regressions are significant at a better than 0.0001 level. The ACT predictor alone accounts for 31.9% of the variance in the grades, the pretest accounts for 27.1%, and the year in school accounts for 2.4%. Thus if we want a single predictor, the ACT does the best job, the pretest is a close second, and the class rank is poor. Interestingly, regression for GRADE versus YEAR has a negative β_1, implying that the average senior does worse than the average freshman. Note that this does not imply that students do worse in algebra the longer they wait to take it, though

Dependent variable is: GRADE
No Selector
R squared = 27.1%   R squared (adjusted) = 26.7%
s = 1.215 with 178 - 2 = 176 degrees of freedom

Source	Sum of Squares	df	Mean Square	F-ratio
Regression	96.6353	1	96.6353	65.4
Residual	259.887	176	1.47663	

Variable	Coefficient	s.e. of Coeff	t-ratio	prob
Constant	-0.802494	0.2543	-3.16	0.0019
ALGEBRA	0.189702	0.0234	8.09	≤ 0.0001

FIGURE 4.59 Regression Table for GRADE vs. ALGEBRA.

4 Empirical Modeling

Dependent variable is: GRADE
No Selector
R squared = 2.4%   R squared (adjusted) = 1.9%
s = 1.406 with 178 - 2 = 176 degrees of freedom

Source	Sum of Squares	df	Mean Square	F-ratio
Regression	8.70906	1	8.70906	4.41
Residual	347.813	176	1.97621	

Variable	Coefficient	s.e. of Coeff	t-ratio	prob
Constant	1.57653	0.2425	6.50	≤0.0001
YEAR	-0.306851	0.1462	-2.10	0.0372

FIGURE 4.60 Regression Table for GRADE vs. YEAR.

this might be true. Other possible explanations are that poorer students put off algebra, or that students who fail algebra are likely to be of a higher rank when they take it again. These issues cannot be resolved with our analysis, and it is incorrect to make any such assertions.

Next let us look at the pairwise regressions. Again all three are significant, this time at a better than 0.01 level, though this is not reported by our program so we must use the F-ratio and either an F-table or a computerized F-test separately. With ACT and PRETEST the amount of variance accounted for rises to 40.3%; with PRETEST and YEAR the amount of variance is 27.9%; and with ACT and YEAR the amount of variance accounted for it 32.6%. Using a two-independent-variable model we conclude that ACT and PRETEST does the best job. Adding YEAR to ACT or PRETEST explains less than one percent additional variance. Further, in the regression models with YEAR, the significance of the YEAR variable is 0.1564 and 0.19734, which is

Dependent variable is: GRADE
No Selector
R squared = 40.3%   R squared (adjusted) = 39.7%
s = 1.102 with 178 - 3 = 175 degrees of freedom

Source	Sum of Squares	df	Mean Square	F-ratio
Regression	143.831	2	71.9154	59.2
Residual	212.692	175	1.21538	

Variable	Coefficient	s.e. of Coeff	t-ratio	prob
Constant	-3.75064	0.5263	-7.13	≤0.0001
ACT	0.191696	0.0308	6.23	≤0.0001
ALGEBRA	0.119652	0.0241	4.97	≤0.0001

FIGURE 4.61 Regression Table for GRADE vs. ACT and ALGEBRA.

Dependent variable is: GRADE
No Selector
R squared = 27.9%   R squared (adjusted) = 27.1%
s = 1.212 with 178 - 3 = 175 degrees of freedom

Source	Sum of Squares	df	Mean Square	F-ratio
Regression	99.6098	2	49.8049	33.9
Residual	256.913	175	1.46807	

Variable	Coefficient	s.e. of Coeff	t-ratio	prob
Constant	-0.489501	0.3356	-1.46	0.1465
ALGEBRA	0.185470	0.0236	7.87	≤0.0001
YEAR	-0.180772	0.1270	-1.42	0.1564

FIGURE 4.62 Regression Table for GRADE vs. ALGEBRA and YEAR.

not significant for most intents and purposes. Finally consider a regression with all three predictors. The regression is significant, and 40.7% of the variance in the data is accounted for; however, this is only slightly better than the 40.3% explained by ACT and PRETEST alone. Further, the predictors PRETEST and ACT are significant at a better than 0.0001 level, while YEAR is significant at a 0.3079 level which is generally unacceptable. Thus we do not consider the three-variable model an improvement over the two-variable model with ACT and PRETEST.

Doing these individual regression analyses and interpreting the individual results is straightforward. Making use of all the regression information is harder. What can we conclude from the regression results? The answer depends on our purpose and our resource constraints. If we want a single best predictor, then ACT is certainly it. Using the ACT and PRETEST predictors explains an additional 10% of the variance

Dependent variable is: GRADE
No Selector
R squared = 32.6%   R squared (adjusted) = 31.8%
s = 1.172 with 178 - 3 = 175 degrees of freedom

Source	Sum of Squares	df	Mean Square	F-ratio
Regression	116.079	2	58.0395	42.2
Residual	240.443	175	1.37396	

Variable	Coefficient	s.e. of Coeff	t-ratio	prob
Constant	-3.56702	0.6160	-5.79	≤0.0001
YEAR	-0.159216	0.1230	-1.29	0.1973
ACT	0.258033	0.0292	8.84	≤0.0001

FIGURE 4.63 Regression Table for GRADE vs. YEAR and ACT.

4 Empirical Modeling

Dependent variable is: GRADE
No Selector
R squared = 40.7%   R squared (adjusted) = 39.7%
s = 1.102 with 178 - 4 = 174 degrees of freedom

Source	Sum of Squares	df	Mean Square	F-ratio
Regression	145.102	3	48.3672	39.8
Residual	211.421	174	1.21506	

Variable	Coefficient	s.e. of Coeff	t-ratio	prob
Constant	-3.50273	0.5794	-6.05	≤0.0001
YEAR	-0.118617	0.1160	-1.02	0.3079
ACT	0.188931	0.0309	6.12	≤0.0001
ALGEBRA	0.117885	0.0241	4.89	≤0.0001

FIGURE 4.64 Regression Table for GRADE vs. YEAR, ACT, and ALGEBRA.

(approximately), so this is better still. On the other hand, the ACT is a test that all students are required to take for admission, while the pretest is a departmental exam which is costly in terms of paper, faculty time, and class time. The question of whether this cost justifies the additional improvement in R^2 is a question of values and cannot be answered by this analysis. This analysis gives information, however, which can be considered in the decision-making process. One of the key things it offers is the question of whether placement decisions can reasonably be made with 60 to 70 percent of the grade variance not accounted for by these exams. One thing this analysis does make clear is that including class rank (YEAR) adds little, and in all but one case, its contribution is not significant at reasonable levels.

Referring to the paragraph preceding this example, we now observe that determining the best model is not straightforward, even given every possible combination of variables. On the topic of every possible combination, notice that for three independent variables there were $2^3 - 1 = 7$ non-trivial models to consider. Take a moment and convince yourself that if there were n independent variables or predictors under consideration, there would be $2^n - 1$ possible models to consider. Hint: remember the size of the power set. Thus for more than a few predictors it becomes impractical to compute all possible models. There are some heuristics for finding good models in these cases, but they are beyond our scope. The interested reader should consult the books on regression listed in this chapter's For Further Reading section.

This example was originally explored as a final project by Cindy Carr, one of our modeling students.

4.13 Curvilinear Regression

Earlier sections of this chapter dealt with fitting both linear and nonlinear curves to data. It was mentioned there that the word *linear* has two meanings. One is that a line is fit to data as is the case in simple regression. The other is that the fitted equation is a linear combination of variables or basis functions. When talking about nonlinear or curvilinear equations, we have two situations: equations which are intrinsically linear (linear combination of the basis functions) and equations which are intrinsically nonlinear (not a linear combination of the basis functions). Fitting $y = Ae^3x + Be^{-x}$ for A and B is intrinsically linear, whereas fitting $y = e^{Ax} + Be^{Bx}$ for A and B is intrinsically nonlinear. In this section, we look at the regression analysis associated with intrinsically linear fits only. We call the functions involved in an intrinsically linear fit *basis functions*.

As with most of this chapter, we assume that the reader has access to a software package that performs curvilinear regression (such as *Mathematica* or a good statistical package). We make no effort to explain how the coefficients are derived and content ourselves with learning to interpret the results of our software tools.

In the multiple regression example, we added independent variables or predictors to the model and observed the improvement in R^2, while taking care to watch significance levels. With intrinsically linear curvilinear fits, we add basis functions and observe the improvement and significance. Care must be taken however. Given n points, an $n-1$ degree polynomial can pass through all of them ($R^2 = 100\%$), but the resulting function may not mean anything at any other point. One strategy is to look at the data and visually observe the patterns and add basis functions which appear to belong. If there is no curve to the data, a line should suffice. Another strategy is to ask oneself what type of behavior is reasonable based on what we know about the situation. The important point here is **do not add basis functions unless you have some justification for doing so**.

4.13.1 Example: A Corn Storage Model

Figure 4.65 shows the stored quantities of corn in Kentucky storehouses quarterly for three years (data from Kentucky Agricultural Statistics [60]). What model should we fit? It is reasonable to fit a line to look for overall upward or downward trends, but there is more structure to this data. There is a periodic pattern to this data suggesting the use of a sine or cosine curve. Based on these observations we make a series of models.

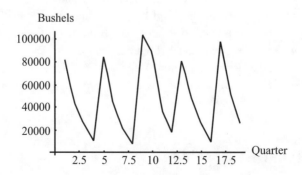

FIGURE 4.65 Corn Data.

4 Empirical Modeling

```
{ParameterTable ->

       Estimate    SE          TStat        PValue
    1  48361.1     15834.1     3.05424      0.00717394

    x  -113.275    1388.74     -0.0815669   0.935944

  RSquared -> 0.000391209, AdjustedRSquared -> -0.0584093,

  EstimatedVariance -> 1.09931 10^9,

  ANOVATable ->

            DoF   SoS            MeanSS         FRatio       PValue

    Model   1     7.31386 10^6   7.31386 10^6   0.00665315   0.935944

    Error   17    1.86882 10^10  1.09931 10^9

    Total   18    1.86955 10^10

}
```

FIGURE 4.66 A Linear Fit to the Corn Storage Data.

Model 1. Linear Fit. Even though we do not expect this to be the final model, it is a good place to start. In particular, we observe whether or not there is a linear trend. This analysis was done in *Mathematica*. The regression table is shown in Figure 4.66.

Several pieces of information are worth noting. First, $R^2 = 0.0004$ which says that very little of the variances is accounted for by this model. Second, the F-ratio has a value of 0.0066 implying a significance at an $\alpha = 0.93$ level which is extremely poor. Finally looking at the individual basis functions of the regression, the constant term is significant at an $\alpha = 0.007$ level and the x term is significant at an $\alpha = 0.936$ level. The implication from this is that there is no significant linear trend to the data. Soon we will add different basis functions to improve this model, but we do not expect the x term to be significant.

Model 2. Sinusoidal fit. As mentioned above, it is poor practice to arbitrarily throw basis functions into a model to reduce R^2. We should have a theory based either on the situation or on the data. We expect storage levels to change during the year with the most grain right after the harvest, and we expect this pattern to repeat year after year. Looking at the data we see a see that it has this periodic nature to it. Prime candidates for fitting periodic data are sine and cosine functions. We will run this model twice, the first time slightly wrong to get a point across and then correct it.

Figure 4.67 shows the regression analysis for basis functions

$$\{1, \sin(\pi x/2)\}.$$

```
{ParameterTable ->

              Estimate    SE        TStat      PValue
    1         47228.4     5197.2    9.08727    0

        Pi x
   Sin[―――]
         2    31576.9     7163.85   4.40781    0.000384682

   RSquared -> 0.533336, AdjustedRSquared -> 0.505886,

                                 8
   EstimatedVariance -> 5.13207 10 ,

   ANOVATable ->

              DoF    SoS                  MeanSS              FRatio     PValue
                                 9                    9
   Model      1      9.97101 10           9.97101 10         19.4288     0.000384682

                                 9                    8
   Error      17     8.72452 10           5.13207 10

                                 10
   Total      18     1.86955 10

}
```

FIGURE 4.67 Corn Data Regression for Basis Functions $\{1, \sin(\pi x/2)\}$.

The $\pi/2$ comes from the fact that the period here is four (x is measured in quarters), so we want $\sin(2\pi)$ when $x = 4$. More precisely, if the period is T, we use $\sin(2\pi x/T)$. A graph of the regression function and the data is shown in Figure 4.68. The regression statistics show $R^2 = 0.53$ and an F-ratio significant at a 0.00038 level, both clear improvements over the linear model. Also both basis functions are significant.

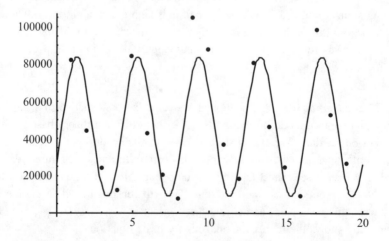

FIGURE 4.68 Corn Data Fit for Basis Functions $\{1, \sin(\pi/2x)\}$.

4 Empirical Modeling

There is a fundamental flaw with the theory of the previous model. Using $\sin(\pi x/2)$ as a basis function implies that the periodic nature of the data is midway between a peak and a valley when the data starts; in other words, the data is in phase with $\sin(\pi x/2)$. There is no reason that it should be. It would be better to fit a function of the form $\sin(\pi x/2+\phi)$. The problem is that the ϕ parameter is unknown, so we need either to eyeball the data for ϕ or fit ϕ to the data. The problem with fitting is that it is an intrinsically nonlinear fit. All this is resolved by including both $\sin(\pi x/2)$ and $\cos(\pi x/2)$. In Exercise 6, the reader is asked to show that, given A, ω, and ϕ, there exist constants C_1 and C_2 such that $A\sin(\omega t + \phi) = C_1 \cos(\omega t) + C_2 \sin(\omega t)$. The implication is that using both sine and cosines allows us to avoid estimating the phase angle ϕ.

Figure 4.69 shows the regression table for basis functions

$$\{1, \sin(\pi x/2), \cos(\pi x/2)\}.$$

Correcting the phase improves the fit from $R^2 = 0.533$ to $R^2 = 0.726$. We further note that the significance of the F-ratio improves from 0.00038 to 0.000031, and that the significance of the sine basis function also improves. Finally, we note that the significance of the cosine basis function is 0.0039 which is also very good.

Model 3. Sinusoidal And Linear Fit. With the results of Model 1, there is no reason to expect a significant improvement by including a linear (x) term into the model constructed

```
{ParameterTable ->

                      Estimate      SE            TStat         PValue
        1             46169.9       4110.24       11.2329       0

            Pi x
        Cos[---]      -20110.9      5972.05       -3.36751      0.00391988
             2

            Pi x
        Sin[---]      31576.9       5648.99       5.58983       0.0000406678
             2

    RSquared -> 0.726899,  AdjustedRSquared -> 0.692761,
                                           8
    EstimatedVariance -> 3.19111 10 ,
    ANOVATable ->
                DoF   SoS           MeanSS        FRatio
                      PValue
                                 10            9
        Model   2     1.35898 10    6.79488 10    21.2931

                      0.0000309451
                                9             8
        Error   16    5.10578 10    3.19111 10

                               10
        Total   18    1.86955 10
```

FIGURE 4.69 Corn Data Fit to Basis Functions $\{1, \sin(\pi /2), \cos(\pi x/2)\}$.

in Model 2, although this could happen. Perhaps after the data is explained by the sinusoids, there is an upward trend on the residuals which was not present earlier. Figure 4.70 shows the regression table using basis functions $\{1, x, \sin(\pi x/2), \cos(\pi x/2)\}$. There is a slight improvement in R^2, but only from 0.727 to 0.733. Further the significance of the x term is 0.569 which is still poor, but a dramatic improvement over 0.936 which was the significance level of the x-term in Model 1. So the inclusion of the sinusoids did increase the significance of the upward trend, which is why Model 3 should be considered. For most criteria, however, the significance level is not enough for Model 3 to be considered an improvement over Model 2.

Further Models. We stop our analysis with these three models, but further improvement is clearly possible. Probably the most notable source of error is that the periodic shape in the data is not sinusoidally shaped. Perhaps we can modify a sinusoid or use some piecewise defined periodic basis function. The theory of Fourier analysis, which is outside of our scope, says that any periodic function can be expressed as a, perhaps infinite, combination of sines and cosines of various periods or frequencies, and that a periodic function is approximated arbitrarily well by a finite sum of sines and cosines. These ideas are explored a little in our present context in Exercise 9.

4.14 For Further Reading

While not a book, an valuable source data and other information can be found on the Internet at Carnegie Mellon University's statistics library at http://lib.stat.cmu.edu.

For a precalculus book discussing both data fitting and data transformation (as well as modeling) see:

Gloria Barrett, et al., The North Carolina School of Science and Mathematics, *Contemporary Precalculus through Applications*, Janson Publications, Inc, Dedham, Mass.,1992.

The modern concept of exploring data as opposed to computing isolated statistical calculations is often attributed to John W. Tukey. His ideas are presented in the classic:

John W. Tukey, *Exploratory Data Analysis*, Addison-Wesley Publishing Company, Reading Mass., 1977.

More discussion on fitting a logistic curve to data can be found in:

Fabio Cavallini, Fitting a Logistic Curve to Data, *The College Mathematics Journal*, Volume 24, Number 3, 1993, 247–253.

For basic statistics background and an introduction to regression, we recommend

Robert V. Hogg and Elliot A. Tanis, *Probability and Statistical Inference* (Fourth Edition), Prentice Hall, Englewood Cliffs, NJ, 1993.

More information on regression analysis can be found in:

Raymond H. Myers, *Classical and Modern Regression with Applications* (Second Edition), Duxbury Press, Belmont, CA, 1990.

```
{ParameterTable ->

                   Estimate      SE          TStat       PValue
        1          41684.1       8778.63     4.74837     0.000258849

        x          448.575       770.984     0.581822    0.569329

           Pi x
        Cos[----]  -20110.9      6099.46     -3.29716    0.00488729
             2

           Pi x
        Sin[----]  32025.5       5820.8      5.5019      0.0000608364
             2

     RSquared -> 0.732926, AdjustedRSquared -> 0.679511,
                                          8
     EstimatedVariance -> 3.32873 10 ,

     ANOVATable ->

                DoF    SoS              MeanSS           FRatio
                       PValue
                                10              9
        Model   3      1.37024 10      4.56748 10       13.7214

                       0.000141937

                                9                 8
        Error   15     4.99309 10       3.32873 10

                                10
        Total   18     1.86955 10
```

FIGURE 4.70 Corn Data Fit to Linear and Sinusoidal Basis Functions.

Leroy A. Franklin, Using Simulation to Study Linear Regression, *The College Mathematics Journal*, Vol. 23, No. 4, September 1992, pp. 290–295. This highly recommended paper discusses learning regression by experimenting with simulated data.

Sam Kosh Kachigan, *Multivariate Statistical Analysis: A Conceptual Introduction* (Second Edition), Radius Press, New York, 1991. A gentle introduction to statistics and regression analysis. Assumes little prior knowledge, yet covers some deep material.

G.W. Snedecor and W.G. Cochran, *Statistical Methods* (Sixth Edition), Iowa State University Press, Ames, Iowa, 1967. A solid presentation of applied statistics and regression.

Elazar J. Pedhazur, *Multiple Regression in Behavior Research* (Second Edition), Holt, Rinehart and Winston, Inc., Fort Worth, 1982. This is a classic in statistical modeling in the behavioral sciences. It is, however, an advanced book and somewhat difficult to read.

4.15 Exercises

1. Verify that the correlation coefficient r satisfies the inequality
$$-1 \leq r \leq 1.$$

2. Verify the equation
$$SS_{\text{Tot}} = SS_{\text{Res}} + SS_{\text{Reg}}.$$

3. By hand (or using only the addition, subtraction, multiplication, division, and square keys of a calculator), use equations (80) and (81) to find the regression line for the following data set:
$$(0, 2.2),\ (1, 5.2),\ (2, 4.8),\ (3, 11.1),\ (4, 13.9),\ \text{and}\ (5, 17.1).$$

4. **a.** Verify that equation (90) has an inflection point when $\exp(rt) = C$.
   b. Put $C = \exp(rt_0)$ into equation (90) to obtain equation (91).
   c. Rewrite equation (92) in vector notation to produce equation (93).
   d. Using equation (93), compute $\partial e/\partial K = 0$ and show that this gives equation (94).
   e. Substitute equation (94) into equation (93) to produce equation (95).

5. Suppose that the true relationship between variables y and x is $y = -4x + 1$, but every measured data point deviates from the true relation by a random amount. Further suppose this random amount follows a normal distribution with mean 0 and standard deviation 0.25. Thus a simulation for this situation would have an equation $y = -4x + 1 + \text{Normal}(0, 0.25)$. Use software to create such a simulation and use it to generate data corresponding to $x = 0, 1, \ldots, 10$. Fit a regression line to the simulation data. Compare the regression equation $\hat{y} = \hat{\beta}_0 + \hat{\beta}_1 x$ you found with the true relation $y = -4x + 1$. Repeat at least 20 times and make histograms of $\hat{\beta}_0$ and $\hat{\beta}_1$. Compare the histograms with $\beta_1 = -4$ and $\beta_0 = 1$.

6. Show that
$$A \sin(\omega x + \phi) = C_1 \cos(\omega t) + C_2 \sin(\omega t)$$
where $A = \sqrt{C_1^2 + C_2^2}$ and $\tan \phi = C_1/C_2$. When determining ϕ, use the signs of C_1 and C_2 to determine which quadrant it lies in. Also consider the special case of $C_2 = 0$.

7. We want to interpolate through the data points $(0,1)$, $(1,0)$, $(2,1)$, and $(3,1)$.

   a. A generic cubic has the form $y = ax^3 + bx^2 + cx + d$. By substituting in the four points listed above, we get four equations in the unknowns a, b, c, and d. Solve this to obtain an interpolating polynomial for this data (a polynomial that passes through all the data points). Plot the polynomial and the data on the same coordinate system.

   b. Use the idea of part **a** to interpolate a quadratic through the first three data points and another through the last three. Plot both together with the points and notice the problem of joining the two together.

   c. As in part **b** interpolate a quadratic through the first three data points (actually, just use the function from part **b**). Next put a quadratic through the last two data points. Notice that so far you have two equations in three unknowns. Add a third equation by

4 Empirical Modeling

requiring that the derivative of the quadratic through the first three points at $(2, 1)$ is equal to the derivative of the quadratic through the last two points at $(2, 1)$. Now plot the two quadratics and the four points. Notice the smooth connection at $(2, 1)$. This is the essence of spline interpolation, quadratic interpolation in particular.

8. Using only the arithmetic features of a handheld calculator, construct a regression table for the data in Table 4.16. In practice you never do this by hand, but it is a useful exercise in understanding the components of the table. Check your answers with a statistical package.

9. This is a detective problem. We created a function by forming a linear combination of functions of the form $\sin(x\pi n)$ and $\cos(x\pi n)$ where $1 \leq n \leq 5$. Some of the values this function has over the domain $[0, 5]$ are listed in Table 4.17. Use curvilinear regression analysis to determine the original function. How confident are you of your answer? (Note: we did not introduce any noise into the data.)

x	1	2	3	4	5	6
y	1	0	2	1	3	1

TABLE 4.16 Data.

x	y	x	y	x	y
0	5	1.7	-3.1171016	3.4	-2.0204683
0.1	-2.9727122	1.8	3.52452539	3.5	-6
0.2	-8.6146953	1.9	7.73549171	3.6	-6.8745703
0.3	-5.8253055	2	5	3.7	-3.1171016
0.4	3.25653629	2.1	-2.9727122	3.8	3.52452539
0.5	10	2.2	-8.6146953	3.9	7.73549171
0.6	8.11063825	2.3	-5.8253055	4.0	5
0.7	-0.1189664	2.4	3.25653629	4.1	-2.9727122
0.8	-6.7605934	2.5	10	4.2	-8.6146953
0.9	-6.4994237	2.6	8.11063825	4.3	-5.8253055
1	-1	2.7	-0.1189664	4.4	3.25653629
1.1	4.2087802	2.8	-6.7605934	4.5	10
1.2	5.37862736	2.9	-6.4994237	4.6	8.11063825
1.3	2.58923753	3.0	-1	4.7	-0.1189664
1.4	-2.0204683	3.1	4.2087802	4.8	-6.7605934
1.5	-6	3.2	5.37862736	4.9	-6.4994237
1.6	-6.8745703	3.3	2.58923753	5.0	-1

TABLE 4.17 Linear Combination of Sinusoids.

Year	Population Size (in millions)
1801	8.89
1811	10.16
1821	12.00
1831	13.90
1841	15.91
1851	17.93
1861	20.07
1871	22.71
1881	25.97
1891	29.00
1901	32.53
1911	36.07
1921	37.89
1931	39.95

TABLE 4.18 Population of England and Wales. Reprinted by permission of Addison Wesley Longman.

4.16 Projects

Project 4.1. England and Wales

In this project, we fit curves to population data from England and Wales from 1801 to 1931. The data set shown in Table 4.18 is the census count of the population (in millions) for England and Wales at every decade from 1801 to 1931 (from Tukey [67]).

0. Enter this data into a suitable software package and create a scatter plot of this data.

1. Fit a line to this data. Plot this line on a graph with the data. Compute R^2 for this data and interpret its meaning. How well does the line model the population?

2. We have often seen that populations tend to grow exponentially. Fit an exponential model to this data by taking the logarithm of the population size. Plot this exponential model along with the data. Compute R^2 for this data and interpret its meaning. How well does the exponential function model the population?

3. Fit a power model to this data by taking the logarithms of both the year and population size. Plot this power model along with the data. Compute R^2 and interpret the result. How well does the power model fit the population?

4. Which is the best model for this data? Is there justification for using a model more complicated than a line?

Project 4.2. A Belly Button Model

The golden ratio 0.618 is the limit of the ratio of successive terms of a Fibonacci sequence presented in Chapter 1. There are numerous claims as to the natural occurrence of this ratio in diverse phenomena ranging from spirals to the great pyramids to rectangles of most pleasing proportion. One claim is that the location of one's belly button or navel

4 Empirical Modeling

is determined by the golden ratio. In particular, the claim is that the ratio of the height of one's navel (distance from navel to the floor) to one's height is 0.618 subject to usual statistical variance. Thus the model is

$$B = 0.618 \times H$$

where H is height and B is distance between navel and the floor. This project tests the validity of this model.

1. Collect data for this model (i.e., actually get out and measure some heights and belly button heights). If possible, collect at least 30 pieces of data.

2. Plot your data with the model line on the same coordinate axes.

3. Compute R^2 for this model. Can you propose a better model for your data? Collect new data and compare how well the golden ratio model and your model (use the regression line if you have no better proposal) explain the new data.

This model is based on a problem in Schnoor [58].

Project 4.3. Bald Eagles

The pesticide DDT was used for many years until it was determined to be responsible for a decline in many raptor populations including eagles and falcons. It was believed that DDT softened the shells of bird eggs which resulted in a lowering of the reproduction rate. DDT was eventually banned on December 31, 1972. In Table 4.19, eagle reproduction from 1966 to 1981 is reported in terms of number of young per unit area (data from [23]).

Year	Young per area
1966	1.26
1967	0.73
1968	0.89
1969	0.84
1970	0.54
1971	0.60
1972	0.54
1973	0.78
1974	0.46
1975	0.77
1976	0.86
1977	0.96
1978	0.82
1979	0.98
1980	1.12
1981	0.82

TABLE 4.19 Number of Bald Eagle Young per Unit Area.

Make a scatter plot of the data. Compute two regression lines, one for the years up to 1972 and one for 1973 on. Create a graph showing the original data and the two regression lines. Comment on the effectiveness of banning DDT on eagle reproduction.

Project 4.4. Logistic *X-Files*

At the end of Section 4.5, we note that the slope of the regression line fit to the second season data is much closer to zero than the slope of the regression line fit to the first season data, suggesting that growth may be slowing down. Based on this observation, it is reasonable to consider a logistic fit to the data.

Using the methods of Section 4.8.1, fit a logistic curve to both seasons of *The X-files* data (Table 4.2). Compute R^2 for this model and compare with the linear models in Section 4.5.

Project 4.5. Alligator Vent Snout Length

In the biological sciences, it is sometimes possible to relate the weight (or volume) of an organism to some linear measurement such as length or weight. If W denotes the weight and ℓ the length, the model

$$W(t) = a_0 \ell^{a_1 t}$$

is known as an *allometric* equation. The model has parameters a_0 and a_1. If we wish to relate the random weight of organisms to observable fixed lengths, we could take logarithms and obtain the linear model

$$\ln W = \ln a_0 + a_1 t \ln \ell.$$

In the following data, W denotes the weight (in pounds) and ℓ the snout vent length (distance in inches from back of head to end of nose) for 15 alligators captured in central Florida. Since ℓ is easier to observe than W for alligators in their natural habitat, we want to construct a model relating weight to snout vent length. Such a model can be used to predict the weight of alligators with particular snout vent lengths. Fit the model

$$\ln W = \ln a_0 + a_1 t \ln \ell$$

to the data in Table 4.20 (data from [42]).

Project 4.6. Interpolating the Population of the United States

United States census data can be found in Table 5.1 in Chapter 5. This project explores this data using interpolating functions.

1. Fit a cubic spline interpolating function to the U.S. Census data and graph. Use it to estimate the population in 1975.

2. Compute and graph the derivative (instantaneous growth rate) of the interpolating function (this should be done with a computer algebra system like *Mathematica* or *Maple*). Identify historical events corresponding to the peaks and valleys of the graph.

3. Compute and graph the percentage growth rate ($\frac{f'(x)}{f(x)} \times 100\%$). Compare this graph with the graph of part **2**, particularly the dips corresponding to the Civil War and the World Wars. Explain the differences.

4 Empirical Modeling

Length ℓ	Weight W
47.94	130.32
36.97	50.91
75.94	639.06
30.88	27.94
45.15	79.84
46.06	109.95
31.82	33.12
42.95	90.02
33.12	35.87
35.87	38.09
66.02	365.03
43.82	83.93
40.85	79.84
41.68	83.10
43.82	70.11

TABLE 4.20 Snout Vent Length and Weights of Florida Alligators. From *Mathematical Statistics with Applications*, by W. Mendenhall, D.D. Wackerly, and R.L. Scheaffer. Copyright ©1994, 1990, 1986, 1981, 1973 Wadsworth Publishing Company. Reprinted by Permission of Brooks/Cole Publishing Company, Pacific Grove, CA, a Division of International Thomson Publishing Inc.

This idea for this project is borrowed from *Calculus&Mathematica* by Davis, Porta, and Uhl [14].

Project 4.7. Black-Capped Chickadees II

This project extends the model in Project 2.3, and the reader should refer there for the data. The reader should also do Project 2.3 first.

If lines are fitted to the Recruitment and Survival Rate data, a trend is observed in the data. This project incorporates the trend into the models of Project 2.3.

1. Fit regression lines to the Recruitment and Survival Rate data and test the lines for significance.

2. Using these functions instead of the constants (means), re-do the deterministic chickadee model (part **2** of Project 2.3). Compare this model to the actual population data and with the original deterministic model.

3. Detrend the Recruitment and Survival Rate data by subtracting the trend lines found in part **1**. Using the detrended data, compute means and standard deviations.

4. Assuming, as before, that these are samples from a normal distribution, create a stochastic model. To do this, at each time step, add a random number from a normal distribution with mean and standard deviation found in part **3** to the appropriate trend line. Thus the recruitment rate will be a function of the form

$$\text{rate}(t) = m\,t + b + \text{Normal}(\bar{x}, s)$$

where $mt + b$ is the trend line, $\bar{x}$ is the sample mean, and s is the sample standard deviation. The survival rate will be similar. Do at least 50 runs over the years 1959 to 1982.

5. Compare the results from **3** with the results of part **3** of Project 2.3. Do the results from this model's improvement justify the complications?

Project 4.8. A Stock Market Model

Conventional wisdom says that over the long term, the stock market exhibits exponential growth. The short-term ups and downs are essentially noise. This project explores this idea. One problem is trying to measure the stock market. We will use one common measurement instrument, the Dow Jones Industrial Average (referred to hereafter as the Dow). Feel free to explore this project using some other measure.

1. Find monthly (at least) data for the Dow for the last 30 years. While this data can be found in the library, there are also online sources for it.

2. Fit an exponential model to this data and plot the data with the model. Comment on how well the model fits the data. Your comments should include a visual inspection, and a discussion of R^2.

3. If you invested $1,000 today and the Dow continued to behave as it had in the past, use your exponential model to predict how much your investment would be worth in 30 years.

4. Compute the residuals for your model. Find their mean and standard deviation. Make a histogram of the residuals. Is it reasonable to assume they are a sample from a population following a normal distribution? If you know how to do a normal probability plot, you may also want to use it to check normality.

5. Assume that the Dow continues to behave in the future as it has in the past. Further assume either that a normal distribution is a reasonable choice for the residuals or that a distribution you think is better is a reasonable choice. Construct a stochastic model as we did in Chapter 2 to obtain a distribution for how much $1,000 invested today will be worth in 30 years. What are its mean and standard deviation? Give a histogram of the distribution. How does it compare with your result from **3**?

Project 4.9. Fitting a Logistic Curve to Data

The purpose of this project is to practice fitting a logistic curve to data. Table 4.21 shows the growth of an algal sample taken from the Adriatic Sea over time. The biomass (amount of algae) is measured as the area of a microscope slide covered by the sample and hence is in units of mm^2.

1. Plot the data in Table 4.21 and verify that a logistic curve is a reasonable model to fit to these data.

2. Use the techniques in Section 4.8.1 to fit a logistic curve to the data and plot the curve with the data.

3. Compute R^2 for this model.

The data in this project originally comes from Zangrandi [73] as reported in Cavallini [10].

Time (days)	Biomass (mm^2)
11	0.00476
15	0.0105
18	0.0207
23	0.0619
26	0.337
31	0.74
39	1.7
44	2.45
54	3.5
64	4.5
74	5.09

TABLE 4.21 Algae Growth Versus Time. Reprinted by permission of The Mathematical Association of America.

Project 4.10. Modeling Increase in Computer Speed

The speed of personal computers to date has been increasing. Research the history of CPU clock speeds and find data for the speed (CPU clock speed) of the newest personal computers actually on the market each year since 1980. Advertisements in personal computer magazines might be one source of data. Using only the data up to five years ago, fit the best model you can with the model's purpose being to predict speed over the next five years. Ideally you should not even know the data for the last five years when you do your model fitting, so put these data aside and do not look at them until you have decided on your best model. From your model would you say that computer clock speeds are growing logistically, exponentially, logarithmically, linearly, quadratically, according to some other power law, or do none of these apply. How well does your model fit the data up to five years ago? How well does your model predict the subsequent five years?

Project 4.11. Modeling Computational Complexity

Even if you have not given much thought to the issue, you are probably aware that numerical solutions take longer to compute as the problems get larger. There are several reasons for this. One is that the number of operations to be performed increases as the size of a problem increases. This is called *algorithmic complexity* and is frequently a topic in discrete mathematics and theoretical computer science courses. A second reason is machine dependent and, hence, much harder to analyze theoretically. Computers have different types of memory (e.g., registers, caches, RAM, hard disks, floppy disks, and tape). Each type operates at its own speed. The list given is in order of speed with registers being the fastest (able to bring data to the CPU in a single clock cycle) and floppy disks and tape being the slowest (further slowed by the potential need for an operator to find and insert/mount the disk/tape). This list is also in order of memory cost with registers being essentially part of the CPU and hence very expensive and few in number; RAM is much cheaper, and much more is available; while hard disks, floppy disks, and tape are so cheap and in such larger quantities they are called mass storage devices. As problems become larger, they require more memory. If all the data/instructions for a problem can be stored in the CPU cache, it will run faster than if it exceeds cache capabilities and must switch information with RAM. Similarly if a problem exceeds RAM size and must be partially stored on a hard disk and intermediate results moved back and forth between RAM and the hard disk, then it will run more slowly than if there is enough RAM to hold the entire problem. For a particular machine, as one increases the problem size, there will be points at which the problem passes over these memory boundaries resulting in much worse performance than an analysis of algorithmic complexity alone would suggests. On the other hand, if algorithmic complexity analysis indicates that a problem completely contained in cache memory will take 100 years to run, then the issue of memory types is irrelevant. Finally, software packages may have memory management techniques which affect speed as problems grow. This project empirically explores the relationship between project size and speed.

While this project can be modified for different types of models and different types of software, as it is written it is assumed that software is available which can find matrix sums, products, and inverses. Some means of accurately timing these procedures is also required.

1. Generate or construct a random 2×2 matrix (entries randomly chosen from $[0, 1]$). Compute the amount of time required to add the matrix to itself. Repeat for a 3×3 matrix, a 4×4, and so on until your computer takes an uncomfortable amount of time (a minute or two) to process. If you have sufficient programming skills, you ought to be able to automate this process. Plot matrix size n (for an $n \times n$ matrix) versus time. Determine if there is any point at which the behavior deviates from the trend (indicating a memory type boundary or other size problem) and fit a polynomial regression curve to the data before this curve. Algorithmic complexity predicts that the polynomial should be a line.

2. Repeat for matrix multiplication and finding matrix inverses. Using algorithmic complexity alone, the regression polynomial should be quadratic for multiplication and cubic for inversion (if Gaussian elimination is used). Use the regression curves to predict how long it will take to compute a product and an inverse for a matrix of size one larger than used in the data. Run them and compare the actual run times with your predicted results. These results should be saved since they are useful when trying to design the largest model that your particular machine can handle in a reasonable time.

Project 4.12. Tridiagonal Matrices and the Thomas Algorithm

In Project 4.11, we observed that the time required to compute the inverse of a matrix by Gaussian elimination grows as the number of rows of the matrix cubed. This rate of growth makes solving systems of linear equations (the matrix equation $AX = R$) by Gaussian elimination impossible for large systems of equations. Some natural problems involve a coefficient matrix with non-zero entries occurring only on the diagonal, subdiagonal, and superdiagonal. Such matrices are called *tridiagonal*. The Thomas algorithm is a method of solving a tridiagonal system which takes advantage of all the zeros and results in significant time savings over Gaussian elimination. In general,

4 Empirical Modeling

matrices which consist mostly of zeros are called *sparse*, and a great deal of study has gone into finding efficient solutions to matrix equations involving sparse matrices. The Thomas algorithm is one useful example of a sparse matrix method.

The Thomas Algorithm. We wish to solve the matrix equation

$$AX = R$$

where A has the form

$$\begin{pmatrix} b_1 & c_1 & 0 & 0 & \cdots & 0 \\ a_2 & b_2 & c_2 & 0 & \ddots & 0 \\ 0 & a_3 & b_3 & c_3 & \ddots & \vdots \\ \vdots & \ddots & \ddots & \ddots & \ddots & 0 \\ \vdots & \ddots & \ddots & a_{n-1} & b_{n-1} & c_{n-1} \\ 0 & \cdots & \cdots & 0 & a_n & b_n \end{pmatrix},$$

X is $(x_1, \ldots, x_n)^T$, and $R = (r_1, \ldots, r_n)^T$.

The Thomas algorithm finds X as follows:

$$b_i = b_i - \frac{a_i}{b_{i-1}} c_{i-1} \forall i = 2, 3, \cdots, n$$

$$r_i = r_i - \frac{a_i}{b_{i-1}} r_{i-1} \forall i = 2, 3, \cdots, n$$

$$x_n = \frac{r_n}{b_n}$$

$$x_j = \frac{r_j - c_j x_{j+1}}{b_j} \forall j = n-1, n-2, \cdots, 1.$$

1. Write a routine to implement the Thomas algorithm for arbitrary n. Ideally you would use a true programming language, but software such as *Mathematica* or *Maple* can be used. A spreadsheet could also be used. The ability to accurately time the algorithm is essential.

Notice that this algorithm makes very efficient use of computer memory since an $n \times n$ matrix requires storing only $3n$ entries (technically only $3n - 2$ entries, but typically A, B, and C arrays are used, all of size n). Notice further that the A array can be used to store the X result, further making efficient use of memory.

2. Consider the problem where all $a_i = -1$, $b_i = 2$, $c_i = -1$, and $r_1 = -1$ and all other $r_i = 0$. Do timing runs solving this problem (see Project 4.11) for different values of n, using Gaussian elimination, and then again using the Thomas algorithm. Plot the results and discuss how they compare. Finally determine the largest matrix that your computer can solve in a reasonable time. If you are using a numerical package (such as *Mathematica* or *Maple*) to do the Gaussian elimination, be sure it is actually using Gaussian elimination. A smart package might recognize a sparse matrix and switch to a faster algorithm, such as the Thomas algorithm, unbeknownst to the user.

Project 4.13. A Body Fat Model

This project uses estimates of the percentage of body fat determined by underwater weighing and various body circumference measurements for 252 men. This data set can be obtained from Carnegie Mellon's statistics library at http://lib.stat.cmu.edu/datasets/bodyfat and is contributed by Roger Johnson at Carleton College and reproduced with his kind permission.

It is costly and complicated to measure body fat accurately. One method is to determine body volume by submersion in water, then compute the percent of body fat using several formulas involving the density of different types of body tissue. This project involves finding a model to estimate body fat using only easily measured attributes such as height, weight, age, and the circumference of various body parts. The data set includes the following variables:

Density determined from underwater weighing
Percent body fat from Siri's [61] equation
Age (years)
Weight (lbs)
Height (inches)
Neck circumference (cm)
Chest circumference (cm)
Abdomen circumference (cm)
Hip circumference (cm)
Thigh circumference (cm)
Knee circumference (cm)
Ankle circumference (cm)
Biceps (extended) circumference (cm)
Forearm circumference (cm)
Wrist circumference (cm)

(Measurement standards are listed in Benhke and Wilmore [2] pp. 45–48 where, for instance, the abdomen circumference is measured "laterally, at the level of the iliac crests, and anteriorly, at the umbilicus.")

For all parts, consider *Percent body fat* to be the dependent variable. **DO NOT USE** the *Density* variable for parts **2**, **3**, and **4**.

1. Do a simple regression analysis (one predictor or independent variable) for each of the predictors. Make a table showing R^2 and levels of significance. Which is the best single predictor model for easy to measure predictors (i.e., not *Density*)?

2. Do a regression on all predictors except *Density*. How much variance is explained by the full set of predictors? Which predictors are significant?

3. The amount of distinct variance an individual predictor explains (i.e., variance not explained by any other predictor) can be computed by removing it from the full model. Using the full model (from **2**), remove each variable, record the change in R^2, and replace it. Make a table and indicate which predictor explains the most distinct variance.

4 Empirical Modeling

4. Find the best three predictor model you can. (We are not asking for the best; just the best you can find. Consider it a competition to see whose model is the best.) Use it to estimate either your body fat, or if you are self-conscious or not male, the percent body fat of a 25-year-old male, who is 69 inches tall, weighs 190 lbs, has a 30-cm neck, a 100-cm chest, a 91-cm abdomen, a 92-cm hip, a 55-cm thigh, a 25-cm knee, a 20-cm ankle, a 30-cm bicep, a 25-cm forearm, and 15 cm-wrist.

5. Find the best model you can with all predictors significant.

These data are used to produce the predictive equations for lean body weight given in the abstract "Generalized body composition prediction equation for men using simple measurement techniques" by K.W. Penrose, A.G. Nelson, and A.G. Fisher [51].

5

Continuous Models

In the introduction to Chapter 1, we discussed how maple trees differ from humans. Specifically a yearly census of trees is generally adequate, but the human population changes from instant to instant. Chapters 1, 2, and 3 were devoted to models and modeling of processes which either changed in discrete time increments or were approximated well by a discrete time model. This chapter is dedicated to processes which change continuously over time, or which are discrete but with time steps small enough that continuous models are acceptable approximations. As mentioned in Chapter 1, continuous models frequently are used on discrete systems and vice versa if it is pragmatic to do so and the resolution of the resulting model is acceptable.

While our emphasis is on models and modeling, it is beneficial to have some knowledge of differential equations and methods for solving special classes of them. This chapter is a blend of models, modeling, and the mathematics of differential equations. As with all mathematics content in this book, the intent is to build understanding and develop the tools needed for the problems addressed here. It is not intended to be a differential equations course. Specifically we primarily address differential equations of the form

$$\frac{dx}{dt} = f(x, t).$$

The only higher-order differential equations addressed are linear with constant coefficients. We also look at systems of first-order differential equations. Many specific models result in special types of differential equations. Some are thoroughly studied, and others are not. Two types, important for the models discussed here, are linear and separable differential equations, and their methods of solution are presented. For other types, the interested reader is encouraged to consult one of the many excellent texts on differential equations, several of which are listed in the For Further Reading section at the end of this chapter.

In Chapter 1, we studied difference equations. One convenient way of writing a difference equation was $x(n+1) - x(n) = f(x(n), \ldots, x(0))$. The left-hand side of the equation expresses the change in the variable x over one time step. The right-hand side gives a function expressing this change in terms of the values of x at earlier times. This way of writing difference equations connects naturally with the way of writing differential equations. Most of the differential equations under consideration have the form $\frac{dx}{dt} = f(x)$ or $\frac{dx}{dt} = f(x,t)$. Now the left-hand side of the equation expresses the instantaneous rate of change of the x variable, and the right-hand side gives a function expressing this change in terms of the x variable and, perhaps, time. This form is useful for our problems since we often know the rate at which a quantity changes. Other applications require other differential equation forms. Physics problems dealing with forces, for example, involve differential equations with second derivatives.

Differential equations are very natural objects arising from everyday phenomena. In spite of this naturalness, differential equations are frequently either hard to solve or, worse, do not have solutions which can be found by analytic integration methods. There are several ways of dealing with this. First there are a number of special classes of differential equations which are easily solved. Frequently problems which cannot be solved are approximated by similar differential equations in one of these classes. In their differential equations book, Hubbard and West [27] liken this situation to the old joke about the man who has lost his car keys at night and is looking for them under a light pole. A stranger offers to help and asks if this is where the man lost his keys. "No," he replies, "I lost them over there, but I am looking for them over here because this is where the light is." In a similar way, frequently we do not work on the real problem, but only an approximation because that is "where the light is" in the sense that we are able to solve it. In this book, we look at two such light poles: separable differential equations and linear differential equations. A typical differential equations course would present many more. A second manner of dealing with difficult or impossible-to-solve differential equations is to look at the nature of the solution curves, and, in particular, asymptotic behavior. We address this issue by looking at fixed points and their stability. There are some geometric approaches to asymptotic behavior that we will study. Finally, even if an analytic solution is not obtainable, computers or calculators numerically approximate solutions, and for modeling purposes this is often sufficient.

5.1 Setting Up The Differential Equation: Compartmental Analysis II

In Chapter 1, we used compartmental analysis to set up certain classes of difference equations and systems of difference equations. The same ideas are used here to set up differential equations. We quickly review the basic concepts.

Compartmental analysis is a conceptual method for setting up the equations which model one or more variables which are increasing or decreasing over time and which may or may not influence one another. Each variable is represented by a box or compartment. Arrows are drawn into, out of, and between boxes. Arrows into a box represent changes which cause the quantity in the box to increase. Arrows out of a box represent changes which cause the quantity in the box to decrease. (More precisely, arrows into a box represent quantities which cause the variable to increase when they are positive and the

5 Continuous Models

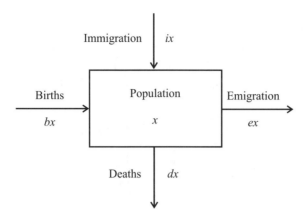

FIGURE 5.1 Compartmental Analysis of a Population.

variable to decrease when they are negative. Arrows out of a box represent quantities which cause the variable to decrease when they are positive and increase when they are negative.) The equations representing these inflows and outflows are written adjacent to each arrow. The rate of change in the variable x (i.e., $\frac{dx}{dt}$) is the sum of the arrows into the box minus the sum of the arrows out of the box. We refer to this as "rate of change equals inflows minus outflows."

Here are a few examples: Figure 5.1 has one box or variable representing population size. There are two inflows representing births and immigration and two outflows representing death and emigration. In this simple example, each of these flows is proportional to the current population size. Putting the pieces together, we have the differential equation

$$\frac{dx}{dt} = bx + ix - dx - ex.$$

The next example is for a logistic equation and is illustrated in Figure 5.2. It has one inflow which represents growth rate. Even though there is only an inflow, when the value of the growth rate is negative it works as an outflow. Some programs that implement this style of modeling require inflows to truly flow in and outflows to truly flow out. Usually the program also has a construction which is a biflow which allows flow both ways, with one direction defined as positive.

Our final example considers the families (income earning units) of a town. They are divided into three income classes: lower, middle, and upper. Each year a certain

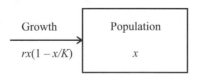

FIGURE 5.2 Compartmental Analysis for a Logistic Model.

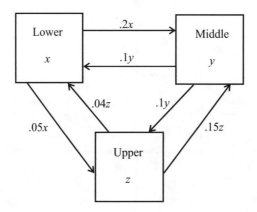

FIGURE 5.3 Income Class Model.

percentage of lower-income families move into the middle and upper classes and a percentage of middle-class families move into the upper class due to raises, promotions, spouses' entering the work force, and so forth. Conversely, certain percentages of the upper two classes move down. In this example, the population flows among the three classes and we will model the annual percentage rates as instantaneous rates. A compartmental diagram for this situation is shown in Figure 5.3, and from it we construct the equations

$$\frac{dx}{dt} = -0.25x + 0.1y + 0.04z$$
$$\frac{dy}{dt} = 0.2x - 0.2y + 0.15z$$
$$\frac{dz}{dt} = 0.05x + 0.1y - 0.19z.$$

Again the value of this technique is that we draw a diagram of a model, discuss it with experts who may not understand differential equations, and formulate the equations easily when the model is conceptually correct.

5.2 Solving Special Classes of Differential Equations
5.2.1 Separable Equations
A differential equation is called *separable* if it can be rewritten in differential form with all occurrences of one variable on the left of the equals sign and all the occurrences of the other variable on the right. For example,

$$\frac{dx}{dt} = xt$$

can be rewritten as

$$\frac{dx}{x} = t\,dt.$$

5 Continuous Models

This rewriting process is called *separating the variables*, and, as the name implies, differential equations which can be separated are called *separable*. After separating the variables, the differential equation is solved by integrating both sides, assuming this integration is possible. The result has a constant of integration. Initial conditions are used to find the constant of integration.

Continuing with the example above

$$\int \frac{dx}{x} = \int t\, dt$$

yields

$$\ln|x| = \frac{t^2}{2} + C \quad \text{or} \quad |x| = e^{\frac{t^2}{2}+C}$$

which is rewritten as

$$x = \tilde{C} e^{\frac{t^2}{2}}.$$

This is known as the *general solution*. If, in addition, we know that $x(0) = 2$, we determine that $\tilde{C} = 2$, and we say that the solution of the *initial value problem* is

$$x = 2e^{\frac{t^2}{2}}.$$

The single most important differential equation in modeling is the *exponential differential equation*

$$\frac{dx}{dt} = rx. \tag{96}$$

In other words, the rate of change of x is proportional to a value of x. This assumption is used in a wide variety of models, including population growth, radioactive decay, and compound interest. Further, as seen in the previous chapter, in the absence of any clear curvilinear trend, a line is the most reasonable function to fit to data. The separation of variables method works on this differential equation giving the solution $x(t) = Ce^{rt}$. It is instructive to compare this result to the result from the analogous difference equation. Recall that if

$$x(n+1) - x(n) = rx(n),$$

then the solution is

$$x(n) = C(1+r)^n.$$

Generally speaking, base e exponential functions arise naturally in continuous models, while other bases arise in discrete models. The base changing formula

$$a^x = e^{\ln a\, x}$$

converts any exponential equation to a base e equation. This is a convenient way of converting a discrete model to a continuous approximation. For example, if a model is discrete with difference equation

$$x(n+1) - x(n) = 0.2x(n)$$

with initial condition

$$x(0) = 1,$$

then the solution is
$$x(n) = 1.2^n.$$
This is analogous to
$$x(t) = e^{\ln 1.2 \, t},$$
which is the solution of
$$\frac{dx}{dt} = \ln 1.2 \, x.$$
Both these models give the same output at integer values of time, and if it suits the purpose, a continuous model might be used to approximate the discrete model.

The *logistic differential equation* is another important separable differential equation. It is the continuous version of the logistic difference equation studied in Chapter 1. The differential equation is
$$\frac{dx}{dt} = rx\left(1 - \frac{x}{K}\right).$$

An important skill is being able to predict the solution curve of a differential equation. This skill aids in both understanding the behavior and in the construction of differential equation models. The expression $\frac{dx}{dt}$ is the rate of change of x; whenever this expression is zero, x is constant or a fixed point; whenever it is positive, x is growing; and whenever it is negative, x is declining. Assuming that r and K are positive constants and that x is very much smaller than K, then $\frac{x}{K}$ is very small and the $rx(1-\frac{x}{K})$ term is approximately rx. Thus for small x, the differential equation is approximately
$$\frac{dx}{dt} = rx,$$
so the solution should look like an exponential function. If x is approximately equal to K, then $\frac{x}{K}$ is approximately equal to 1, and $rx(1-\frac{x}{K})$ is approximately equal to zero. Thus when x is near K, the differential equation is approximately
$$\frac{dx}{dt} = 0,$$
so x is approximately a constant function. From this discussion, we conclude that if $x(t)$ is small relative to K, then $x(t)$ exhibits nearly exponential growth. When $x(t)$ is near K, it levels out to be nearly constant. The behavior in between is harder to predict, but it would seem reasonable that there is a smooth transition from exponential growth to no growth. If x is greater than K, the results are similar except the derivative is negative, so $x(t)$ starts off decaying exponentially until it nears K where it levels out. These behaviors are illustrated in Figure 5.4.

The constant K is called the *carrying capacity* since if x represents a population, the population rises or decays until it reaches this level. As with the discrete case, the logistic model is useful for situations that begin with exponential growth until they reach some saturation point after which the growth levels off. One example of a logistic model is the black bear age-versus-weight model from Section 4.8.1.

5 Continuous Models

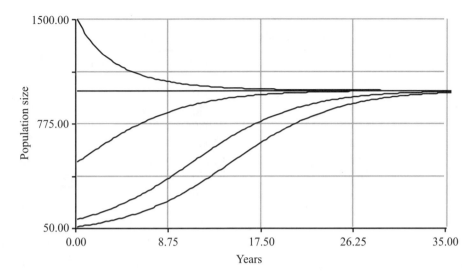

FIGURE 5.4 Logistic Curves for Different Initial Values.

The solution of the logistic differential equation is
$$\frac{K}{1+Ce^{-rt}}.$$
We leave the actual solving as an exercise (see Exercise 2). Again notice that the decaying exponential causes this function to approach the carrying capacity asymptotically.

In the case of the exponential models, except for domain issues, continuous and discrete models give essentially the same behavior. For logistic models the situation is quite different. With the continuous logistic model, the behavior is always as we described, namely either smoothly rising or falling to the carrying capacity. Recall that in the case of the discrete logistic, the behavior was dependent on the growth rate. Certain growth rates resulted in periodic, quasi-periodic, or chaotic behavior. None of this behavior is exhibited in the continuous case.

5.2.2 Example and Method: U.S. Population Growth and Using Curves Fitted to Data

This example does double duty. It uses a separable differential equation and shows how a curve fit to data is used in a differential equation model.

Throughout this text, we have seen that a first modeling assumption for the growth of a population is to assume that the population grows at a rate proportional to the population size. This assumption, in the continuous time case, leads to the differential equation

$$\frac{dP(t)}{dt} = rP(t)$$

which we have just seen has an exponential solution. In this context, r is a constant. Later in this chapter we study r factors which are not constants, yielding the logistic and

Year	Population Size (in millions)	Year	Population Size (in millions)
1790	3.9	1900	76.2
1800	5.3	1910	92.2
1810	7.24	1920	106.0
1820	9.60	1930	123.2
1830	12.86	1940	132.2
1840	17.0	1950	151.3
1850	23.2	1960	179.3
1860	31.4	1970	203.3
1870	38.5	1980	226.5
1880	50.2	1990	248.7
1890	63.0		

TABLE 5.1 Population of the United States. Data are from Davis, Porta, Uhl, *Calculus& Mathematica: Derivatives* ©1994 Addison Wesley Longman Inc. Reprinted by Permission of Addison Wesley Longman.

Allee effect models. Here we determine an appropriate functional relation for r given data for P as a function of time.

The official census data of United States population counts is given in Table 5.1. The figures are in millions for the corresponding year and can be found in *Calculus& Mathematica: Derivatives*[14] or the 1996 World Almanac [52].

Figure 5.5 shows a scatter plot of this data. At first glance, the data appears to exhibit exponential growth. As was done in Example 4.7, we fit an exponential function to the data by first taking the natural logarithm of the population size. This logged data is given

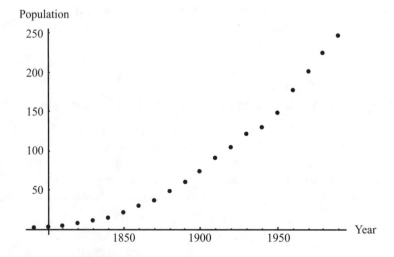

FIGURE 5.5 United States Population (in Millions).

5 Continuous Models

Year	Logarithm of Population
1790	1.36098
1800	1.66771
1810	1.97962
1820	2.26176
1830	2.55412
1840	2.83321
1850	3.14415
1860	3.44681
1870	3.65066
1880	3.91602
1890	4.14313

Year	Logarithm of Population
1900	4.33336
1910	4.52396
1920	4.66344
1930	4.81381
1940	4.88432
1950	5.01926
1960	5.18906
1970	5.31468
1980	5.42274
1990	5.51625

TABLE 5.2 Logarithm of Population of the United States.

in Table 5.2. A plot of this data is shown in Figure 5.6. Fitting a line to this data gives

$$\ln P(t) = -35.4471 + 0.0207868t.$$

Solving this equation for $P(t)$ gives the exponential model

$$P(t) = \exp(-35.4471 + 0.0207868t) \tag{97}$$

which is graphed in Figure 5.7 along with the data. The fit starts out very good, but becomes worse as time goes on. Using (97), SS_{Res} for this model is $25{,}196.57$.

Upon closer inspection, we see that the growth rate r is not a constant. If r were constant, the time for the population size to double would be constant. In 1840, the

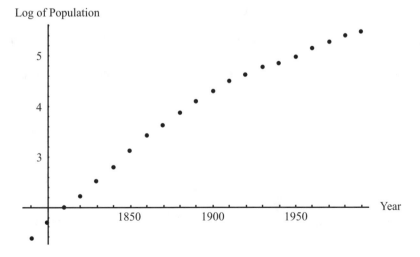

FIGURE 5.6 Log of Population.

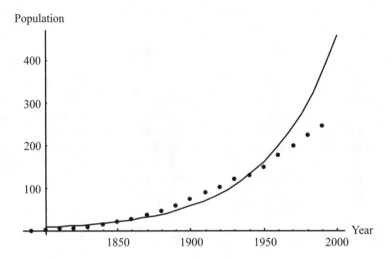

FIGURE 5.7 Exponential Model Fit to U.S. Population.

population was 17 million and by 1870 it had more than doubled to 38.5 million. This doubling time was between 20 and 30 years. The population again doubled between 1900 and 1910, shorter than 30 years. Thus, the doubling time is not constant, but, in fact, is decreasing. We can improve our model by making r a function of t.

We are able to determine r from the data in Table 5.1. We assume that $P(t)$ is effectively modeled by an equation of the form

$$\frac{dP(t)}{dt} = r(t)P(t).$$

If this expression is rearranged, we obtain

$$r(t) = \frac{1}{P(t)} \frac{dP(t)}{dt}; \tag{98}$$

thus $r(t)$ is the amount of population growth per unit population. This agrees with our intuition. Equation (98) is used to approximate values of $r(t)$. Since $dP(t)/dt$ represents the slope of the tangent line to $P(t)$, we approximate $dP(t)/dt$ by values from the data in Table 5.1. More specifically, we use a secant line approximation

$$\frac{dP(t)}{dt} \approx \frac{P(t+h) - P(t-h)}{2h}. \tag{99}$$

Using this expression in (98) gives an approximation for $r(t)$ as

$$r(t) \approx \frac{1}{P(t)} \left(\frac{P(t+h) - P(t-h)}{2h} \right), \tag{100}$$

from which approximate values of $r(t)$ are calculated. Using the data in Table 5.1,

$$r(1800) \approx \frac{1}{P(1800)} \left(\frac{P(1810) - P(1790)}{20} \right)$$

5 Continuous Models

Year	Growth Rate
1800	0.0315094
1810	0.0296961
1820	0.0292708
1830	0.0287714
1840	0.0304118
1850	0.0310345
1860	0.0243631
1870	0.0244156
1880	0.0244024
1890	0.0206349
1900	0.0191601
1910	0.0161605
1920	0.0146226
1930	0.0106331
1940	0.0106278
1950	0.0155651
1960	0.0145008
1970	0.0116085
1980	0.0100221

TABLE 5.3 Approximate Rate of Population Growth.

$$= \frac{1}{5.3}\left(\frac{7.24 - 3.9}{20}\right)$$
$$= 0.0315094$$

The remaining values of $r(t)$ are calculated similarly and are given in Table 5.3. A plot of this growth rate versus time is shown in Figure 5.8. This plot shows the decline in the growth rate for the U.S. population that was observed earlier.

This trend is modeled using a regression line fit to this growth rate data. In general, if the data does not exhibit a clear functional relation, but a trend upwards or downwards is observable, then the simplest way to model this trend is a straight line

$$r(t) = \beta_0 + \beta_1 t.$$

Before we estimate the values of the least-squares coefficients β_0 and β_1, let us examine the nature of the model

$$\frac{dP(t)}{dt} = (\beta_0 + \beta_1 t)P(t).$$

This is a separable differential equation with solution

$$\ln P(t) = \frac{1}{2}\beta_1 t^2 + \beta_0 t + C \tag{101}$$

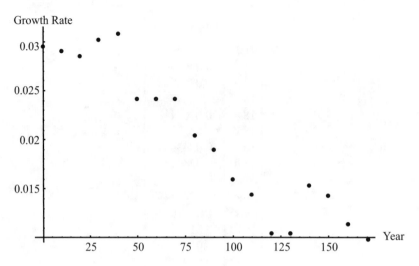

FIGURE 5.8 Population Growth Rate.

where C is a constant to be determined from initial conditions. To simplify calculations, let $t = 0$ correspond to the year 1800 so that equation (101) gives

$$C = \ln P(0).$$

This assumption together with equation (101) implies

$$P(t) = P(0)e^{\frac{1}{2}\beta_1 t^2 + \beta_0 t}. \tag{102}$$

Here the least-squares coefficients β_0 and β_1 are for the growth rate data with the year 1800 taken as time $t = 0$. Using the least-squares equations for the rescaled data gives

$$\hat{\beta}_0 = 0.0317591 \text{ and } \hat{\beta}_1 = -0.000134485 \tag{103}$$

so that

$$r(t) = 0.0317591 - 0.000134485t.$$

This regression line is plotted together with the rescaled growth rate data in Figure 5.9. The decreasing rate of growth is reflected in the line having negative slope.

Using equations (103) in (102) along with $P(0) = 5.3$ gives the model

$$P(t) = 5.3 \exp(0.0317591t - 0.0000672425t^2). \tag{104}$$

This equation is plotted in Figure 5.10 together with the actual data. This model is visually much better than the exponential model. Only the last few data points appear not to match. The SS_{Res} for this model is 3,507, vastly better than the 25,196 obtained for the exponential model. This model is a substantial improvement over equation (97). It has $R^2 = 0.91276$, so this model accounts for 91.3% of the variance in the data. In Section 2.4.1 we saw that this model was significant (up until 1960) at a $\alpha = 0.1$ level, while the exponential model was not.

Some other models using this data are examined in Projects 4.6 and 5.11.

5 Continuous Models

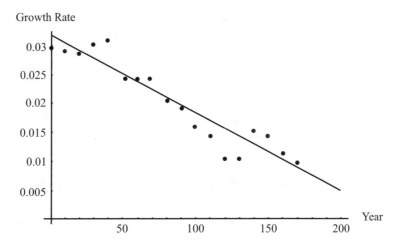

FIGURE 5.9 Regression Line through Growth Rate Data.

5.2.3 Linear Differential Equations

A linear differential equation is a differential equation that can be written in the form

$$a_0(t)\frac{d^n x}{dt^n} + a_1(t)\frac{d^{n-1} x}{dt^{n-1}} + \cdots + a_{n-1}(t)\frac{dx}{dt} + a_n(t)x = f(t).$$

Here the functions $a_i(t)$ are any functions of t, linear or not. At first some differential equations classified as linear (such as $\sin(t)\frac{dx}{dt} + e^t x = \sqrt{t}$) may contradict your intuitive notion of linearity. What could be more nonlinear than trigonometric functions, exponentials, and square roots? The idea of linearity used here, however, is that the

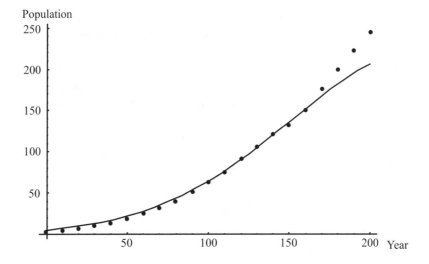

FIGURE 5.10 Model of the Population of the United States.

differential equation is a linear combination of x, and the derivatives of x, and the coefficients need to be constants as far as x is concerned and hence can be functions of t.

In this section, we address only first-order differential equations, namely those which have the form $a_0(t)\frac{dx}{dt} + a_1(t)x = f(t)$. These can be and usually are written in the form $\frac{dx}{dt} + P(t)x = Q(t)$. The function $Q(t)$ is called a *forcing term*. In any differential equations book, equations of this type are shown to be convertible into a form called *exact* by use of a function called an *integrating factor*. Since we do not have occasion to discuss exact equations or integrating factors further, we merely present the method and do not discuss its derivation.

A Method of Solving Linear Differential Equations.

1. Write the linear differential equation in the form

$$\frac{dx}{dt} + P(t)x = Q(t).$$

2. Define the function

$$\mu(t) = \exp \int P(t)\,dt.$$

The function $\mu(t)$ is called an *integrating factor*.

3. The general solution of the differential equation is

$$x(t)\mu(t) = \int \mu(t)Q(t)\,dt + C \quad \text{or} \quad x(t) = \frac{1}{\mu(t)}(\int \mu(t)Q(t)\,dt + C).$$

While we do not verify why this works, the reader should always check a solution obtained through this method by differentiating or implicitly differentiating to see that, indeed, it is a solution.

5.2.4 Example: Drugs in the Body

The body processes drugs in either a linear or exponential manner (the rate being constant or linear) for most substances. An accepted mechanism for this process is the Michaelis-Menton equation which is studied in Projects 5.1 and 5.6. For this example, we assume that the body is processing a particular drug with a rate of change (derivative) which is linear, hence with a solution curve which is exponential. Further assume that the drug is provided continuously for a period of 12 hours intravenously and then stopped. We are interested in finding the amount of the drug in the body at any given time.

Using the compartmental analysis idea that rate of change is "inflow minus outflow," we have

$$\frac{dA}{dt} = D(t) - P(A)$$

where A is the amount of the drug in milligrams, $D(t)$ the rate at which the drug is administered, and $P(A)$ is the processing rate. Let us assume that $D(t)$ is 10 mg/hour for the first 12 hours and then zero afterwards. Thus

$$D(t) = \begin{cases} 10 & \text{if } 0 \leq t \leq 12 \\ 0 & \text{if } t > 12. \end{cases}$$

5 Continuous Models

We are assuming that drug processing is a linear relation; specifically let us assume that it follows the equation $P(A) = 0.1A$. Further assume that there is no drug in the system (body) at time $t = 0$.

It is probably easiest to write this differential equation in terms of two differential equations; one with time domain $[0, 12]$, the other with domain $(12, \infty)$, with care taken to match the solutions at 12.

On $[0, 12]$ the differential equation is

$$\frac{dA}{dt} = 10 - 0.1A$$

which is written in standard linear form as

$$\frac{dA}{dt} + 0.1A = 10.$$

Here $P(t) = 0.1$ and $Q(t) = 10$. The integrating factor is

$$\mu(t) = e^{0.1t},$$

which implies that the solution is

$$A(t)e^{0.1t} = \int 10e^{0.1t}\, dt + C.$$

Thus

$$A(t)e^{0.1t} = 10/0.1\, e^{0.1t} + C, \quad \text{or} \quad A(t) = 100 + Ce^{-0.1t}.$$

Using the initial information gives $C = -100$ so the solution is

$$A(t) = 100 - 100e^{-0.1t}.$$

This solution starts at 0 mg and asymptotically approaches 100 mg. After 12 hours there are $A(12) = 69.88$ milligrams of the drug in the system. We use this number as the initial condition for the second part of the time domain. For time greater than 12 hours, the system follows the differential equation $\frac{dA}{dt} = -0.1A$. This is an exponential differential equation. Appealing to prior work, the general solution is

$$A(t) = Ce^{-0.1t},$$

and the solution of the initial value problem is

$$A(t) = 69.88e^{-0.1t}.$$

Combining our two answers (and shifting the second appropriately) gives the solution over the whole domain

$$A(t) = \begin{cases} 100 - 100e^{-0.1t} & \text{if } 0 \leq t \leq 12 \\ 69.88e^{-0.1(t-12)} & \text{if } t > 12. \end{cases}$$

This solution is plotted in Figure 5.11.

5.2.5 Linear Differential Equations with Constant Coefficients

The most important fact about differentiable functions of one variable is that they are approximated arbitrarily well by lines over small enough intervals. The approximating

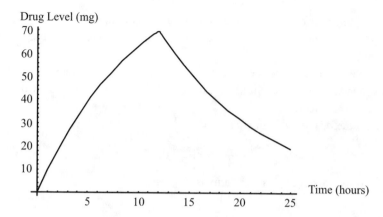

FIGURE 5.11 Drug Levels in the Body.

line is, of course, the tangent line studied in first-semester calculus. Often students have the idea that the important thing about the tangent line is that it touches the curve locally at one point. While this is true for nonlinear functions, it creates the impression that if one were to zoom in close enough, the tangent line would balance on this point as though it were the tip of a pin. This is, however, the wrong image. In fact if we use a computer or calculator to zoom in on a curve and its tangent line at the point of tangency, it is possible to zoom in close enough that the two cannot be distinguished from one another to within the resolution of the computer or calculator screen. If the reader finds this idea unfamiliar, it would be highly instructive to verify it on a computer or calculator. It is the defining concept of differentiability, and similar ideas involving tangent planes or hyperplanes are used for functions of more than one variable.

Linear differential equations with constant coefficients are important for several reasons. First, they are very well understood and easy to solve. Second, since differential functions are approximated locally by lines, linear differential equations with constant coefficients are frequently used to locally approximate linear differential equations with nonconstant coefficients or even nonlinear differential equations. Third, as seen in the next section, systems of linear differential equations can be converted into linear differential equations of one variable. Finally, in many modeling applications, data is scarce and the best that one is able to do is to estimate constant growth rates which result in linear models with constant coefficients.

A *homogeneous linear differential equation with constant coefficients* is a differential equation of the form

$$a_0 \frac{d^n x}{dt^n} + a_1 \frac{d^{n-1} x}{dt^{n-1}} + \cdots + a_{n-1} \frac{dx}{dt} + a_n x = 0.$$

The term homogeneous refers to the zero on the right-hand side of the equals sign. Notice that the equation $\frac{dx}{dt} = rx$ studied in Section 5.2.1 can be rearranged to have the form recognizable as a homogeneous linear differential equation. The idea behind the method is that since the exponential differential equation has a solution of the form $x(t) = e^{rt} x(0)$, perhaps a linear homogeneous equation of higher order has a similar solution. Note that

5 Continuous Models

we are not solving this differential equation in the sense of integrating; rather we are assuming the form of the solution and merely finding the parameters. Sometimes this is called *formally* solving a differential equation since we are guessing solutions based on the equation's form.

The method. Consider the equation

$$a_0 \frac{d^n x}{dt^n} + a_1 \frac{d^{n-1} x}{dt^{n-1}} + \cdots + a_{n-1} \frac{dx}{dt} + a_n x = 0. \tag{105}$$

1. Assume a solution of the form $x(t) = Ce^{\lambda t}$.

2. Substitute $x(t) = Ce^{\lambda t}$ into equation (105) to get

$$a_0 C \lambda^n e^{\lambda t} + a_1 C \lambda^{n-1} e^{\lambda t} + a_2 C \lambda^{n-2} e^{\lambda t} + \cdots + a_{n-1} C \lambda e^{\lambda t} + a_n C e^{\lambda t} = 0.$$

3. Factor out the common factors of $Ce^{\lambda t}$ and λ^{n-m} (where a_m is the last non-zero coefficient). Since we are not interested in the cases $C = 0$ and $\lambda = 0$ (these give a constant function which is fine but uninteresting), we set the rest equal to zero. The result is called the *characteristic* or *auxiliary equation*

$$a_0 \lambda^m + a_1 \lambda^{m-1} + a_2 \lambda^{m-2} + \cdots + a_{m-1} \lambda + a_m = 0.$$

4. Solve the characteristic equation for λ. The characteristic equation has m roots; some may be repeated and some may be complex. These roots are called *characteristic roots* or *eigenvalues*, and they are the growth rates of the system. They are denoted by $\lambda_1, \ldots, \lambda_m$. In practice, we can solve second-degree polynomials using the quadratic formula by hand, but some type of equation solver should probably be used for higher-degree equations.

5. Each equation of the form $x(t) = Ce^{\lambda_i t}$ is a solution of the original equation. The *general solution* is the linear combination of these solutions, namely

$$x(t) = C_1 e^{\lambda_1 t} + C_2 e^{\lambda_2 t} + \cdots + C_m e^{\lambda_m t}.$$

6. The C_i's can be found, provided enough initial information is given. If there are m constants, then it is necessary to know $x(t)$ at m values of t or some equivalent amount of information such as the values of m derivatives of x at $t = 0$. Usually this information results in m equations in m unknowns, which are solved by any standard method such as Gaussian elimination. Again computers or calculators should be used if the m is large.

This method yields a closed-form solution

$$x(t) = C_1 e^{\lambda_1 t} + C_2 e^{\lambda_2 t} + \cdots + C_m e^{\lambda_m t}.$$

Since each term is exponential, we analyze its pieces using the next proposition.

Proposition 5.1. *If $x(t) = e^{\lambda t}$ and the eigenvalue λ is such that*

a. $\text{Im}(\lambda) = 0$, *and*

  i. $\text{Re}(\lambda) > 0$, *then $x(t)$ exhibits exponential growth;*

  ii. $\text{Re}(\lambda) < 0$, *then $x(t)$ exhibits exponential decay;*

    iii. $\text{Re}(\lambda) = 0$, *then $x(t)$ is a constant;*

b. $\text{Im}(\lambda) \neq 0$ *and*

    i. $\text{Re}(\lambda) > 0$, *then both the real and complex parts of $x(t)$ exhibit undamped oscillation;*

    ii. $\text{Re}(\lambda) < 0$, *then both the real and complex parts of $x(t)$ exhibit damped oscillation;*

    iii. $\text{Re}(\lambda) = 0$, *then both the real and complex parts of $x(t)$ exhibit pure oscillatory behavior.*

In our setting, the complex eigenvalues always occur in complex conjugate pairs. With real initial conditions, the complex parts of terms due to complex conjugate eigenvalues cancel, leaving a pure real solution.

Over the short term the various eigenvalues or growth constants can contribute significantly to $x(t)$, but eventually the contribution due to the eigenvalue which has the largest real part (remember the eigenvalues can be complex) dominates the contribution of the others. The eigenvalue λ_d which has the largest real part, that is $\text{Re}(\lambda_d) > \text{Re}(\lambda_j)$ for all $j \neq d$, is called the *dominant eigenvalue*. Thus just by looking at the list of eigenvalues for a differential equation, finding the dominant eigenvalue, and using Proposition 5.1, we determine the nature of its long-term behavior. In general, the real part $\text{Re}(\lambda_d)$ determines long-term growth or decay, and the imaginary part $\text{Im}(\lambda_d)$ determines whether or not the long-term behavior exhibits oscillations.

One point should be emphasized here. While it is mathematically possible for a linear system to have a stable non-zero solution (i.e., $\text{Re}(\lambda_d) = 0$), for any real (nonclosed) system it is impossible. What we mean by this is that the dominant eigenvalue must have a real part which is exactly zero in order to be stable. If $\lambda_d = -0.0000000001$, the system approaches zero eventually. If $\lambda_d = 0.000000001$, the system ultimately exhibits exponential growth. There are two issues here. The first is that the probability of picking the number zero from any real interval containing zero under any continuous probability distribution is zero. Thus zero will not be chosen unless there is some natural law in place. The second is that even if there is some natural law that says that $\text{Re}(\lambda_d) = 0$, the law probably holds only under some highly idealized set of circumstances (e.g., point masses in a vacuum attached to a massless spring which has no thickness). In most practical situations, these idealized assumptions fail, and even if they fail only slightly, our eigenvalue changes from zero by some minuscule amount. Thus, in order to have a non-zero stable solution, we must walk a tightrope which is one point thick, and the noise of most any realistic situation will cause us to fall. One notable exception is seen in closed systems where nothing enters or leaves the system.

Let us look at several examples to illustrate this method.

Consider the differential equation

$$\frac{d^2x}{dt^2} - \frac{dx}{dt} - 6x = 0,$$

5 Continuous Models

with initial values $x(0) = 1$ and $x'(0) = 1$. Substituting $x(t) = Ce^{\lambda t}$ into this equation gives the characteristic or auxiliary equation

$$\lambda^2 - \lambda - 6 = 0$$

which factors to give

$$(\lambda - 3)(\lambda + 2) = 0.$$

The eigenvalues are $\lambda = 3$ and $\lambda = -2$. The general solution of this differential equation is

$$x(t) = C_1 e^{3t} + C_2 e^{-2t}.$$

The dominant eigenvalue is 3, so we know that the system ultimately exhibits exponential growth. Since we know that $x(0) = 0$ and $x'(0) = 1$, then

$$C_1 + C_2 = 0 \quad \text{and} \quad 3C_1 - 2C_2 = 1.$$

Thus $C_1 = 1/5$, and $C_2 = -1/5$. With these initial conditions, the solution of the differential equation is

$$x(t) = \frac{1}{5} e^{3t} - \frac{1}{5} e^{-2t}.$$

For another example consider the initial value problem

$$\frac{d^2 x}{dt^2} + \frac{dx}{dt} + x = 0,$$

subject to the conditions that $x(0) = 2$ and $x(1) = 3$. Substituting $x(t) = Ce^{\lambda t}$ into this equation and simplifying gives

$$\lambda^2 + \lambda + 1 = 0.$$

Using the quadratic formula gives

$$\lambda = \frac{-1 \pm i\sqrt{3}}{2},$$

and the general solution is

$$x(t) = e^{\frac{(-1-i\sqrt{3})t}{2}} C_1 + e^{\frac{(-1+i\sqrt{3})t}{2}} C_2.$$

Now using the initial information we find that

$$C_1 = 2 + \frac{2 - 3e^{\frac{1+i\sqrt{3}}{2}}}{-1 + e^{i\sqrt{3}}} \quad \text{and} \quad C_2 = \frac{-2 + 3e^{\frac{1+i\sqrt{3}}{2}}}{-1 + e^{i\sqrt{3}}}$$

or $C_1 = 1.0 + 2.39606\, i$ and $C_2 = 1.0 - 2.39606\, i$. The solution to the boundary value problem is then

$$x(t) = (1.0 + 2.39606\, i) e^{\frac{(-1-i\sqrt{3})t}{2}} + (1.0 - 2.39606\, i) e^{\frac{(-1+i\sqrt{3})t}{2}}.$$

The eigenvalues both have negative real components, and so the solution tends asymptotically to zero. The presence of the imaginary part of the solution indicates that the solution oscillates. Thus the solution is a damped oscillation. The solution is graphed in Figure 5.12.

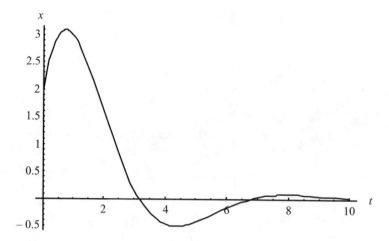

FIGURE 5.12 Solution of $x'' + x' + x = 0$.

We should note that while the work just shown follows the prescribed method exactly, working with complex coefficients is a little nasty and in most applications the imaginary parts cancel out. Because of this, if we have a pair of eigenvalues which are complex conjugates ($\lambda_1 = a + ib$ and $\lambda_2 = a - ib$), then rather than writing the general solution as $C_1 e^{(a+ib)t} + C_2 e^{(a-ib)t}$, we write it as $C_1 e^{at} \cos(bt) + C_2 e^{at} \sin(bt)$. From here we solve for C_1 and C_2 without complex arithmetic. This form also makes clear the oscillatory nature of the solution and whether the solution is growing, stable, or damped.

For our example $\lambda = -\frac{1}{2} \pm \frac{i\sqrt{3}}{2}$, so

$$x(t) = C_1 e^{-\frac{1}{2}t} \cos(\frac{\sqrt{3}}{2}t) + C_2 e^{-\frac{1}{2}t} \sin(\frac{\sqrt{3}}{2}t).$$

From this form it is clear that the behavior is oscillatory and the oscillations are decaying. Using the conditions that $x(0) = 2$ and $x(1) = 3$, we solve for C_1 and C_2. In this case, $C_1 = 2$ and $C_2 = 4.79212$. Thus an equivalent formulation of our answer is

$$x(t) = 2e^{-\frac{1}{2}t} \cos(\frac{\sqrt{3}}{2}t) + 4.79212 e^{-\frac{1}{2}t} \sin(\frac{\sqrt{3}}{2}t).$$

For details about why this procedure is valid, see Exercise 10. Of related interest is Exercise 6 of Chapter 4.

5.2.6 Systems of First-Order Homogeneous Linear Differential Equations With Constant Coefficients

Systems of differential equations arise whenever there is more than one dependent variable for an independent variable. For example, one might consider a system with two or more interacting animal species with the population sizes changing over time. The more interesting systems have the rate of change of one variable dependent on several of the other variables. In a predator-prey system, for example, the rate at which the prey grow depends on the number of predators which kill the prey. Similarly the rate at which the

5 Continuous Models

predators grow depends on the size of their food supply, namely the prey population. In general these conditions produce nonlinear equations which are very difficult to solve analytically. In Section 5.3.5 below, we discuss methods for finding fixed points of these systems. Of particular interest is the behavior near the fixed points. To analyze behavior close to a fixed point, we "linearize" or take a linear approximation to the differential equation. The result is a system of linear differential equations which we solve by the methods of this section.

The objects of study of this section are equations of the form

$$\frac{dx_1}{dt} = a_{11}x_1 + a_{12}x_2 + \cdots + a_{1n}x_n$$
$$\frac{dx_2}{dt} = a_{21}x_1 + a_{22}x_2 + \cdots + a_{2n}x_n$$
$$\vdots \qquad \vdots \qquad \vdots \qquad \ddots \qquad \vdots$$
$$\frac{dx_n}{dt} = a_{n1}x_1 + a_{n2}x_2 + \cdots + a_{nn}x_n.$$

Systems of two homogeneous first-order differential equations are easily solved by hand, and the behaviors of their solutions are readily categorized. For these reasons, we discuss these systems first in full detail. We then discuss systems of arbitrary size which usually require technological assistance and which have solutions which are harder to classify.

It is left as an exercise (Exercise 4) to show that

$$\frac{dx}{dt} = ax + by$$
$$\frac{dy}{dt} = cx + dy$$

can be rewritten as

$$\frac{d^2x}{dt^2} - \beta\frac{dx}{dt} + \gamma x = 0$$

where $\beta = a + d$ and $\gamma = ad - cd$. Notice that this system can be written in matrix form as

$$\begin{pmatrix} \frac{dx}{dt} \\ \frac{dy}{dt} \end{pmatrix} = \begin{pmatrix} a & b \\ c & d \end{pmatrix} \begin{pmatrix} x \\ y \end{pmatrix}.$$

With this matrix notation, β is the trace of the coefficient matrix and γ is the value of the determinant. The characteristic equation of the second-order differential equation is $\lambda^2 - \beta\lambda + \gamma = 0$. Using the quadratic formula we have

$$\lambda = \frac{\beta \pm \sqrt{\beta^2 - 4\gamma}}{2}.$$

With $x(t)$ found, we differentiate to find $x'(t)$ and substitute both x and x' into $\frac{dx}{dt} = ax + by$ to find $y(t)$. Note that y has the same eigenvalues as x did. Thus there are only two (possibly repeated) eigenvalues for this system.

This system has fixed points or equilibrium points whenever there is no change, hence, whenever both derivatives are zero. As long as the system is non-degenerate, the only fixed point is the origin $(0,0)$. As discussed in Section 5.2.5 on linear equations, the long-term behavior of x and y is determined by the dominant eigenvalue. If the dominant eigenvalue is positive, both x and y grow exponentially. If these are plotted on an xy coordinate system, the effect is a trajectory away from the origin (equilibrium point). Conversely if both are negative, the result is a trajectory towards the origin (equilibrium point) as both the x and y values decay. If the eigenvalues have imaginary parts, then there are inward or outward spirals or stable orbits about an equilibrium point. The possibilities seem daunting, but they can be organized in a succinct manner.

If all the trajectories move toward an equilibrium point, it is called a *sink*; if they all move away, it is called a *source*; if some trajectories move towards and others move away, it is called a *saddle*; and if all trajectories orbit around the equilibrium point, it is called a *center*. Further sinks or sources may exhibit spiraling behaviors in which case they are called *spiral sinks* or *spiral sources*. Sometimes we will use the word *node* (e.g., source node or sink node) to indicate that no spiraling is occurring. These results are summarized in the following two theorems.

Theorem 5.2. *If λ_1 and λ_2 are the two eigenvalues of a system of two homogeneous linear first-order differential equations with constant coefficients and*
   a. $\operatorname{Re}(\lambda_1) < 0$ *and* $\operatorname{Re}(\lambda_2) < 0$, *then the fixed point is a sink;*
   b. $\operatorname{Re}(\lambda_1) < 0$ *and* $\operatorname{Re}(\lambda_2) > 0$, *then the fixed point is a saddle;*
   c. $\operatorname{Re}(\lambda_1) > 0$ *and* $\operatorname{Re}(\lambda_2) > 0$, *then the fixed point is a source;*
   d. *If λ_1 and λ_2 are complex conjugates of each other, then a sink or source is a spiral;*
   e. *If $\operatorname{Re}(\lambda_1) = 0$ and $\operatorname{Re}(\lambda_2) = 0$, then the fixed point is a center.*

While it is not hard to consider the signs of the real parts of λ_1 and λ_2, it is often convenient to look at β and γ.

Theorem 5.3. *Let $\beta = a + d$ and $\gamma = ad - bc$; then*
   a. *If $\gamma < 0$, then the fixed point is a saddle.*
   b. *If $\gamma > 0$ and $\beta < 0$, then the fixed point is a sink.*
   c. *If $\gamma > 0$ and $\beta > 0$, then the fixed point is a source.*
   d. *If $\gamma > 0$ and $\beta = 0$, then the fixed point is a center.*
   e. *If $\beta^2 - 4\gamma < 0$, then a sink or source is a spiral.*

Figure 5.13 summarizes these behaviors, based on the values of β and γ.

Often our primary interest is in the behaviors of solutions, rather than the solutions themselves. From the above discussion, notice that by looking at the system of equations and evaluating β and γ, we are able to predict the behavior of the system.

For example, consider the system:

$$\frac{dx}{dt} = 2x + 1y$$
$$\frac{dy}{dt} = -3x + 4y.$$

5 Continuous Models

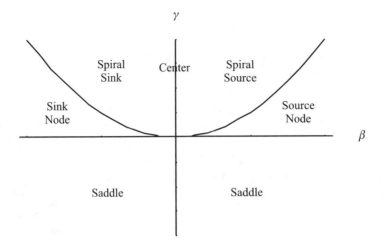

FIGURE 5.13 The Beta-Gamma Plane.

Here $\beta = 6$ and $\gamma = 8 + 3 = 11$, so $\beta^2 - 4\gamma = 36 - 44 = -8$. The origin is an equilibrium point. From Figure 5.13 we note that since β and γ are both positive, we have either a source or outward spiral and since $\beta^2 - 4\gamma < 0$, the trajectories spiral away from the fixed point. Sample trajectories are shown in Figure 5.14.

The following fact should be obvious from the above discussion, but is of sufficient importance that we emphasize it by making it a proposition.

Proposition 5.4. *The long-term behavior of a system of linear equations is such that each variable either goes to zero or becomes arbitrarily large with the exception of a few special cases that essentially never occur. If an actual nonclosed system exhibits robust stable behavior then a model reflecting this behavior must be nonlinear.*

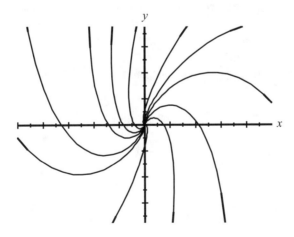

FIGURE 5.14 Spiral Trajectories.

We clarify this with several examples. In Example 5.2.7 below, we analyze the income bracket model first discussed in the compartmental analysis section. In it a population of a town moves among three income brackets. The analysis below shows that it has a non-zero stable solution which seems to contradict the assertion made in the previous proposition. Further, this solution is robust in the sense that changing the percentages moving back and forth changes only the equilibrium numbers, but not the fact that the system stabilizes. The reason, however, is that we assume in this model that we have a closed population in the sense that no families enter, leave, die, are created, or split apart. If we start with 30,000 families and insist that we always have 30,000 families, then clearly no demographic sector explodes, nor do all go to zero. But how realistic is this assumption? If we are interested in tracking a fixed portion of the population (as is done in Markov models) and making a state for deaths or for families otherwise disappearing, then this could be acceptable. Example 5.2.7 concludes by expanding the model to allow additions to and deletions from each group. The subsequent analysis shows that the population will either explode or crash. As a final example of this issue, see Exercise 8, where we ask the reader to set up a program on a computer to solve a system of three or more linear equations. We then invite you to arbitrarily pick coefficients and try to obtain a stable non-zero solution.

Next we look at a general method for solving systems of first-order homogeneous linear differential equations with constant coefficients. It ties the material on linear equations in this chapter with material in both Chapter 1 and Chapter 3. It also provides a theoretical framework and makes one aware of what type of solutions are expected from such a linear system. Practically speaking, however, solving systems of three or more equations involves solving of polynomials of degree 3 or higher. Except in special cases, we need to use some technological tool to solve them. If we are using a computer algebra system, then we might as well just use it to solve the system and skip the method of this section. This is especially true if the graph of the solution is the only item of interest. On the other hand, if one is interested in the form of the solution equations, frequently the form of the solution given by a computer algebra system is large, cumbersome, and hard to study. Even simplification routines may not help much. In this case, the method of this section is valuable, even if we are using a machine to perform the individual steps. From a philosophical point of view, it is important that we know how to solve a problem, even though we may let the machine do the work. This is much different from the discussion in Section 5.4 below where we let the machine do the work with no clue as to what the machine is doing.

The method. We are solving equations of the form

$$\mathbf{X}' = \mathbf{A}\mathbf{X} \tag{106}$$

where

$$\mathbf{X}^T = (x_1, x_2, \ldots, x_n),$$
$$(\mathbf{X}')^T = \left(\frac{dx_1}{dt}, \frac{dx_2}{dt}, \ldots, \frac{dx_n}{dt}\right),$$

5 Continuous Models

and

$$A = \begin{pmatrix} a_{11} & a_{12} & \cdots & a_{1n} \\ a_{21} & a_{22} & \cdots & a_{2n} \\ \vdots & \vdots & \vdots & \vdots \\ a_{n1} & a_{n2} & \cdots & a_{nn} \end{pmatrix}.$$

As with solving a higher-order linear homogeneous differential equation of one variable, we assume a solution of the form

$$\mathbf{X}(t) = C\mathbf{v}e^{\lambda t}.$$

Substituting this into equation (106), we get

$$\lambda C\mathbf{v}e^{\lambda t} = AC\mathbf{v}e^{\lambda t}.$$

Rearranging gives

$$AC\mathbf{v}e^{\lambda t} - \lambda C\mathbf{v}e^{\lambda t} = \mathbf{0}.$$

Factoring and inserting an identity matrix to maintain proper matrix form yields

$$(A - \lambda I)C\mathbf{v}e^{\lambda t} = \mathbf{0}. \tag{107}$$

Now we are in the realm of the eigenvalue/eigenvector problem from Chapter 3. In particular, the only way to have solutions other than $\mathbf{v} = \mathbf{0}$ is to have the matrix $A - \lambda I$ be singular which means that $\det(A - \lambda I) = 0$. Evaluating this determinant yields an nth degree polynomial in λ which has n, possibly repeating and possibly complex, solutions $\lambda_1, \ldots, \lambda_n$ called eigenvalues. Each of these eigenvalues is substituted for λ in equation (107), which in turn is solved for $\mathbf{v}_i$ (called the *eigenvector associated with* λ_i). Thus one solution is

$$\mathbf{X}(t) = \mathbf{v}_i e^{\lambda_i t} C_i$$

where C_i is an arbitrary constant. Since the set of such solutions is linearly independent, the general solution of $\mathbf{X}' = \mathbf{AX}$ is

$$\mathbf{X}(t) = \mathbf{v}_1 e^{\lambda_1 t} C_1 + \mathbf{v}_2 e^{\lambda_2 t} C_2 + \cdots + \mathbf{v}_n e^{\lambda_n t} C_n.$$

If

$$\mathbf{X}(0) = \mathbf{X}_0,$$

then we set up a system of n equations in the unknowns $C_1, \ldots, C_n$ and solve for the C_i. Equivalently, the general solution is sometimes written as

$$\mathbf{X}(t) = \Phi(t)\mathbf{C}$$

where

$$\Phi(t) = \left(\mathbf{v}_1 e^{\lambda_1 t} | \mathbf{v}_2 e^{\lambda_2 t} | \ldots | \mathbf{v}_n e^{\lambda_n t}\right) \quad \text{and} \quad \mathbf{C} = \{C_1, C_2, \ldots, C_n\}.$$

(Here the vertical lines separate the columns.) Thus given the initial condition $\mathbf{X}(0) = \mathbf{X}_0$, we write the matrix equation $\Phi(0)\mathbf{C} = \mathbf{X}_0$ which has a solution

$$\mathbf{C} = \Phi^{-1}(0)\mathbf{X}_0.$$

5.2.7 Example: Social Mobility

Consider a town of 30,000 families (economic units) which have been divided by the local chamber of commerce for planning purposes into three economic brackets: lower, middle, and upper. Each year 20% of the lower move into the middle, and 10% of the middle move back to the lower; 10% of the middle move to the upper, and 15% of the upper move down to the middle; finally 5% of the lower move directly to the upper, and 4% of the upper move down to the lower. These transitions occur continuously throughout the year as people are hired and fired, promoted and demoted, retire, and change careers. In fact, this is a continuous version of the Markov chain discussed in Chapter 3, only here the columns of the matrix sum to 0 instead of 1. This difference we have seen before in another guise. In the discrete case, a stable exponential has growth rate 1, while in the continuous case the stable growth rate is 0. From a more descriptive point of view, in the discrete Markov case the columns summed to 1 because the matrix entries represented the probability of moving into a state and everything had to move somewhere. Thus the sum of the probabilities of moving to each state is 1. In the continuous case, the entries of a matrix column represent where items in a class are going. If the system is closed, then any members entering another class must have left their own; which is to say the numbers must sum to zero.

Assume that initially there are 12,000 lower, 10,000 middle, and 8,000 upper income families. A compartmental diagram for this situation is shown in Figure 5.3 in Section 5.1. There the system of equations was determined to be

$$\frac{dx}{dt} = -0.25x + 0.1y + 0.04z$$

$$\frac{dy}{dt} = 0.2x - 0.2y + 0.15z$$

$$\frac{dz}{dt} = 0.05x + 0.1y - 0.19z.$$

Thus

$$A = \begin{pmatrix} -0.25 & 0.1 & 0.04 \\ 0.2 & -0.2 & 0.15 \\ 0.05 & 0.1 & -0.19 \end{pmatrix},$$

which has eigenvalues $-0.38245, -0.25755, 0$ with associated eigenvectors

$$\mathbf{v}_1 = (0.522233, -0.804675, 0.282441),$$

$$\mathbf{v}_2 = (-0.522233, -0.282441, 0.804676), \text{ and}$$

$$\mathbf{v}_3 = (0.388813, 0.769173, 0.507147),$$

respectively. With two negative eigenvalues and one zero eigenvalue, we know that the system eventually stabilizes (in proportions of the eigenvector associated with the 0 eigenvalue). The matrix $\Phi(t)$ is

$$\Phi(t) = \begin{pmatrix} 0.522233e^{-0.38245t} & -0.522233e^{-0.25755t} & 0.388813 \\ -0.804675e^{-0.38245t} & -0.282441e^{-0.25755t} & 0.769173 \\ 0.282441e^{-0.38245t} & 0.804676e^{-0.25755t} & 0.507147 \end{pmatrix}.$$

5 Continuous Models

The general solution is $\Phi(t)\mathbf{C}$ or

$$x(t) = C_1 0.522233 e^{-0.38245t} + C_2(-0.522233)e^{-0.25755t} + C_3 0.388813,$$

$$y(t) = C_1(-0.804675)e^{-0.38245t} + C_2(-0.282441)e^{-0.25755t} + C_3 0.769173,$$

$$z(t) = C_1 0.282441 e^{-0.38245t} + C_2 0.804676 e^{-0.25755t} + C_3 0.507147.$$

The initial conditions are $\mathbf{X}(0) = (12{,}000, 10{,}000, 8{,}000)^T$, so we find the C_i's by computing $\Phi^{-1}(0)\mathbf{X}(0)$ which gives $\mathbf{C} = (6{,}033.67, -3{,}530.88, 18{,}016.6)^T$. Our solution to the initial value problem is

$$x(t) = 3150.98 e^{-0.38245t} + 1843.94 e^{-0.25755t} + 7005.08$$

$$y(t) = -4855.13 e^{-0.38245t} + 997.255 e^{-0.25755t} + 13857$$

$$z(t) = 1704.15 e^{-0.38245t} - 2841.21 e^{-0.25755t} + 9137.06. \quad (108)$$

Looking at the decaying exponentials in these solutions, it is evident that the system stabilizes with 7,005 lower income families, 13,857 middle income families, and 9,137 upper income families. This type of information is useful to builders trying to determine the number of houses to build in certain price ranges or to businesses determining whether or not there is a sufficient market to locate in an area. A graph of the solution curves is exhibited in Figure 5.15.

Perturbations to the Basic Model. Next let us consider the effect of slightly perturbing the above model to include some movement into and out of the system. We consider two scenarios. The first involves a one-percent net inflow to the lower income bracket, a one-percent net loss from the upper income bracket, and an assumption that flows into the middle bracket from outside equal the flows out of the middle bracket to outside the system. This is illustrated by the compartmental model shown in Figure 5.16. The second scenario is identical except that the flows are five percent instead of one percent for both the lower and higher income brackets. Naively it might seem like "x percent in equals

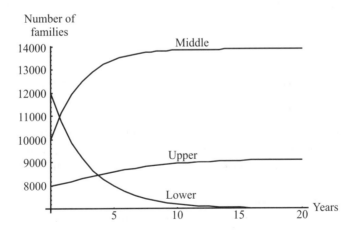

FIGURE 5.15 Income Model.

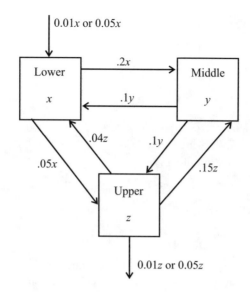

FIGURE 5.16 Income Model for a Nonclosed System.

x percent out" results in another equilibrium situation. More careful thought reveals that removing one percent from a large group is different from adding one percent of a small group. More interesting is that this analysis gives an overall decline for the one-percent model and an overall growth for the five-percent model. Again it is all in the eigenvalues. The results of these two cases are illustrated in Figures 5.17 and 5.18. Notice that even though we now have population growth and decline, it starts out being very gentle so that by the time the overwhelming nature of exponential growth or decay occurs, the model probably is no longer valid (in particular the constants probably have changed). Perhaps the most important point to note is that the system quickly moves into a ratio prescribed by the eigenvectors. Thus for commerce decisions, the ratio of lower:middle:higher is robust and dependable, even though growth or decline of the overall population is fairly sensitive.

5.3 Geometric Analysis and Nonlinear Equations

5.3.1 Phase Line Analysis

Phase line analysis is a term coined by the authors because it is essentially phase plane analysis (discussed below) performed on one dependent variable. This is an important method for understanding nonlinear equations of the form $\frac{dx}{dt} = f(x)$, and it is convenient to have a name to refer to it.

The key to phase line analysis comes from differential calculus. If a function has a positive first derivative at a point, then the function is increasing at that point; if a function has a negative first derivative at a point, then the function is decreasing at that point; and if a function has a first derivative of zero at a point, the function is neither

5 Continuous Models

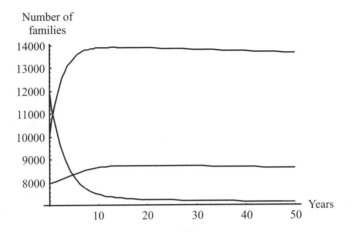

FIGURE 5.17 Income Model: One Percent Perturbation.

increasing nor decreasing there. In differential calculus, however, one usually is interested in equations of the form $x = F(t)$ and hence in derivatives of the form $\frac{dx}{dt} = F'(t)$. In this section, we are interested in differential equations of the form $\frac{dx}{dt} = f(x)$, with a function involving the dependent variable on the right-hand side. When there is no explicit time dependence, a differential equation is called *autonomous*. Our goals are slightly different here. In calculus, we wanted to plot x versus t on a plane. To do this we constructed a sign chart for the first derivative. Here we are trying to understand x without the t axis, and the phase line plays the role of the sign chart with arrows drawn instead of signs. To help, we construct a plane with the vertical axis representing either $f(x)$ or some functions that build up $f(x)$. Technically this is not part of the phase line, but it helps make clear what is happening on the line. For this reason, we leave the vertical axis unlabeled and understand it to be the range of whatever functions are of interest to us.

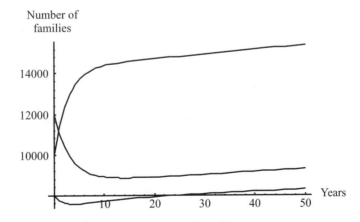

FIGURE 5.18 Income Model: Five Percent Perturbation.

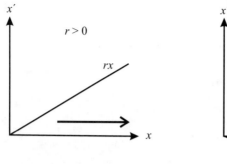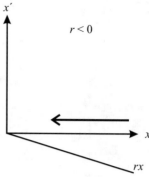

FIGURE 5.19 Simple Phase Line Diagrams.

We begin with some simple models and successively build our way up to more complex models.

Our simplest model is $\frac{dx}{dt} = rx$, the exponential growth model discussed above. Phase line diagrams are shown in Figure 5.19 for two cases: $r > 0$ and $r < 0$. The horizontal axis is the x-axis and the vertical axis represents $\frac{dx}{dt}$ which is rx. In one case, the derivative is always positive, implying that x is growing. This is indicated by an arrow pointing to the right on the phase line (right being the direction of increase in x). In the other case, the arrow points to the left because the derivative is always negative. More can be said although it does not show in the arrows. In the first diagram, the derivative function is growing, which indicates not only that x is growing, but it is growing more and more rapidly.

As discussed earlier, the exponential model is a growth model with $\frac{dx}{dt}$ given by a linear equation. In Exercise 1, models with constant $\frac{dx}{dt}$ are explored and found to have solutions which are lines. Working our way up in complexity, the next model should have $\frac{dx}{dt}$ expressed as a quadratic, which is what the logistic equation is. Figure 5.20 shows a phase line diagram for $\frac{dx}{dt} = rx(1 - x/K)$. From precalculus, we know that $f(x) = rx(1 - x/K)$ is a parabola, opening downward with zeros at $x = 0$ and $x = K$. With the parabola drawn, the phase line arrows can be determined by inspection. Whenever the graph is above the x-axis, the derivative is positive and the arrow points right, and, analogously, when the graph is below the x-axis, the arrows point to the left.

The use of phase line analysis for linear or exponential models is mostly for the illustration of the concepts. With the quadratic or logistic model, the value of this method becomes clear. Namely it is a conceptual way to identify fixed points and test their stability. In continuous models, fixed points occur when $\frac{dx}{dt} = 0$. Graphically this occurs where the graph of the derivative function crosses the x-axis. With the logistic model, this happens at $x = 0$ and $x = K$. Looking at the arrows on the phase plane, we observe that if x is perturbed from the fixed point at K, the growth rates as indicated by the arrows tend to push x back towards K. This is a stable steady state. There is another steady state at $x = 0$; observe that a slight perturbation from zero causes x to move away from zero. (For population models, we are interested only in positive x, but

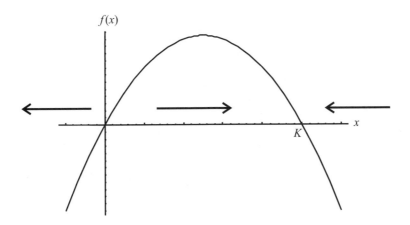

FIGURE 5.20 Phase Line Diagrams for the Logistic Model.

this statement is true for negative x as well.) This is actually a weakness in the logistic model. It suggests that a non-zero population, no matter how small, will recover. There are several resolutions to this problem. One is to put restrictions on the range of model validity. One might say that this model holds only for populations above the size of, say, 100. On the other hand, one might want a model which tends towards zero if the population gets too small. One such model is an *Allee effect model*.

An Allee effect model is an improvement over the logistic model in that if a population gets too small the growth rate becomes negative. This is used to handle a number of complex processes that occur with small populations, such as difficulties finding mates or inbreeding effects, without complicating an otherwise simple model. The Allee effect model is also the next natural step in our progression for it involves $\frac{dx}{dt}$ with a cubic equation. For simplicity, an *Allee effect model* is any model that can be written in the form $\frac{dx}{dt} = a_3 x^3 + a_2 x^2 + a_1 x + a_0$. Several other forms are $\frac{dx}{dt} = d(x-a)(x-b)(x-c)$ which is useful if the fixed points are known, and $\frac{dx}{dt} = x(r - a(x-b)^2)$ if one is interested in the form $xr(x)$. A typical Allee effect growth curve and phase line are presented in Figure 5.21. There are three fixed points: one corresponding to the carrying capacity, one corresponding to zero population, and one at the threshold between "large" and "small" population sizes. With this model, both the zero population and the carrying capacity are stable fixed points, which generally corresponds to intuition. The "large-small" threshold point is an unstable fixed point. Populations which start out above this value tend towards the carrying capacity; populations below this point tend to extinction.

We have now worked through analysis of three specific cases: linear, quadratic, and cubic. From this foundation, generalizations to other functions are obvious. The analysis performed for the logistic model is essentially identical for any continuous function which has two zero crossings and which is positive between the zero crossing. The sine function on a restricted domain of $[0, \frac{3\pi}{2}]$, for example, has identical steady state behavior although the time-dependent solutions are slightly different (see Exercise 3). These differences, of course, could be important to how well a model reflects reality.

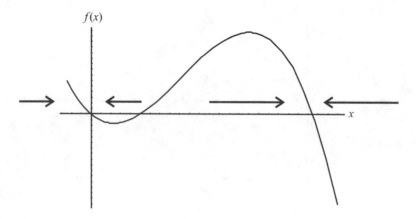

FIGURE 5.21 Phase Line Diagrams for Allee Effect Model.

One family of curves that recently has been used in population models is the θ-logistic. The θ-logistic curve is the solution of the differential equation

$$\frac{dx}{dt} = rx\left(1 - \left(\frac{x}{K}\right)^\theta\right).$$

Figure 5.22 shows the shape of this family for different values of θ. Since the θ-logistic is of increasing importance and is relatively unknown in mathematical modeling books, we will explore it more fully before looking at applications of phase line analysis. The elk harvesting model will make use of the θ-logistic.

5.3.2 Gilpin and Ayala's θ-logistic Model

The θ-logistic model is a variant of the logistic model, and its development follows a number of good modeling practices. Gilpin and Ayala [20] studied a logistic predator-prey

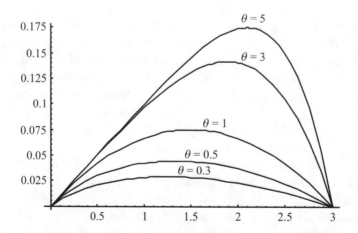

FIGURE 5.22 θ-Logistic Model $r = 0.1$ and $K = 3$.

model developed by Lotka and Volterra, namely

$$\frac{dN_1}{dt} = r_1 N_1 (1 - N_1/K_1 - \alpha_1 N_2/K_1)$$

$$\frac{dN_2}{dt} = r_2 N_2 (1 - N_2/K_2 - \alpha_2 N_1/K_2)$$

where for species $i = 1$ and 2, N_i is the population or population density, r_i is the intrinsic growth rate, α_i is an interaction parameter, and K_i is the carrying capacity. Predator-prey models are studied later in this chapter, but for now it suffices to observe that the model is a pair of simultaneous differential equations. Also observe the logistic differential equation as a building block for these equations (noncompetitive part).

Gilpin and Ayala noted that Volterra proposed this equation as a global "law" without any data as a basis. Lotka suspected a more complicated interaction term, but derived this model taking the first term of a Taylor's series expansion. Some experimental verification has been done with competing microorganisms. Gilpin and Ayala sought to test this model with experimental data from more advanced organisms and also to examine the question of whether a slightly more complicated model would yield significantly more accurate results. They performed a series of experiments with competing fly (*Drosophila*) strains. Adult flies in desired proportions were placed in culture bottles for a week. At that time survivors of each strain were counted and removed. Four weeks later offspring were counted. These "survivors + recruits" were used to compute the population change ΔN for a change in time Δt. The experiment was repeated for many initial proportions of each strain yielding an number of initial conditions representable in a phase plane (N_1 versus N_2) format.

Twenty different competition models were fit to the data. The adequacy of a model was determined according to the following criteria:

1. Simplicity. The model should have the least number of parameters necessary to account for the observed results. Models were all required to be statistically significant at an $\alpha = 0.05$ (95% confidence) level. Within this confidence level, a three-parameter model was considered superior to a four-parameter model.

2. Reality. Parameters should have biological interpretations.

3. Generality. The model should be as general as possible. Since the Lotka-Volterra equations adequately explain some microorganism competitions, it was desirable to have these equations as a special case.

4. Accuracy. Explained variances (R^2) should be as close to 100% as possible. Predicted isoclines should visually match observed isoclines (isoclines are defined in the predator-prey section below).

As modelers these are good goals for us to strive for in our work as well.

After analyzing the 20 models, two models satisfied the above criteria.

$$(A) \quad \frac{dN_i}{dt} = r_i N_i (1 - (N_i/K_i)^{\theta_i} - \alpha_i N_j/K_i)$$

for $i, j = 1, 2$ and

$$(B) \quad \frac{dN_i}{dt} = r_i N_i (1 - N_i/K_i - \alpha_i N_j/K_i - \beta_i N_i^2/K_i$$

for $i, j = 1, 2$. Note that these are still pairs of equations.

Both models contain four parameters (equally simple), are general, and contain the Lotka-Volterra equation (equally general), and there is no statistically significant difference in their very good explanation of the data (equally accurate). On the basis of reality, however, negative parameters occurring in model (B) indicate cooperation between strains which does not match the known behavior of the species. Thus model (A) was considered to be the superior model.

The first part of model (A) is

$$\frac{dN_i}{dt} = r_i N_i (1 - (N_i/K_i)^{\theta_i})$$

which is the noncompetitive (single species behavior) part of the model. This is what we have been calling the θ-logistic model.

5.3.3 Example: Harvesting Models

The examples and methods in Section 5.3.1 on phase line analysis are interesting and important in and of themselves, but they are also building blocks and tools for constructing and analyzing new models. In this section, we examine the effects of introducing harvesting into several of the models studied above. Much of this discussion follows the first chapter of the classic book *Mathematical Bioeconomics* [12] by Colin Clark. By *harvesting* we mean the removal of animals or plants from a population by any means, including fishing, hunting, culling, or lumbering. The elk herd in the Grand Teton National Park-Jackson Hole area, for example, is harvested by hunting. At one time, the elk herd in Yellowstone National Part was harvested by ranger slaughter, though currently management is by natural means such as predation, disease, and winter starvation. By contrast, the national bison herd in Moiese, Montana, is harvested by selling surplus, aged, or diseased animals at auction. All of these various methods are considered harvesting.

Constant Harvesting. One of the simplest harvesting models is constant harvesting. For example, suppose that the managers of a hunting region allow 10,000 deer to be killed by hunting per year. If fewer than 10,000 deer are around, then the entire population is killed. Again this is a simple model, and refinements are discussed later.

Let us assume that an unharvested population of some animal grows according to the logistic differential equation $\frac{dx}{dt} = rx(1 - \frac{x}{K})$ and each year h animals are killed. One very elegant aspect of phase line analysis is that it can be performed for general parameters r, K, and h, and hence the results hold for many situations and are highly robust. We define two functions, a population growth function $F(x) = rx(1 - \frac{x}{K})$ and a harvest function $H(x) = h$. Using the idea of inflow minus outflow, the model for the harvested population is

$$\frac{dx}{dt} = F(x) - H(x)$$

or in this case

$$\frac{dx}{dt} = rx\left(1 - \frac{x}{K}\right) - h.$$

5 Continuous Models

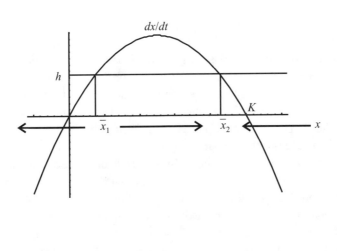

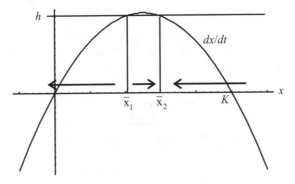

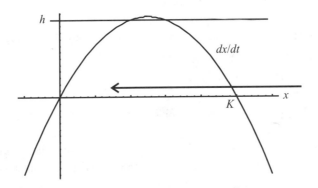

FIGURE 5.23 Constant Harvesting.

The phase line analysis is shown in Figure 5.23. Notice that now we are plotting $F(x)$ and $H(x)$ on the vertical axis so the fixed points are now the intersection points of $F(x)$ and $H(x)$, not zero crossings. Also, right arrows or increasing populations correspond to intervals where $F(x) > H(x)$, and left arrows or decreasing populations correspond to intervals where $F(x) < H(x)$. There are several important features to this analysis. First, the fixed points move as h changes. In particular, they move closer together as h increases. Although zero does not appear as a fixed point, it must be considered to be one as it is at the edge of the domain. The upper fixed point is not the carrying capacity. Thus if an unmanaged herd were in equilibrium, and some reasonable but constant level of hunting were allowed, the population's equilibrium size would drop due to the nature of the system and this should not be a cause for concern. The other equilibrium point is unstable, and should a population drop below it, the population will crash, unlike a population governed by the logistic equation alone. Second, as the harvesting levels increase, the two fixed points move closer together and touch when the harvesting level reaches the vertex of the logistic's quadratic curve. This harvesting level is called the Maximum Sustainable Yield (MSY). Going beyond it results in a population crash which is irreversible without a nearly total cessation of harvesting. Commercial harvesting efforts, such as fisheries, often play a dangerous game of harvesting as close to the MSY as possible. As seen in Chapters 2 and 4, these curves are models, and the observed data follows some distribution about the curve due to various random factors (see Figure 5.28). Thus treating the value of the MSY as a fixed number for a population and managing accordingly can lead to catastrophic results.

Harvesting Effort. A second harvesting model takes effort into account and is useful for harvesting methods such as fishing. If there are few fish around, it is hard to catch fish. If there are many around, it is easier. Many functions can be used to describe this situation and the simplest is a linear function. In this model $H(x) = hx$, a line with slope h and intercept zero. For a given value of x, the larger h is the greater the harvest. Thus the parameter h is a measure of the effort put into the harvesting process. Again we consider unharvested growth to be logistic in nature, and the overall model is

$$\frac{dx}{dt} = F(x) - H(x) \quad \text{or} \quad \frac{dx}{dt} = rx\left(1 - \frac{x}{K}\right) - hx.$$

Figure 5.24 shows phase line analysis for several values of h. When $h = 0$, this is, of course, just the logistic model, and its analysis holds. Figure 5.24 shows the phase line analysis for small h values ($h < F''(0)$). There are two equilibria values, and again the non-zero equilibrium point is less than the carrying capacity. An interesting phenomena is observed with this model. Increasing effort h causes an increase in harvest (the y value at the intersection of the two curves) to a point. Beyond that, increasing effort results in a decreased harvest, even though the population is completely stable. This is often illustrated with a *yield-effort* curve such as the one shown in Figure 5.25. The size of the population is shown on the x-axis, and the size of the harvest or yield is indicated on the y-axis. Once again there is a MSY although increasing effort does not cause a population crash, rather only a decrease in the harvest. If this model accurately describes the population that one is harvesting, then understanding the MSY is of great

5 Continuous Models

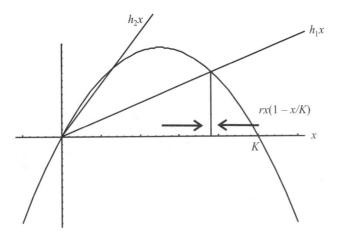

FIGURE 5.24 Harvesting a Logistic Model.

consequence. There is no need to expend more effort (money, time, energy) if the result is a decrease in yield. On the other hand, if the harvesting has been going on for years but no population analysis has been done, it may be possible to decrease effort and increase yield.

Next suppose that the logistic's quadratic curve is replaced with the Allee effect curve illustrated in Figure 5.26. This model is similar to the logistic, except that there are two non-zero equilibria, one stable and the other unstable. If the effort is increased to the point that the two equilibria coincide, the result is an immediate tendency towards extinction, as opposed to just a lowering of yield. The yield-effort curve is shown in Figure 5.27. Notice that if the effort causes the equilibria to coincide and the population starts to drop, there is a point beyond which the population never recovers even if all harvesting is permanently terminated.

The methods of this section are simple and elegant and provide tremendous information to someone trying to make management decisions. On the other hand, it should be clear that a slightly incorrect model could lead to poor decisions which have undesirable

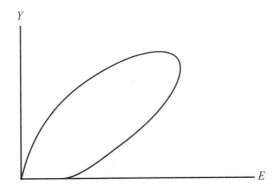

FIGURE 5.25 Yield-Effort Curve.

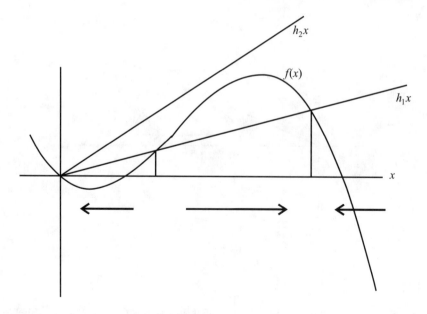

FIGURE 5.26 Harvesting an Allee Effect Model.

consequences. In general it is difficult to have sufficient data on a wild population to accurately determine the $F(x)$ function. Following are some examples which illustrate use of these ideas in practice.

Example: Jackson Elk Herd. The elk herd of Grand Teton National Park has been studied extensively (see Boyce [5]). In Chapter 4, we discussed fitting curves to data. In this case a θ-logistic curve was fit to the census data for the summer population of elk in the central valley of the national park. This fit resulted in a model with parameters $K = 2043$, $r = 0.881$, and $\theta = 3.5$. Thus our model is

$$\frac{dx}{dt} = 0.881x\left(1 - \left(\frac{x}{2043}\right)^{3.5}\right).$$

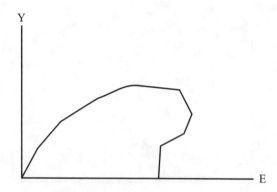

FIGURE 5.27 Yield-Effort Curve.

5 Continuous Models

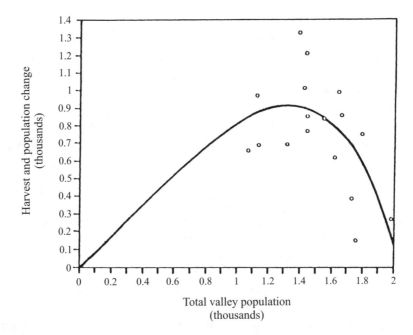

FIGURE 5.28 Elk in Grand Teton National Park's Central Valley. Used by Permission of Cambridge University Press.

By taking derivatives of the right-hand side to find relative extrema, we find that $F(x)$ has a maximum of 910 at 1,329. What this implies is that a yearly harvest of 910 elk maintains the population at 1,329. A plot of valley population versus population change (adjusted for hunting) is shown in Figure 5.28. This model suggests that with no hunting the population eventually stabilizes at approximately 2,000 elk in the central valley.

5.3.4 Example: Spruce Budworm

A modeling success story is the spruce-budworm model of Ludwig, Jones, and Holling [41]. The spruce budworm is a parasite which defoliates and kills Canadian Balsam forests. The budworm always exists in the forests, but at an equilibrium level where the budworms and the forest coexist. Occasionally there are outbreaks during which the budworm population soars, leading to defoliation of whole forests. During these outbreaks, management efforts do little to control the budworm's population explosion. The model in this section effectively describes the dynamics of these outbreaks and suggests management schemes to combat these problems. This is also an excellent example of modeling methods, and we will go through the steps of this model in some detail.

The main players in this scenario are the budworms, the trees whose foliage they eat, and birds which prey upon the budworms. Several simplifications are made to start the modeling process (Occam's razor). First the budworm's life cycle is relatively short, and a population can increase several hundredfold in a few years. Trees operate on a much slower time scale, and the time interval for trees to completely replace their foliage is 7–10 years. In the absence of budworms, these trees live 100–150 years. In assigning time scales to processes, we observe that the budworm shows measurable change over time increments of months, while trees show measurable change over years. Our first simplification is to assume the budworms are modeled accurately by a continuous model and that the tree growth is so much slower that their contribution is governed by a parameter with no time dependence. Rather than allowing the tree parameters to vary, we fix the tree parameter, allow the model to stabilize, then set the tree parameter again and allow the model to stabilize again.

Our second assumption is that in the absence of predation the budworm population obeys a logistic model

$$\frac{dB}{dt} = rB\left(1 - \frac{B}{K}\right).$$

Here B is the population density of the budworm, r is the density independent or "intrinsic" growth rate (i.e., when $B = 0$), and K is the carrying capacity which depends on the amount of foliage available. Due to our first assumption, we assume K is a constant since it changes so much more slowly than the budworm's density. As discussed in the phase line section, other functions could be used here as long as there is a self-limiting density dependence.

The third assumption is that although birds like to eat budworms, if budworms are difficult to find, they merely switch to some other food. So we do not need a bird population component to this model. There are three aspects to the birds contribution: 1) If budworms are available, they become a favorite food; 2) if budworms are scarce, much fewer get eaten, for the birds go after the easy food; and 3) if budworms are abundant, the birds eat more and more budworms up to a saturation point. We assume this behavior is adequately described by the function

$$g(B) = \beta\left(\frac{B^2}{\alpha^2 + B^2}\right),$$

where α and β are constants or parameters. A graph of this function is shown in Figure 5.29. For small B, this term behaves like a quadratic which starts with a slow growth. The growth becomes more rapid with an inflection point at $(\frac{\alpha}{\sqrt{3}}, \frac{\beta}{4})$. As B continues to grow, it asymptotically approaches β, which is the saturation point.

The model for the budworm is then the intrinsic growth less the predation (or inflow minus outflow), and the equation is

$$\frac{dB}{dt} = rB\left(1 - \frac{B}{K}\right) - \beta\left(\frac{B^2}{\alpha^2 + B^2}\right).$$

The next step in the model analysis is very important. Currently there are two variables (B and t) and four parameters (r, K, α, β). Trying to understand the effects of six changing or changeable quantities at once is quite daunting. Frequently, however, parameters are present just because of the choice of scale. Anyone with experience working with temperatures will attest that working with a Celsius scale gives cleaner equations than a Fahrenheit scale, and often a Kelvin scale is the cleanest of all. Similarly we rescale our variables using a method called *dimensionless analysis*. The idea is to replace our current variables with variables which are dimensionless. We replace B by the dimensionless variable $x = B/\hat{B}$ where $\hat{B}$ is a scalar or number with the same

5 Continuous Models

dimensions as B. Similarly we replace t by the dimensionless variable $\tau = t/\hat{t}$. The symbols $\hat{B}$ and $\hat{t}$ are left undefined for the time being. We may assign any number or scale that we desire, and later we will make the assignment that will be the most beneficial.

Thus we have $B = \hat{B}x$ and $t = \hat{t}\tau$, and so $dB = \hat{B}dx$ and $dt = \hat{t}d\tau$. Making the appropriate substitutions

$$\frac{dB}{dt} = \frac{\hat{B}}{\hat{t}}\frac{dx}{d\tau} = r\hat{B}x\left(1 - \frac{\hat{B}x}{K}\right) - \beta\left(\frac{(\hat{B}x)^2}{\alpha^2 + (\hat{B}x)^2}\right).$$

Simplifying yields

$$\frac{dx}{d\tau} = rx\hat{t}\left(1 - \frac{\hat{B}x}{K}\right) - \frac{\beta\hat{t}}{\hat{B}}\left(\frac{x^2}{\left(\frac{\alpha}{\hat{B}}\right)^2 + x^2}\right).$$

Now we define $\hat{t}$. Notice that the coefficient on the second term would be one if $\hat{t} = \hat{B}/\beta$. After making this substitution, making the assignment $\hat{B} = \alpha$ leads to

$$\frac{dx}{d\tau} = \frac{rx\hat{B}}{\beta}\left(1 - \frac{\hat{B}x}{K}\right) - \left(\frac{x^2}{1 + x^2}\right).$$

Next we combine several of the old parameters to make two new ones, namely

$$R = \frac{r\hat{B}}{\beta} \quad \text{and} \quad Q = \frac{K}{\hat{B}}.$$

After making these substitutions, we have

$$\frac{dx}{d\tau} = Rx\left(1 - \frac{x}{Q}\right) - \frac{x^2}{1 + x^2}.$$

The key achievement is that we have reduced our equation from four parameters down to two. Although the new parameters R and Q are not the growth rate and carrying capacity

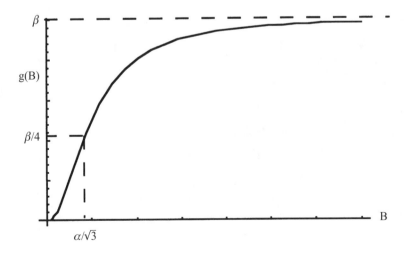

FIGURE 5.29 Predation Function.

in the old dimensions, they behave similarly and may be thought of as a growth rate and carrying capacity in the dimensionless coordinate system.

Factoring the right-hand side we have

$$\frac{dx}{d\tau} = x\left(R\left(1-\frac{x}{Q}\right) - \left(\frac{x}{1+x^2}\right)\right)$$

which has the form

$$\frac{dx}{d\tau} = x(F(x) - G(x)).$$

For positive x, we need to consider only the sign of $F(x) - G(x)$ to perform the phase line analysis. The term $F(x) = R(1-\frac{x}{Q})$ is a line with intercepts $(0, R)$ and $(Q, 0)$. The function $G(x) = x/(1+x^2)$ is zero at the origin, has a maximum at $(1, \frac{1}{2})$, an inflection point at $(\sqrt{3}, \sqrt{3}/4)$, is concave down on $(0, \sqrt{3})$ and concave up on $(\sqrt{3}, \infty)$, and has the line $y = 0$ as a horizontal asymptote. The graphs of $F(x)$ (for several values of R and Q) and $G(x)$ are shown in Figure 5.30. There are three cases with one, two, or three equilibrium points. Phase line analysis shows that if there is a single equilibrium, it is stable. If there are two, one is stable and the other is unstable. There are lines which will make either one stable. Finally, if there are three equilibrium points, the largest and smallest are stable, but the middle one is unstable.

Now recall that R and Q are not constants, but, rather, are very slowly changing parameters which are assumed constant for the sake of finding equilibrium values. In particular as the forest matures and foliage increases, the carrying capacity of the budworms rises. The value of R also rises with more favorable conditions, or is lowered by management efforts such as pesticides.

Consider Figures 5.31 through 5.34. We assume R is held constant and let Q change slowly enough that the system is always in equilibrium. At first, there is one equilibrium point. This is the normal endemic population of budworm in the forest which is at a

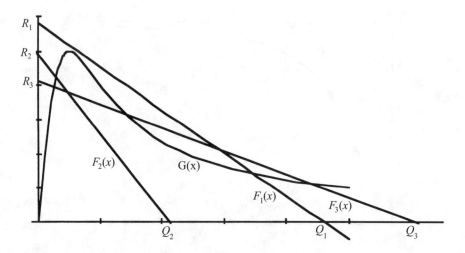

FIGURE 5.30 Phase Line for Different Values of R and Q.

5 Continuous Models

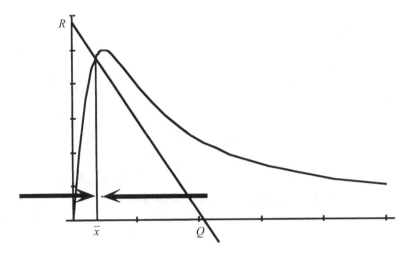

FIGURE 5.31 Spruce Budworm Phase Plane Analysis.

low enough level that both the insects and the trees coexist. As the forest matures, Q increases until there are two, then three equilibrium points. At the moment when there are two equilibrium points, the insect population is at the lower point, which is unstable, and hence the system stabilizes at the upper equilibrium. This is the population outbreak. Notice that even drastic reductions in R have little effect on the population size. Naturally the carrying capacity begins to fall, but there is little change until the carrying capacity drops back to the level with one equilibrium point. Of course this may result in significant forest damage or loss.

This model is a theoretical model for which we used no data or even values for parameters. Its purpose is to explain behavior and suggest control methods. We conclude

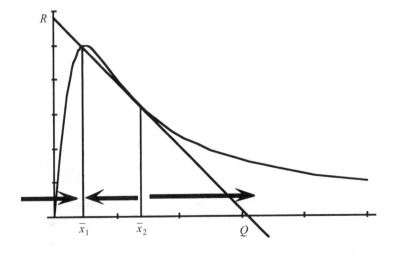

FIGURE 5.32 Spruce Budworm Phase Plane Analysis.

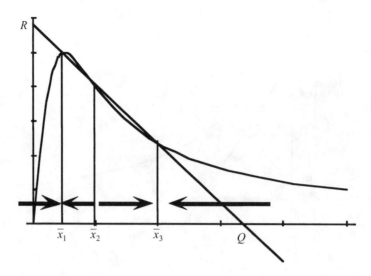

FIGURE 5.33 Spruce Budworm Phase Plane Analysis.

this discussion by looking at what control methods might be effective. One common method is spraying pesticides. As long as this is done continuously and keeps R low enough, the budworms always are held in check. Spraying is, however, expensive and often has undesirable consequences. If spraying keeps R down while Q gets large, then even a short period of not spraying (as might happen during a period of economic hardship) would result in an immediate rise in R and hence an outbreak. An alternative method is to try to lower Q. This can be done through pruning and thinning and removing unwanted foliage that might contribute to the carrying capacity.

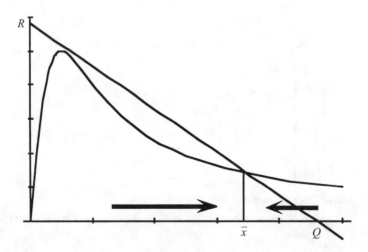

FIGURE 5.34 Spruce Budworm Phase Plane Analysis.

5 Continuous Models

This is a model worth studying carefully. It has a purpose which is met. Reality has been simplified through a series of well-stated assumptions to give a model which we are able to analyze. True, the effects of winters or rainfall may play some role, but for the purpose of understanding the behavior they are not needed and are left out. The model is robust in several senses. The model's predictions can be made for ranges of parameter values. Even if these functions are slightly off, it is easy to see that many similarly shaped functions would give the same predictions. The model also makes use of dimensionless analysis and phase line analysis and illustrates the usefulness of each. Finally, the understanding given by this model leads to suggestions of management strategies.

5.3.5 Phase Plane Analysis

Phase plane analysis is a two-dimensional version of phase line analysis. Recall that we coined the name phase line analysis because it is a one-dimensional version of phase plane analysis. If we have systems with two variables x and y changing over time, a phase plane diagram is a plot of this behavior over time on an xy coordinate system. As a simple example, suppose $x(t) = t$ and $y(t) = t^2$. Then if we plot the points $(x, y) = (t, t^2)$ for the desired values of t, the result is a parabola (or portion thereof) in the xy phase plane.

To do phase plane analysis we need a system of two differential equations of the form

$$\frac{dx}{dt} = f(x, y)$$
$$\frac{dy}{dt} = g(x, y).$$

Since there is no explicit dependence on time, such a system is called *autonomous*. This is crucial for the type of analysis we are going to do as we will be marking regions with certain flow properties with the assumption that, once marked, a region's behavior remains the same through time. These flows help determine asymptotic behaviors.

We track the following example through this discussion:

$$\frac{dx}{dt} = y - x^2 \qquad \frac{dy}{dt} = y - x.$$

Fixed points. We begin our analysis by finding the fixed points of an autonomous system of first-order differential equations. The idea is intuitive. When is a point said to be fixed? Clearly when it does not move. How do we know when a point does not move? When its rates of change are zero. More precisely, when $\frac{dx}{dt} = 0$ and $\frac{dy}{dt} = 0$. In order to find fixed points, we need to consider the curves $f(x, y) = 0$ and $g(x, y) = 0$ in the phase plane. Any curve of the form $h(x, y) = k$ is called an *isocline* or *level curve* of the function h. In other words, a curve is an isocline or level curve of a function if the function takes the same value at every point on the curve. The isoclines $f(x, y) = 0$ and $g(x, y) = 0$ are called *nullclines* since they take the value zero. The intersection point of the two nullclines is, necessarily, a fixed point. These are found by solving the two equations simultaneously, or they can be approximated from their phase plane graphs.

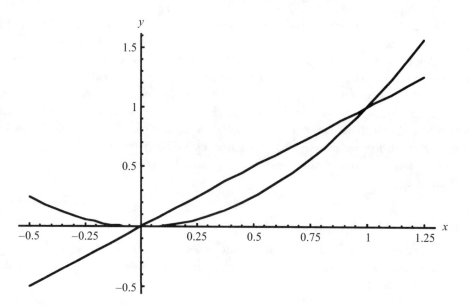

FIGURE 5.35 Nullclines.

In our example, the nullclines are $y = x^2$ and $y = x$, and the fixed points are $(0, 0)$ and $(1, 1)$.

Directions of flow. From calculus of a single variable, we know that a function changes sign only at zero crossings and discontinuities. This is why we mark zeros and discontinuities when beginning sign charts. Similar ideas hold in two dimensions, and we assume that our functions are continuous so we only need to worry about zeros. A continuous surface, such as $z = f(x, y)$, changes sign only at places where z is zero, which is a nullcline. So by marking the f nullclines we have divided the plane up into regions where $\frac{dx}{dt}$ is positive and negative and similarly for g and $\frac{dy}{dt}$.

Using these ideas we mark arrows to indicate the direction of derivatives. In each region where $\frac{dx}{dt}$ is positive we mark an arrow to the right (direction of growth in x), and where $\frac{dx}{dt}$ is negative we mark an arrow to the left. Similarly for y, only now arrows are up for positive $\frac{dy}{dt}$ (direction of increasing y) and down for negative. These arrows indicate the direction a trajectory moves. One may think of the phase plane as a flowing body of water. If we drop a stick in the water, it moves in a path determined by the flow vectors. While our drawing is a little crude, we know that a point moves up and left or, perhaps, down and right. A computer-generated phase plane might have arrows of different lengths giving a more accurate sense of which direction and how fast, but, for an understanding, our crude picture is very informative. As a point crosses nullclines, the direction it moves changes. Consider our example. The arrows drawn by hand are shown in Figure 5.36. Hand-drawn trajectories are also shown and are discussed next.

From a phase plane drawing, saddle points are evident; however, it is difficult to tell a spiral sink from a center or a spiral source. We discuss two methods. The first is to use a differential-equations solver to find either exact or approximate solutions for initial points

5 Continuous Models

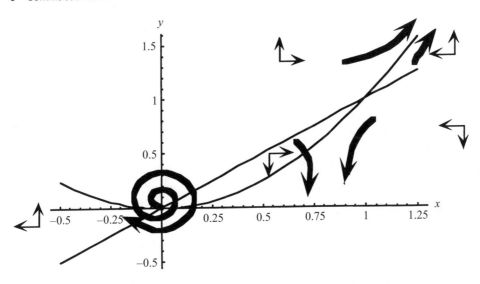

FIGURE 5.36 Hand Drawn Phase Plane with Trajectories.

in various locations and observe which behavior they exhibit. This is explored more fully in the section on solvers below. A second method, discussed here, is to linearize the differential equations.

Linearization. If the reader has not read Section 5.2.6, now would be a very good time to do so. Linearization of a differential equation is the replacement of the equation of a curve or surface by the appropriate tangent line or tangent plane. This is usually done to study the behavior of a system near a point, and the tangent line or plane is centered at this point. After this is done, we have a system of linear differential equations, and we use the methods and analysis of Section 5.2.6 to determine the behavior close to the point of interest.

While linearization can be done at any point, we linearize only around fixed points. Let $(\bar{x}, \bar{y})$ be fixed points of the system (do not confuse fixed point notation with the sample mean)

$$\frac{dx}{dt} = f(x, y)$$

$$\frac{dy}{dt} = g(x, y).$$

Note that this means $f(\bar{x}, \bar{y}) = 0$ and $g(\bar{x}, \bar{y}) = 0$. We use the notation $f_x(a, b)$ to denote the partial derivative of f with respect to x evaluated at (a, b), and $f_y(a, b)$ is the partial of f with respect to y. Similar notation is used for g. Recall that a partial derivative with respect to x gives the slope in the x direction; the partial derivative with respect to y gives the slope in the y direction.

The reader should recall from calculus or the discussion in Section 5.2.5 that a differentiable function $F(x)$ can be approximated near a point by its tangent line. That is, $F(x) \approx F(a) + F'(a)(x - a)$. In calculus of several variables, this is extended to a plane tangent to a surface at a point. Even if this is unfamiliar to the reader, the form should be a plausible extension of the tangent line equation:

$$\frac{dx}{dt} = f(x,y) \approx f(\bar{x},\bar{y}) + f_x(\bar{x},\bar{y})(x-\bar{x}) + f_y(\bar{x},\bar{y})(y-\bar{y})$$

$$\frac{dy}{dt} = g(x,y) \approx g(\bar{x},\bar{y}) + g_x(\bar{x},\bar{y})(x-\bar{x}) + g_y(\bar{x},\bar{y})(y-\bar{y}).$$

Notice $f(\bar{x},\bar{y}) = 0$ and $g(\bar{x},\bar{y}) = 0$. Also

$$\frac{d(x-\bar{x})}{dt} = \frac{dx}{dt} \quad \text{and} \quad \frac{d(y-\bar{y})}{dt} = \frac{dy}{dt}$$

so we have the following pair of equations:

$$\frac{d(x-\bar{x})}{dt} \approx f_x(\bar{x},\bar{y})(x-\bar{x}) + f_y(\bar{x},\bar{y})(y-\bar{y}).$$

$$\frac{d(y-\bar{y})}{dt} \approx g_x(\bar{x},\bar{y})(x-\bar{x}) + g_y(\bar{x},\bar{y})(y-\bar{y}).$$

It is simpler to change coordinate systems and translate the fixed point to the origin. Let $u = x - \bar{x}$ and $v = y - \bar{y}$. Then we have

$$\frac{du}{dt} \approx f_x(\bar{x},\bar{y})u + f_y(\bar{x},\bar{y})v$$

$$\frac{dv}{dt} \approx g_x(\bar{x},\bar{y})u + g_y(\bar{x},\bar{y})v.$$

This is a linear system which we rewrite as

$$\begin{pmatrix} \frac{du}{dt} \\ \frac{dv}{dt} \end{pmatrix} = \begin{pmatrix} f_x(\bar{x},\bar{y}) & f_y(\bar{x},\bar{y}) \\ g_x(\bar{x},\bar{y}) & g_y(\bar{x},\bar{y}) \end{pmatrix} \begin{pmatrix} u \\ v \end{pmatrix}.$$

The matrix of partial derivatives is sometimes called the *Jacobian matrix*, and we use the notation

$$J(a,b) = \begin{pmatrix} f_x(a,b) & f_y(a,b) \\ g_x(a,b) & g_y(a,b) \end{pmatrix}.$$

This matrix is used with Theorem 5.3 to determine the nature of the solution.

This linearization process may seem involved, but much of the complication is in the development of the equations. In practice it is only necessary to compute the four partial derivatives at the fixed point and then use Theorem 5.3 to determine the nature of the solution. Applying the technique to our example yields the Jacobian matrix:

$$J(0,0) = \begin{pmatrix} 0 & 1 \\ -1 & 1 \end{pmatrix},$$

so $\beta = Tr(J(0,0)) = 1$ and $\gamma = \det(J(0,0)) = 1$ so $(0,0)$ is a spiral source.

$$J(1,1) = \begin{pmatrix} -2 & 1 \\ -1 & 1 \end{pmatrix},$$

so $\beta = Tr(J(1,1)) = -1$ and $\gamma = \det(J(1,1)) = -1$; thus $(1,1)$ is a saddle point.

5.3.6 The Classical Predator-Prey Model

This section deals with a classical model describing the interaction between two species: a predator and its prey. This model was originally proposed by A. J. Lotka and V. Volterra in the 1920s. Although this model is simplistic and it is easy to find weaknesses in it, it is still a valuable example to study. Many wild populations oscillate as do a number of other phenomena such as the auto industry (in fact there is a whole industrial sector called cyclicals). The predator-prey model is a theoretical model in the sense that while it may not predict correct numbers, it proves that conditions can be formulated which lead to stable oscillatory behaviors.

The model assumptions, simplifications, and notation.

1. There are two species interacting: a prey species x and a predator species y. For the purposes of this model no other species interact with these two.

2. In the absence of the predator, the prey exhibits pure exponential growth. In particular $\frac{dx}{dt} = ax$ where $a > 0$. Implicit in this assumption is that there is sufficient food and space to allow the prey species to grow indefinitely.

3. In the absence of the prey, the predator species dies out exponentially. In particular, $\frac{dy}{dt} = -dy$ where $d > 0$. Thus although it is not explicitly mentioned, there is other food for the predators, but not enough to sustain the population. Thus it dies out over several years rather than over a month or two.

4. When the two species are in the presence of each other, the predators kill the prey in such a way that the predator population increases at a rate proportional to the product of the number of predators and the number of prey (i.e., xy). Similarly the prey population is decreased by an amount proportional to the product of the population sizes.

This xy term perhaps deserves some further discussion. Naively it makes sense that if either the number of predators or the number of prey increase, the number of interactions and hence deaths increase also. But why xy and not, say $x^2\sqrt{y}$, or some other such function with the same properties? Originally this term was borrowed from chemistry models of rates of reactions where molecules in solution interacted by randomly bumping into one another. In her book, *Mathematical Models in Biology*, Leah Edelstein-Keshet [15] writes:

> "The term xy approximates the likelihood that an encounter takes place between predators and prey given that both species move about randomly and are uniformly distributed over their habitat.... The form of this encounter rate is derived from the *law of mass action* that, in its original context, states that the rate of molecular collisions of two chemical species in a dilute gas or solution is proportional to the product of the two concentrations. We should bear in mind that this simple relationship may be inaccurate in describing the subtle interactions and motions of organisms."

Perhaps the biggest flaw with the term xy is that the predators never become satiated. If a pride of lions kills one gazelle over some time period when there are 1,000 gazelles, the same pride probably kills one gazelle over that same time period even if there are 100,000 gazelles. While it is easy to pick this assumption apart, there may be situations where it is reasonable, at least over some range of values. The following discussion is intended to indicate conditions under which the xy term might be reasonable.

We count the interactions between the species, some proportion of which result in a lethal encounter. What is meant by an interaction? Clearly one species bumping into the other is an interaction, but seeing, smelling, or hearing one another could be called an interaction as well. To keep things simple, we assume that the predator species has some territory and any prey individual that lies within that territory constitutes an interaction. Some percent of these interactions result in a prey death by whatever means of hunting the predator employs. Secondly we require that all the animals reside in some fixed region. Doubling the prey over an infinite extent would not have any effect on the number of interactions. Thirdly, we assume that for the population sizes under consideration the predators never become satiated (i.e., more prey always involves more interactions, hence prey deaths).

With these three assumptions, it should be clear that if the predator's territories do not intersect and the prey are uniformly distributed over the region, then doubling the number of prey doubles the number of interactions on average. Doubling the number of prey also doubles the number of interactions. The next issue is what happens if the predator regions are allowed to intersect. The answer depends on what assumptions are made concerning the prey lying in the regions of intersection. There are two extreme cases. The first is that the prey lying in the two regions are twice as likely to be eaten as prey lying in only one region. If there are plenty of prey or if prey are difficult to locate, then two predators are likely to eat twice as much as one. On the other hand if prey is scarce and easy to locate, then one predator might kill all the prey in its territory so two predators do not double the chances of being killed. After all, dead is dead. Of course there are many in-between situations.

If a prey is twice as likely to be eaten when it is in the territory of two predators and 100 times as likely if it is in the territory of 100 predators, then it really does not matter if the regions intersect or not. Since each territory a prey is in counts for an interaction, there is mathematically no difference between one prey in n territories or n prey in n separate nonintersecting territories. With this assumption in force, the xy term is valid. With the "dead is dead" assumption, every prey counts only once. With a fixed number of prey as the number of predators increases, the number of interactions must asymptotically approach the number of prey. The xy term is not valid in this case, nor in the "in between" cases.

Thus we have a fourth assumption, which is the key one. If a prey lies in the territory of n predators, then it is n times as likely to be killed.

We mentioned predators' and prey's being uniformly distributed earlier, but due to the additive property of the expected value (i.e., $E(x+y) = E(x) + E(y)$), one can position the predators and the prey according to any random distribution, and the average number of interactions over all such positionings is xy. Figure 5.37 shows the positioning of predators and prey according to normal distributions in both the horizontal and vertical directions with the predator and prey species centered around different points (predators

5 Continuous Models

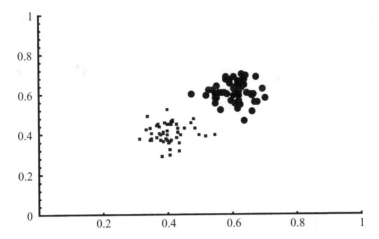

FIGURE 5.37 Predators and Prey Normally Distributed about Different Points.

at $(0.4, 0.4)$, prey at $(0.6, 0.6)$, and $\sigma = 0.05$). The territory size and standard deviation of the normals are set so that the midpoint between the two centers is approximately three standard deviations. The territories of predators near the edge of the 3σ region contain some of the prey closest to the edge of the prey's 3σ region. Figure 5.38 shows the average number of interactions found in simulations with prey (x) held fixed and the predator size (y) varied. It is evident that this function is linear (i.e., xy with x constant), but with noise due to the randomness. This noise is ignored in these analytic models, but could be dealt with in a simulation model.

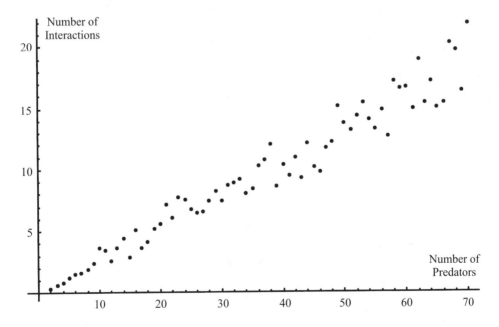

FIGURE 5.38 Average Number of Interactions as a Function of Number of Predators.

The model equations and analysis. The predator-prey assumptions yield the following system of equations:

$$\frac{dx}{dt} = a\,x - b\,xy$$

$$\frac{dy}{dt} = c\,xy - d\,y.$$

Factoring gives

$$\frac{dx}{dt} = x(a - b\,y)$$

$$\frac{dy}{dt} = y(c\,x - d).$$

The nullclines for $\frac{dx}{dt} = 0$ are $x = 0$ and $y = a/b$. The nullclines for $\frac{dy}{dt} = 0$ are $y = 0$ and $x = d/c$. Graphing these in the phase plane reveals two fixed points $(0,0)$ and $(d/c, a/b)$. (The points $(d/c, 0)$ and $(0, a/b)$ are not points of intersection of different nullclines.) A hand-drawn phase plane is shown in Figure 5.39. Naturally, we are not interested in the $(0,0)$ solution. The $(d/c, a/b)$ solution has trajectories circling about it, though it is not clear from this analysis if the solutions are spiraling in or out, or if they are circling in pure oscillatory motion. Jacobian analysis determines this.

Here $f(x, y) = ax - bxy$ and $g(x, y) = cxy - dy$, and the Jacobian matrix is

$$J(x,y) = \begin{pmatrix} a - by & -bx \\ cy & cx - d \end{pmatrix}.$$

While we are at it, we compute

$$J(0,0) = \begin{pmatrix} a & 0 \\ 0 & -d \end{pmatrix}.$$

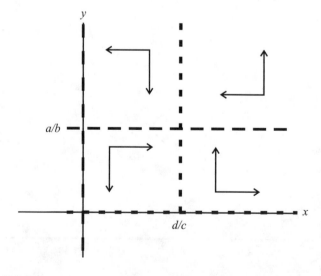

FIGURE 5.39 Phase Plane and Nullcline Diagram for Predator-Prey Model.

5 Continuous Models

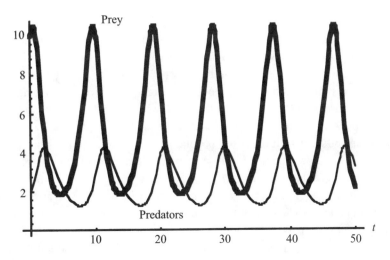

FIGURE 5.40 Predator and Prey Populations as a Function of Time.

Here $Tr(J(0,0)) = a - d$ and $Det(J(0,0)) = -ad$. Since a and b are both positive, $Det(J(0,0)) < 0$, and hence $(0,0)$ is a saddle point. At $(d/c, a/b)$ we have

$$J(c/d, a/b) = \begin{pmatrix} 0 & \frac{-bd}{c} \\ \frac{ca}{b} & 0 \end{pmatrix}.$$

Thus $Tr(J(d/c, a/b)) = 0$ and $Det(J(d/c, a/b)) = ad > 0$. Referring to Theorem 5.3, we have $\beta = 0$ and $\gamma > 0$ which indicates that this fixed point is a center.

In Section 5.4, we discuss finding and plotting solution curves using a differential equation solver. Figures 5.40 and 5.41 show the output from *Mathematica* for a predator-prey system with $a = 1$, $b = 0.4$, $c = 0.5$, and $d = 0.1$. As this is a theoretical model, these are more or less arbitrarily chosen parameters. Figure 5.40 shows the predator and

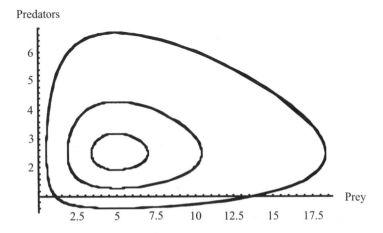

FIGURE 5.41 Predator-Prey Solutions Plotted in a Phase Plane.

prey solution curves plotted as a function of time. Figure 5.41 plots these solutions on an xy phase plane. The phase plane diagram verifies that the fixed point is a center. Observe the following behavior on both graphs. When the predator population is low, the prey population starts to rise. Soon the predator population rises to the point that the prey population starts to drop. Even though the prey population is dropping, for a time the prey population is still high enough that the predator population continues to rise. At a point, the prey population drops so low that the predator population can no longer be maintained and it starts to drop. Both populations drop until the predator population is low enough that the prey population begins to grow again and the cycle repeats. From the graphs, one can even determine the predicted period of the cycle.

The work up to now has been a general technique which applies to any system of two differential equations. The analysis from this point on depends on the particular form of these equations. In particular, the linearized equations are combined to form a single separable differential equation. This is used to analytically determine the shape of the trajectories near a center for several important cases.

Going back to the information on linearization in Section 5.3.5, we see that if we translate the coordinate system, so that a fixed point $(\bar{x}, \bar{y})$ is moved to the origin using the equations $u = x - \bar{x}$ and $v = y - \bar{y}$, the system behaves near the origin according to the linearized system of equations. In particular,

$$\frac{du}{dt} = f_x(\bar{x}, \bar{y})u + f_y(\bar{x}, \bar{y})v$$

$$\frac{dv}{dt} = g_x(\bar{x}, \bar{y})u + g_y(\bar{x}, \bar{y})v.$$

In the case of the predator-prey model, $u = x - d/c$, $v = y - a/b$ and

$$\frac{du}{dt} = -\left(\frac{bd}{c}\right)v,$$

$$\frac{dv}{dt} = \left(\frac{ac}{b}\right)u.$$

The chain rule from calculus implies that

$$\frac{du}{dv} = \frac{du}{dt}\frac{dt}{dv};$$

thus

$$\frac{du}{dv} = -\frac{b^2 d}{ac^2}\frac{v}{u}.$$

The key to this analysis is that this is a separable differentiable equation. Thus

$$ac^2 u\, du = -db^2 v\, dv.$$

Integrating both sides yields

$$\frac{ac^2 u^2}{2} = -\frac{b^2 dv^2}{2} + C.$$

5 Continuous Models

Rearranging gives

$$\frac{ac^2 u^2}{2} + \frac{b^2 dv^2}{2} = C \quad \text{or} \quad \frac{u^2}{b^2 d} + \frac{v^2}{ac^2} = C.$$

Substituting to get back our to our original variables yields

$$\frac{(x - \frac{d}{c})^2}{b^2 d} + \frac{(y - \frac{a}{b})^2}{ac^2} = C.$$

This is, of course, the equation for an ellipse centered at the fixed point with axes parallel to the coordinate axes. The implication is that close to the fixed point, the trajectories look like ellipses; which confirms our prior determination that the fixed point was a center. Figure 5.41 was found using a computer to plot trajectories for several initial conditions. Indeed, close to the fixed point the trajectories are ellipses, further away they still oscillate about the fixed point, but with a more complicated shape. Interestingly the original nonlinear predator-prey equations can be solved in closed form to give the implicit solution $(y^a e^{-by})(x^c e^{-dx}) = K$, and the reader is referred to Olinick's book [48] for a derivation. Olinick's book was written in 1978, before computers became commonplace, and there are some interesting discussions about how to graph solutions such as $(y^a e^{-by})(x^c e^{-dx}) = K$ using drafting techniques.

Discussion. Many modeling books report a famed data set of Canadian lynxes and hares that appear to support this model. The graph of this data is shown in Figure 5.42. There has been much controversy over this data set, however. The data represents the trapping for fur, and there is a question about how well trapping data represents the true census (trends in fur prices might play a role in the number of animals trapped). An observation that at one point the rise and fall in the lynx population preceded the rise and fall in the hare population resulted in a paper by Gilpin entitled *Do Hares Eat Lynx?* [19]. Hall

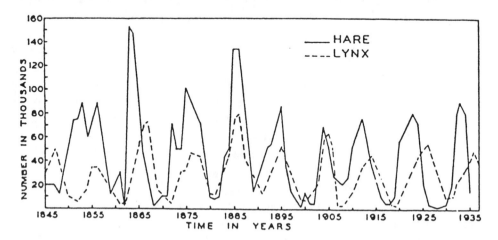

FIGURE 5.42 Lynx-Hare Fur Returns. Figure from *Fundamentals of Ecology,* Second Edition, by Eugene P. Odum and Howard T. Odum, copyright ©1959 by Saunders College Publishing and renewed 1987 by Eugene P. Odum and Howard T. Odum. Reproduced by Permission of the Publisher.

points out, in a paper [24] that should be read by all aspiring modelers, that the data set represents hares from eastern Canada and lynx from western Canada (lynx furs were worth shipping but hare furs were not). In fact these populations were not interacting at all. Further, hare populations were seen to oscillate on Anticosti Island in a manner similar to hares on the mainland, but no lynx lived on the island. Great care must be taken when modeling if you are using data you have not collected yourself.

In natural settings, it is very hard to observe a true predator-prey relationship due to other factors that affect the populations. There are, however, examples of the predator-prey phenomena observed in experimental settings. One performed in 1957 by Huffaker studied two species of mite: one fed on oranges, the other fed on the other species of mite. The results agree favorably with the predictions of the predator-prey model.

5.4 Differential Equation Solvers

If we run across a differential equation while model building and cannot solve it, there are two possibilities: 1) we do not know how to solve it, but someone else does; 2) we do not know how to solve it, and no one else does either. A computer package called a differential equation solver (DES) is useful for both of these cases, although we may get different types of solutions for the two cases. Differential equation solvers built into computer algebra systems such as *Mathematica* or *Maple* have libraries of differential equation forms and their solutions. A differential equation for which a method of solution is known is likely to be in this library (or will be in later versions as the software becomes more comprehensive), and we can use the DES to find the solution even if we have no idea how the solution is obtained. If the method of solution is unknown to the mathematics community or if an elementary solution cannot be found, then DES can be used to find approximate numerical solutions using some form of numerical integration. This section overviews some of the important ideas involved in using differential equation solvers with examples from several current packages. It is not intended to teach the reader how to use any one package in particular. The *Mathematica* appendix demonstrates the use of *Mathematica* to solve some of the examples of this chapter. Readers interested in using *Mathematica* to solve differential equations are directed to the book by Ross [56]. We recommend that the reader obtain and have ready a differential equations solver. While reading this section, the reader's assignment is to figure out how to get one's chosen solver to do the things mentioned in this section.

Before beginning, here is a short sermon. Ecologist, applied mathematician, and modeler Anthony Starfield (who has had profound impact on the authors) related to the authors the story of a model he was developing that involved a differential equation that he could not solve. He asked a colleague who did not know either, but agreed to consider it. A year or so later the colleague came back saying that he still did not know how to solve the differential equation Starfield had asked about, but had proven that the solution of it was equivalent to finding the solution of a different problem. Starfield replied that the colleague should stop working on the problem since the differential equation he had originally been interested in had changed as the model developed and he no longer cared about its solution. The moral of the story is that, historically, differential equations have been linked to problems in physics, and the differential equation was, in essence, a law which remained unchanged unless someone like Einstein changed the whole paradigm under which one worked. Thus, once a differential equation was formulated, it made sense to spend years exploring it and its solutions. In modeling situations other than the hard sciences (or even in the hard sciences where the situation is complicated), however, models are more theory than law. Theories tend to change rapidly and their models with them. Often people engage in "what if" scenarios in which pieces of a model or theory are switched with other pieces. The result is that one may be interested in a particular differential equation for only a few weeks, or even a few minutes. In this book, we frequently speculate on how much effect refining a model has. In this context, spending years trying to solve a differential equation is unreasonable and a waste of time. Differential equation solvers, however, allow us to consider a differential equation for a model, analyze it in minutes, and decide, based upon this analysis, whether to accept it or reject it. Consequently, this section is probably the most important in this chapter. There is a downside though. In the sections on special differential equation forms, phase line, and phase plane analysis, we were able to develop a tremendous understanding of a model, usually with parameters unspecified except for being positive or negative or within some interval. With numerical analysis much of our understanding is lost. We can learn only that with particular parameter values, the model behaves in a particular way. Many runs must be performed with different parameter values in order to determine whether some observation is a general phenomena or an isolated occurrence. Finally, as we discuss below, certain occurrences fool a differential equations solver and give useless results.

5.4.1 The Answer Is Known, But Not By You

Knowing a closed-form solution to a differential equation is always preferable to a numerical approximate solution. The differential equation capabilities of a computer algebra system may give you a closed-form solution. Solvers are not perfect, however, and may not give an answer when one is known, or may not recognize the differential equation as being known if it were written in an unexpected form. Unless you have made it a priority to know how to solve as many differential equation types as possible, computer algebra systems are probably better at solving differential equation than you are. How do we know whether a computer algebra system will work or not? Try it. There is no penalty except for a few lost minutes if a solver tries to analytically solve a differential equation and cannot.

In Exercise 2 the reader is asked to compute the solution of the logistic differential equation. Regardless of whether the reader has done or can do this exercise, we can find the solution. Here we use the **DSolve** command of *Mathematica*.

Entering the following command in *Mathematica*

$$\textbf{DSolve}[\textbf{x}'[\textbf{t}] == \textbf{r}\,\textbf{x}[\textbf{t}](1 - \textbf{x}[\textbf{t}]/\textbf{K}), \textbf{x}[\textbf{t}], \textbf{t}]$$

results in the solution

$$\left\{ \left\{ x(t) \to \frac{e^{rt} K}{e^{rt} + K\,\mathrm{C}(1)} \right\}, \{x(t) \to 0\} \right\}$$

indicating that there are two solutions, one of which is the zero solution, the other the logistic curve. *Mathematica* expresses the constant of integration as C(1). This constant is found using the initial information. Notice that we can find C(1) either by hand or by using a computer algebra system. By using a computer algebra system, we either find the constant the way we would by hand, simply using the machine to do the dirty work, or the initial conditions are specified in the **DSolve** command. Suppose that the initial population was 1,000. We enter this as

$$\textbf{DSolve}[\textbf{x}'[\textbf{t}] == \textbf{rx}[\textbf{t}](1 - \textbf{x}[\textbf{t}]/\textbf{K}), \textbf{x}[0] == 1000, \textbf{x}[\textbf{t}], \textbf{t}]$$

with solution

$$\left\{\left\{x(t) \to \frac{e^{rt}K}{e^{rt} + \left(\frac{1}{1000} - \frac{1}{K}\right)K}\right\}\right\}.$$

In theory we should be able to use this method for multiple constants of integration, but in practice this does not always work and we may have to find other constants the traditional way with the machine's assistance.

There are two points to notice. First, the solutions are expressed in terms of arbitrary parameters r and K, so we maintain complete generality in our solutions. Second, the form in which a computer algebra system presents the answer may not be the form we are used to or even a form we consider to be "clean." We may need to do some algebra (which a computer algebra system might help with) to change the answer to a more appropriate looking form. Sometimes this simplification is a very complicated endeavor.

Next, we modify the model to include a varying carrying capacity. To be concrete, this time suppose that the growth rate r is 1.0 and the carrying capacity is initially 3,000, but drops by 100 each year due to habitat destruction. Now K becomes $K(t) = 3,000 - 100t$. Our differential equation becomes $\frac{dx}{dt} = 1.0x(1 - \frac{x}{3000-100t})$ and we assume $x(0) = 1,000$. Offhand this is not a differential equation form we recognize, perhaps, so we either try to solve it manually and search the differential equations literature, or see if it is in the library of a computer algebra system. Entering the code

$$\textbf{DSolve}[\{\textbf{x}'[\textbf{t}] == 1.0\textbf{x}[\textbf{t}](1 - \textbf{x}[\textbf{t}]/(3000 - 100\textbf{t})), \textbf{x}[0] == 1000\}, \textbf{x}[\textbf{t}], \textbf{t}]$$

results in the solution

$$\left\{\left\{x(t) \to \frac{e^{1.t}}{0.0006771 - 1.06865\ 10^{11}\ \text{ExpIntegralEi}(-30. + 1.0t)}\right\}\right\}.$$

By $ExpIntegralEi[z]$, *Mathematica* means the exponential integral special function $Ei(z)$, which average undergraduate mathematics students probably have not encountered unless they have taken courses in applied mathematics or partial differential equations. It is not an elementary function in the sense that it can be expressed as a finite number of algebraic operations on the basic functions from precalculus. It is an example of what are known as *special functions* which have definitions involving power series or integrals. It does not matter if we are familiar with Ei; *Mathematica* uses it like any other function. Figure 5.43 is a plot of this solution over a 10-year period. It looks much like we would expect. The population rises rapidly to the current carrying capacity and then drops linearly with the carrying capacity.

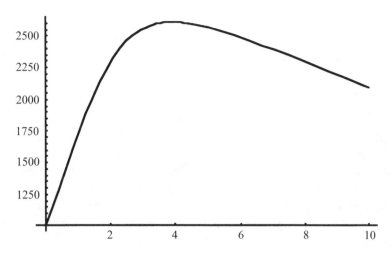

FIGURE 5.43 Decreasing Carrying Capacity.

The use of special functions in a problem raises the question of what it means to "solve" a differential equation. Transcendental functions are functions which cannot be expressed in terms of a finite number of algebraic operations on x. Two familiar examples are the natural log function and the sine function. Definitions of these functions are

$$\ln(x) = \int_1^x \frac{1}{t} dt \text{ for } x > 0$$

$$\sin(x) = \sum_0^\infty \frac{x^{2n+1}}{(2n+1)!}.$$

One is defined in terms of an integral, and the other in terms of a power series. Sometimes problems have solutions which cannot be expressed in terms of a finite number of algebraic operations on x, nor in terms of the familiar transcendental functions from calculus such as logs, exponents, the trigonometric functions, and hyperbolic trigonometric functions. These problems are said to be not solvable in terms of elementary functions. Some of these problems arise from modeling problems in physics and engineering and have been studied extensively. Names have been given to their solutions, the important properties of the solutions have been determined, and graphs of the solutions have been drawn. These named solution functions are called the special functions. Several examples are Erf, Ei, and the Bessel functions. Definitions for these (actually just one of the Bessel functions) are

$$\text{Erf}(x) = \frac{1}{\sqrt{2\pi}} \int_{-\infty}^x e^{\frac{-x^2}{2}} dx,$$

$$Ei(x) = \int_{-\infty}^x \frac{e^s}{s} ds,$$

$$J_0(x) = \sum_{k=0}^\infty \frac{(-1)^k x^{2k}}{(k!)^2 2^{2k}}.$$

Again some are defined in terms of integrals and some in terms of infinite sums. $\text{Erf}(x)$ is important since it is the integral of the normal probability density function discussed in Chapter 2. Notice that these are a special class of the transcendental functions.

Here is a dilemma. Suppose we have the differential equation

$$\frac{dx}{dt} = \frac{1}{\sqrt{2\pi}} e^{\frac{-t^2}{2}}.$$

One could claim the solution is $x(t) = \text{Erf}(t) + C$. This, however, is a circular argument. In particular, since $\text{Erf}(t)$ is defined to be the solution of this differential equation, it is not really solving it. On the other hand, once one analyzes the solution of this differential equation, determines its properties, and graphs it, then the solution becomes familiar and it is reasonable to give it a name (i.e., Erf). Then saying we have solved the differential equation by calling the name the solution has been given some meaning. It means we can graph the solution and enumerate some of its properties. The situation is similar to being asked to solve $\int_1^x \frac{1}{t} dt$ and writing the answer $\ln(x)$ and being satisfied with this answer. Its real importance comes in solving differential equations like

$$\frac{dx}{dt} = \frac{2}{\sqrt{2\pi}} e^{\frac{-t^2}{2}}$$

and calling the answer $2\,\text{Erf}(t)$. What we are saying is that the solution is two times the solution of this other problem which cannot be solved using elementary functions, but which is well understood.

One of the goals of this book is to instill in the reader the sense that models are not static objects etched in stone. Rather most of the models we present are building blocks to be used as beginnings for a model. Just now we used the building blocks of the logistic differential equation and a linear equation (declining carrying capacity) and combined them to model a specific situation. A differential equation solver helped us evaluate an integral that was beyond our knowledge. With this success under our belt, suppose that further analysis of our situation reveals that we should be using a θ-logistic with $\theta = 0.5$ instead of a logistic. Rerunning our *Mathematica* analysis with the appropriate elements changed gives us the following result.

DSolve[{x'[t] == 1.0x[t](1 − (x[t]/(3000 − 100t))^(1/2)), x[0] == 1000}, x[t], t]

$$\text{DSolve}[\{x'[t] == 1.0x[t](1 - \text{Sqrt}[\frac{x[t]}{3000 - 100t}]), x[0] == 1000\}, x[t], t]$$

This time our solver just spit back the input. This means that the form of this equation was not found in the solver's library. This is not a disaster. We can analyze it numerically.

5.4.2 Numerical Solvers When No One Can Solve the Differential Equation

Most differential equations one runs across cannot be solved. For these equations a numerical differential equation solver is a powerful tool to have in our modeling kit. Before looking at some of these solvers in action, we briefly discuss how they work.

Direction Fields Or Slope Fields. We are looking at equations of the form $\frac{dx}{dt} = f(x,t)$ so we know the derivative of the function everywhere and at all times. Recalling that

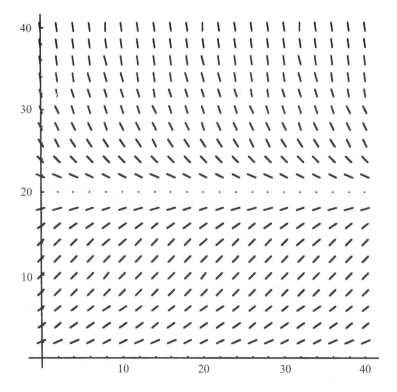

FIGURE 5.44 Direction Field for the Logistic Differential Equation.

the derivative is the slope of the tangent line to the solution curve, we know the slope of the tangent line at all points in space (1-dim) and time which is called the *slope field* or *direction field*. We draw a representation of the direction field by selecting a grid of points and drawing a short tangent line at each point. For example, the slope field of the logistic differential equation $\frac{dx}{dt} = rx\left(1 - x/K\right)$ is shown in Figure 5.44.

With a slope field and the knowledge that the slopes are slopes of a solution, we draw by hand an approximation to a solution starting at any point in time and space. For example, see Figure 5.45.

Some solvers use color or arrows to help determine which way the solution is moving at any time.

Several points should be obvious. 1) The finer the mesh of the grid for the representation of the slope field, the better the approximate solution curve which we are able to draw. This is the same as saying the tangent line is a good approximation to a curve close to the point of tangency. 2) A computer ought to be able to do this better than a human.

Euler's Method. Euler's method of numerically solving a differential equation of the form $\frac{dx}{dt} = f(x, t)$ is essentially an algorithm that formalizes the method of solution we found by drawing the slope field. This is the way the method works. Pick an initial value, say $x(0) = a$. Compute the tangent line at this point, namely $x_0(t) = a + f(a, 0)t$. This

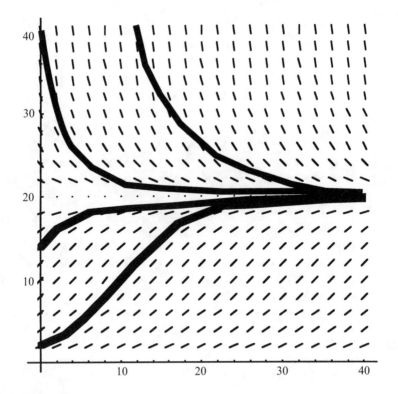

FIGURE 5.45 Hand-Drawn Trajectories on a Direction Field.

approximate solution is good for a "little bit." Euler's method is ignorant and requires the user to specify what a "little bit" is. This value (called the *step size*) h tells the algorithm how far to go in time before recomputing a tangent line. In essence we are choosing a mesh size although the mesh is uniform on the t axis, but not the x axis. So the next piece of the solution is $x_1(t) = x_0(h) + f(x_0(h), h)(t - h)$. The process is continued until the last desired t value is reached. The smaller h is, the smaller we expect the error to be. Usually when people use Euler's method, they do not write down the equations of all the lines; rather they give the series of points that the lines connect (linear interpolation). The formula for this algorithm is:

Specify (t_0, x_0)
$t(n) = t_0 + nh$
$x(n) = x(n-1) + f(x(n-1), t(n-1))h.$

Euler's method is easy to implement on a spreadsheet, calculator, or with any programming language geared towards mathematical computations.

There are many methods considered to be superior to Euler's method for accuracy and speed, but Euler's method has the advantage of simplicity, intuition, and clarity. It is also an important method to understand because when $h = 1$, it treats differential equations as difference equations. In particular, we can use a numerical differential equation solver to give numerical solutions to difference equations provided we specify that Euler's method be used and that a step size $h = 1$ is always used.

In Exercise 9, the reader investigates the effects of different step sizes on Euler's method. There one sees that with some curves large step sizes give slightly wrong numbers, but the correct basic behavior. With others, however, too large step sizes result in strange oscillations and other aberrant behaviors which disappear with smaller step sizes. As a rule of thumb when choosing step sizes, continue to decrease the step size until these oscillations smooth out.

Other Methods. There are numerous other methods. Two that are frequently encountered are Runge-Kutta 2nd order and Runge-Kutta 4th order. These work in manners similar to Euler's method except that instead of using first-order approximations (lines) they use second-order (parabolas) and fourth order (quartics) curves, although they do it in ways which avoid the computation of higher-order derivatives. We do not discuss how to implement these methods, but just understand that they generally give more accurate results than Euler's method. They are easy to program and hence are commonly used. More information on algorithms for numerical solving can be found in most books on differential equations or numerical analysis. Table 5.4 compares the results of Euler, second-order and fourth order Runge Kutta (all with step size 0.25) with the solution of the logistic differential equation $\frac{dx}{dt} = 1.2x(1 - \frac{x}{20})$ with initial value $x(0) = 1$.

Many differential equation solvers let the user specify the desired method and step size. Others, such as *Mathematica*, do not, which gives the solver great flexibility. For example, it may choose large step sizes when the solution is fairly straightforward and small step sizes when the solution changes rapidly. It may switch between one method and another. It may use state-of-the-art algorithms of which the reader may be unaware, unless one regularly reads the appropriate research journals. In other words, by taking control of method away from the user, exceptional results can be achieved. As an example, consider the differential equation $\frac{dx}{dt} = -e^x$. Figure 5.46 shows a graph of the solution. Figure 5.47 shows an approximate solution generated by a fourth-order Runge-Kutta with $h = 0.01$. This solution is clearly unusable. Figure 5.48 shows an approximate solution again using a fourth order Runge-Kutta with $h = 0.001$. While this is an improvement over the previous solution, it also requires the use of 5,000 steps. While decreasing the step size further would improve the solution, one can run into problems with machine precision and round-off errors and with machine speed and capacity. Ideally one would continue to reduce the step size until the solution curves converge, but this is often not practical. Finally Figure 5.49 shows the approximate solution generated by *Mathematica*'s numerical differential equation solver (**NDSolve**) which is a "black box" solver where the algorithm used is unknown to the user. There are some parameters that the user can set, but this graph was created using the default parameter values. Clearly the results favor the complicated package even though the user may feel a loss of control in the solution process. In addition to feeling out of control, if one is interested in analyzing and bounding the error in the approximate solution, one needs to know the solution algorithm.

We need a fine print statement at this point so that we do not convey a false impression. A fair statement is that "black box" solvers (and even differential equation solvers) are excellent exploratory tools, but research-grade final results require knowledge of the

Comparison of several numerical differential equation solving techniques

Time	Euler	Runge-Kutta 2	Runge-Kutta 4	Solution
0.00	1.00	1.00	1.00	1.00
0.25	1.28	1.32	1.33	1.33
0.50	1.65	1.74	1.75	1.75
0.75	2.10	2.27	2.29	2.29
1.00	2.66	2.95	2.97	2.97
1.25	3.35	3.78	3.82	3.82
1.50	4.19	4.77	4.83	4.83
1.75	5.19	5.94	6.01	6.01
2.00	6.34	7.26	7.34	7.34
2.25	7.64	8.69	8.78	8.78
2.50	9.05	10.18	10.28	10.28
2.75	10.54	11.65	11.76	11.76
3.00	12.04	13.06	13.16	13.17
3.25	13.47	14.34	14.44	14.44
3.50	14.79	15.47	15.56	15.57
3.75	15.95	16.43	16.51	16.51
4.00	16.92	17.22	17.30	17.30
4.25	17.70	17.85	17.92	17.92
4.50	18.31	18.36	18.42	18.42
4.75	18.77	18.75	18.80	18.80
5.00	19.12	19.06	19.10	19.10
5.25	19.37	19.29	19.33	19.33
5.50	19.55	19.47	19.50	19.50
5.75	19.69	19.60	19.62	19.62
6.00	19.78	19.70	19.72	19.72
6.25	19.84	19.78	19.79	19.79
6.50	19.89	19.83	19.85	19.85
6.75	19.92	19.87	19.89	19.89
7.00	19.95	19.91	19.91	19.91
7.25	19.96	19.93	19.94	19.94
7.50	19.97	19.95	19.95	19.95
7.75	19.98	19.96	19.97	19.97
8.00	19.99	19.97	19.97	19.97
8.25	19.99	19.98	19.98	19.98
8.50	19.99	19.98	19.99	19.99
8.75	20.00	19.99	19.99	19.99
9.00	20.00	19.99	19.99	19.99
9.25	20.00	19.99	19.99	19.99
9.50	20.00	20.00	20.00	20.00
9.75	20.00	20.00	20.00	20.00

TABLE 5.4

algorithm used and detailed knowledge of the algorithm's implementation. Again the purpose of the model is key. In this book, we consider our models to be exploratory and avoid the issues of numerical analysis which are essential for many applications.

Mathematica uses the command **NDSolve** to solve differential equations numerically. As just discussed, it is a "black box" command in the sense that the user has no idea what

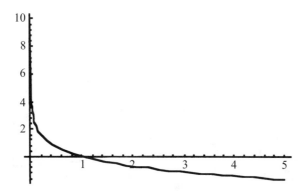

FIGURE 5.46 Comparison: Exact.

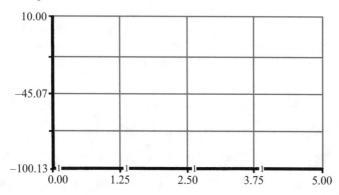

FIGURE 5.47 Runge-Kutta 4: $h = 0.01$.

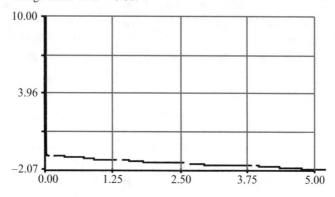

FIGURE 5.48 Runge-Kutta 4: $h = 0.001$.

method is being used, but is confident that the results are better than using a standard method alone. One can, however, set parameters determining the working precision and the precision and accuracy of the answer.

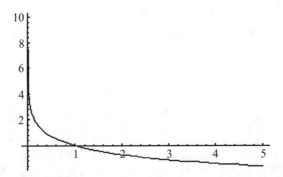

FIGURE 5.49 *Mathematica*'s **NDSolve** command.

For our first example, we use **NDSolve** to solve the logistic differential equation $\frac{dx}{dt} = rx(1 - \frac{x}{K})$ when $r = 1.0$, $K = 3{,}000$, and $x(0) = 1{,}000$. Since we can also solve this differential equation exactly, we compare the approximation to the exact solution.

The command

NDSolve[{x'[t] == 1.0x[t](1 − x[t]/3000), x[0] == 1000}, x[t], {t, 0, 5}]

numerically solves this differential equation, and the solution is

$$\{\{x[t] -> \text{InterpolatingFunction}[0., 5., <>][t]\}\}.$$

Notice that the answer does not look like any standard functions. *Mathematica* gives the answer as a function interpolated through a list of points. The list is suppressed and abbreviated with the $<>$ symbol. While this form of the answer is of dubious value, we can use the interpolation function to do the various things we do with a closed-form solution function. (For details on interpolation see Section 4.9 of Chapter 4.) In the table below, we compare the approximate solution with the exact (closed-form) solution for a period of five years.

t	*Exact*	*Approximate*
0.0	1000.	1000.
0.5	1355.588285632818132	1355.588300824846135
1.0	1728.35065429748733	1728.352961138969948
1.5	2074.315362108682793	2074.315990251557731
2.0	2360.958126484795497	2360.957699634676264
2.5	2576.943236032769598	2576.942561955190417
3.0	2728.328995538225606	2728.329171434152039
3.5	2829.135048381036479	2829.135794499951915
4.0	2893.989467915711705	2893.988458211615755
4.5	2934.79475055134679	2934.794165209813031
5.0	2960.109873126803976	2960.109527758743317

Notice that the approximate values are sufficiently close to the true values that a graph of each would be identical on any reasonable scale. Also, if we are modeling animal populations, errors in the fractional animals are irrelevant.

5 Continuous Models

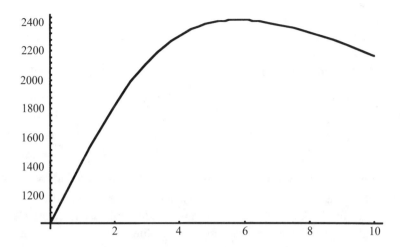

FIGURE 5.50 Numerical Solution of θ-logistic with Declining Carrying Capacity.

Earlier in this section, we tried using *DSolve* on the differential equation

$$\frac{dx}{dt} = 1.0x \left(1 - \left(\frac{x}{3000 - 100t}\right)^{\frac{1}{2}}\right)$$

with $x(0) = 1,000$. *Mathematica* was unable to solve this differential equation exactly. Now let us use **NDSolve**.

The command

NDSolve[x$'$[t] == 1.0x[t](1−(x[t]/(3000−100t))^(1/2)), x[0] == 1000, x[t], {t, 0, 10}]

returns the solution

$$\{\{x[t]-> \text{InterpolatingFunction}[\{0., 10.\}, <>][t]\}\}.$$

Again the solution is an interpolating function, but plotting this function (Figure 5.50) makes it understandable.

In conclusion, it is always better to have a closed-form solution. Its results provide theoretical insights, and there are no problems with errors associated with approximate methods. Unfortunately most of the differential equations one constructs cannot be solved. Numerical solvers allow us to proceed with our analysis, but are limited in that one must explicitly express all parameters as numbers and give all necessary initial conditions, and there are always errors in making approximations.

5.5 For Further Reading

The following pair of papers (appearing sequentially in the journal) present two sides of a debate over the validity of several classical models. We make reading and discussing them part of our course requirements:

Charles A.S. Hall, An Assessment of Several of the Historically Most Influential Theoretical Models Used in Ecology And of the Data Provided in Their Support, *Ecological Modelling*, 43,1988, 5–31.

Hal Caswell, Theory and Models in Ecology, *Ecological Modelling*, 43,1988, 33–44.

Those interested in using *Mathematica* when working with differential equations should investigate:

Clay C. Ross, *Differential Equations: An Introduction with Mathematica*, Springer-Verlag, New York, 1995.

The following books are good sources of modeling with differential equations:

Michael Olinick, *An Introduction to Mathematical Models in the Social and Life Sciences*, Addison-Wesley Publishing Company, Reading, Massachusetts, 1978.
A very good presentation of models used in the life sciences. It predates desktop computers and, consequently, has some interesting methods of avoiding highly numerical computations such as plotting certain curves or fitting special curves to data. Deals with probabilistic aspects of modeling, but not statistical aspects.

Colin Clark, *Mathematical Bioeconomics: The Optimal Management of Renewable Resources*, John Wiley & Sons, Inc., New York, Second Edition, 1990.
The second edition of a highly influential classic. Incorporates economics into biological models. Our discussion of harvesting models roughly follows the first chapter of this book.

Leah Edelstein-Keshet, *Mathematical Models in Biology*, McGraw Hill, Inc., Birkhäuser Mathematics Series, New York,1988.
As discussed in the For Further Reading section of Chapter 1, this is an outstanding book for anyone interested in applications of difference and differential equations to biology. The book also discusses partial differential equation models. It is highly recommended.

A.M. Starfield and A.L. Bleloch, *Building Models for Conservation and Wildlife Management*, Burgess International Group, Inc., Edina, MN, Second Edition, 1991.
A nuts and bolts approach to model building that stresses realism over mathematical elegance. Modeling focuses on African game parks and savanna.

The next book is a somewhat controversial calculus book which teaches calculus using *Mathematica*. The book contains some modeling examples and projects which are gems:

Bill Davis, Horacio Porta, and Jerry Uhl, *Calculus&Mathematica*, Addison-Wesley, Reading, Massachusetts, 1994.

The remaining books are differential equations texts:

W.E. Boyce and R.C. DiPrima, *Elementary Differential Equations*, John Wiley & Sons, Inc., New York, Fourth Edition, 1986.
The classic differential equations text which was used by several generations of science and engineering students.

Martha L. Abell and James P. Braselton, *Modern Differential Equations: Theory, Applications, Technology*, Saunders College Publishing, Fort Worth, 1996.

A good general purpose differential equations text.

J.H. Hubbard and B.W. West, *Differential Equations: A Dynamical Systems Approach, Part I, One Dimensional Equations*, Springer-Verlag, New York, 1990.
This differential equations text stresses visualization and qualitative analysis.

5.6 Exercises

1. Use the separation of variables technique to solve
$$\frac{dx}{dt} = k.$$

2. Use the separation of variables technique to solve the logistic differential equation
$$\frac{dx}{dt} = rx\left(1 - \frac{x}{K}\right).$$

3. Use separation of variables to solve
$$\frac{dx}{dt} = \sin(x)$$
and, using a graphing program, compare this solution with the solution of the logistic equation
$$\frac{dx}{dt} = rx(1 - \frac{x}{\pi})$$
for several different values of r and $x(0) = 1$.

4. Show that
$$\frac{dx}{dt} = ax + by$$
$$\frac{dy}{dt} = cx + dy$$
can be rewritten as
$$\frac{d^2x}{dt^2} - \beta\frac{dx}{dt} + \gamma x = 0.$$

5. Find values for a, b, c, and d so that the fixed point (the origin) of
$$\frac{dx}{dt} = ax + by$$
$$\frac{dy}{dt} = cx + dy$$
is a
a. source node
b. sink node
c. spiral source
d. spiral sink
e. saddle point.

6. Consider the system of equations

$$\frac{dx}{dt} = y^2 - x$$
$$\frac{dy}{dt} = x^2 - 8y.$$

a. Find the fixed points of this system (there are two).
b. Use Jacobian analysis to determine the nature of the fixed points (i.e., source, sink, saddle, spiral, etc.)
c. Graph the nullclines on a phase plane.
d. Use phase plane analysis to mark the flow arrows. Do the directions of flow correspond to the results of **b**?
e. Use a differential equation solver capable of drawing phase plane plots to generate several trajectories (at least one from each region formed by the nullclines). How do these trajectories compare with what you expected from the previous analysis by hand?

7. Consider the system of equations

$$\frac{dx}{dt} = \sin(x) - y$$
$$\frac{dy}{dt} = y - \frac{1}{2}.$$

a. Find the fixed points and determine their nature using Jacobian analysis.
b. Draw the nullclines and flow arrows on a phase plane and compare with the behaviors suggested by the Jacobian analysis.

8. For a final project, a student of ours was attempting to build a model for an ecosystem with a number of interacting variables changing over time. He set the equations as a system of first-order homogeneous linear equations with constant coefficients. Without consulting us, he was arbitrarily assigning values to the coefficients trying to find a set of coefficients that would result in a stable ecosystem.

Try to help him. Set up a software model that solves (numerically or symbolically) a system of three first-order linear homogeneous differential equations with constant coefficients. Arbitrarily (and without mathematical analysis) assign coefficients and initial conditions to try to get a system with solution curves that level out and are not all zero. Keep trying. It can be done.

It can be done, but probably not by arbitrarily picking numbers. Try again, this time carefully choosing numbers and, perhaps, referring to some of the discussions on these types of systems in the text.

If this student wanted a stable model, what advice would you give him? Also feel free to advise about randomly choosing parameters which are supposed to have some meaning in an ecosystem.

9. Use Euler's method (either directly or with a solver that uses Euler's method) to solve numerically the following two logistic differential equations for time t between 0 and 10. Assume that $x(0) = 100$. Solve both using the following h values: 1, 0.5, 0.25, and 0.1. Compare the solutions of the first equation for the different h values. Explain what is happening as h gets small. Repeat the comparison for the second equation. Compare the solutions of the two equations to each other for the different h values. Explain why two differential equations which are identical except for the parameter values exhibit different behaviors while h is in the process of "getting small."

a. $\dfrac{dx}{dt} = 1.7x\left(1 - \dfrac{x}{1000}\right)$

b. $\dfrac{dx}{dt} = 2.7x\left(1 - \dfrac{x}{1000}\right)$

10. Recall DeMoivre's Theorem from trigonometry. It states that
$$e^{i\theta} = \cos\theta + i\sin\theta.$$

a. Show that $e^{-i\theta} = \cos\theta - i\sin\theta$.

b. Use DeMoivre's Theorem and the result of part **a** to show that given C_1 and C_2, there are D_1 and D_2 such that
$$C_1 e^{(a+ib)t} + C_2 e^{(a-ib)t} = D_1 e^{at}\cos bt + D_2 e^{at}\sin bt.$$

5.7 Projects

Project 5.1. The Michaelis-Menton Equation I

The Michaelis-Menton Equation is useful in biological kinetics. As a pharmacokinetic (rate of processing drugs) law, it provides the basis for the equations used in the Blood Alcohol Model (in Project 1.3 in Chapter 1 and in Project 5.5 below) and in the Drugs in the Body Model in the text of this chapter (Section 5.2.4.) This project explores the properties of the equation, while Project 5.6 derives the Michaelis-Menton equation for a particular setting.

In the setting of pharmacokinetics, the Michaelis-Menton equation governs the rate at which the body processes a drug. In this application, the form of the equation is

$$\dfrac{dx}{dt} = \dfrac{-Kx}{A+x} \tag{109}$$

where x is the concentration of a drug in the body at time t, and K and A are positive constants. This differential equation, is frequently approximated by simpler differential equations and this is explored in parts **1** and **2** below.

1. For many drugs (cocaine for example) the A in equation (109) is very much larger than $x(t)$ at all times. An example in Davis, Porta, and Uhl's *Calculus&Mathematica* [14] gives values for cocaine use of $A = 6$ and $x(0) = 0.0025$. Using the idea that, if $A >> x(t)$ then $A + x(t) \approx A$, rewrite the Michaelis-Menton equation and solve it analytically in terms of $x(0)$. Notice that this is the equation discussed in Section 3.5 on drugs in the body.

2. For other drugs (alcohol for example) $x(t) >> A$. In *Calculus&Mathematica*, values for alcohol are given as $A = 0.005$ and $x(0) = 0.025$. Using the idea that if $A << x(t)$ then $A + x(t) \approx x(t)$, rewrite and analytically solve the Michaelis-Menton equation in terms of $x(0)$. Notice that with this simplification the processing rate of alcohol should be approximately constant, which is what we assumed in our blood alcohol model.

3. Assuming that $K = 1$ and using the A and $x(0)$ values in parts **1** and **2**, graph solution curves for these two approximations.

4. Of course, these approximations are not always valid, and we may need to use (109) directly (if A is of a similar magnitude to $x(t)$). Solve the Michaelis-Menton equation analytically in terms of $x(0)$ (an implicit solution is fine). Letting $K = 1$, $A = 0.025$, and $x(0) = 0.025$, plot the solution curve. Compare the curves in part **3**.

Project 5.2. Richardson's Arms Race Model

In this project, we consider several models for an arms race. Consider two economically competing nations which we call Purple and Green. Both nations desire peace and hope to avoid war, but they are not pacifistic. They will not go out of their way to launch aggression, but they will not sit idly by if their country is attacked. They believe in self-defense and will fight to protect their nation and their way of life. Both nations feel that the maintenance of a large army and the stockpiling of weapons are purely "defensive" gestures when they do it, but at least somewhat "offensive" when the other side does it.

Since the two nations are in competition, there is an underlying sense of "mutual fear." The more one nation arms, the more the other nation is spurred to arm.

1. A Mutual Fear Model. Let $x(t)$ and $y(t)$ represent the yearly rates of armament expenditures of the two nations in some standardized monetary unit. To develop a model of mutual fear, we assume that each country adjusts the rate of increase or decrease of its armaments in response to the level of the other's. The simplest assumption is that each nation's rate is directly proportional to the expenditure of the other nation. That is,

$$\frac{dx}{dt} = ay \tag{110}$$

$$\frac{dy}{dt} = bx \tag{111}$$

where a and b are positive constants. Suppose the initial (at some arbitrary time $t = 0$) armament expenditures of the two nations are x_0 and y_0 respectively.

a. Use the chain rule

$$\frac{dy}{dx} = \frac{dy/dt}{dx/dt}$$

to obtain

$$\frac{dy}{dx} = \frac{bx}{ay}. \tag{112}$$

Solve equation (112) using the initial conditions. Show that your answer can be written as the equation for a hyperbola:

$$y^2 - \left(\frac{b}{a}\right) x^2 = y_0^2 - \left(\frac{b}{a}\right) x_0^2. \tag{113}$$

b. Suppose at some time t_s, the point $(x(t_s), y(t_s))$ lies on the upper branch of the hyperbola, show that $y(t_s) > \sqrt{b/a}\, x(t_s)$. Using this inequality and equation (110), show

5 Continuous Models

that
$$\frac{dx}{dt} > \sqrt{ab}\, x(t).$$

Thus, as the x-coordinate increases, so does the velocity of the horizontal motion. Using this observation, show that
$$\lim_{t\to\infty} x(t) = \lim_{t\to\infty} y(t) = \infty.$$

The mutual fear model predicts that as time increases, the armaments of both nations continue to increase without bound. Even if we assume that countries do not have infinite resources and hence this model cannot continue forever, in the short term it suggests a runaway arms race.

c. Using a differential equation solver and several sets of arbitrarily chosen values of a and b and initial values x_0 and y_0, verify numerically the results obtained in part **b**. Graphically display the results as trajectories on a phase plane.

2. The Richardson Model. We now present a refinement of the mutual fear model. The mutual fear model produced a "runaway" arms race with unlimited expenditures. To prevent unlimited expenditures, we assume that excessive armament expenditures present a drag on the nation's economy so that the actual level of expenditure reduces the rate of change of the expenditure. The simplest way to model this is to assume that the rate of change for a nation is directly and negatively proportional to its own expenditure. This refines the mutual fear model and gives equations (110) and (111), as

$$\frac{dx}{dt} = ay - mx \tag{114}$$

$$\frac{dy}{dt} = bx - ny \tag{115}$$

where a, b, m, and n are positive constants.

Before proceeding with a mathematical analysis of this model, we introduce a further refinement. This refinement models any underlying grievances of each country toward the other. To model this, we introduce two additional constant terms, r and s, to the equations (114) and (115) and obtain

$$\frac{dx}{dt} = ay - mx + r \tag{116}$$

$$\frac{dy}{dt} = bx - ny + s. \tag{117}$$

A positive value of r or s indicates that there is a grievance of one country toward the other which causes an increase in the rate of arms expenditures. If r or s is negative, then there is an underlying feeling of good will, so there is a decrease in the rate of arms expenditures.

Equations (116) and (117) are called the *Richardson's Arms Race model* in honor of Lewis F. Richardson [54], who considered this model in 1939 for the combatants of World War I.

a. We use phase plane analysis to investigate the behavior of the Richardson's arms race model. First set the left-hand side of equations (116) and (117) to zero to determine the isoclines of the Richardson's model. Observe that these isoclines are straight lines. Show that, if $mn - ab \neq 0$, these isoclines intersect in the fixed point $\overline{x} = (as + nr)/(mn - ab)$, $\overline{y} = (ms + br)/(mn - ab)$. The behavior of the arms race depends upon the values of a, b, m, n, r, and s.

b. Mutual Grievances. We first investigate the case where each side has a permanent underlying grievance toward the other side. In this case the parameters r and s are positive. Show that, depending upon the sign of $mn - ab$, the fixed point $(\overline{x}, \overline{y})$ lies in either the first or third quadrant. Using phase plane analysis, show that if the fixed point lies in the first quadrant, the arms race stabilizes at the fixed point $(\overline{x}, \overline{y})$. Show that if the fixed point lies in the third quadrant, the Richardson's model represents a "runaway" arms race in the sense that $(\overline{x}, \overline{y})$ is an unstable fixed point.

c. The Effects of Good Will. Feelings of good will are represented by one of the grievance terms' r or s being negative. If both r and s are negative, show that the fixed point is unstable. Further, show that if $mn - ab$ is positive, the arms race results in total disarmament of both sides. If $mn - ab$ is negative, show that there will be either disarmament or a runaway arms race, depending upon the initial level of expenditure.

d. Pick representative values of a, b, m, n, r, and s for both the mutual grievances and the goodwill scenarios. Numerically analyze the differential equations using initial conditions from the regions formed by the nullclines. Graphically represent the trajectories on a phase plane plot and verify the numerical results.

3. Open-Ended Extensions of the Richardson Model. This part offers some suggested further extensions of the Richardson model. Some of these extensions are much more involved than the work in parts **1** and **2**.

There are several directions one could choose to extend the Richardson model. Here are two. The first extension concerns a modification of the mutual fear term discussed in part **1**. If we assume that there is an inherent limit to the amount a nation can spend on armaments and let K_p be the maximum expenditure of the Purple nation and K_g be the maximum expenditure of the Green nation, then the proportionality constant can be replaced by an expenditure-dependent rate. A simple way to do this is to replace exponential factors with logistic factors. Thus our arms race model becomes

$$\frac{dx}{dt} = a\left(1 - \frac{x}{K_p}\right)y - mx + r \tag{118}$$

$$\frac{dy}{dt} = b\left(1 - \frac{y}{K_g}\right)x - ny + s. \tag{119}$$

Examine the stability of this "logistic" Richardson's model by performing an analysis similar to part **2**.

A second extension considers three mutually fearful nations. Each nation is spurred to arm by the expenditures of the other two. Build a Richardson's model for three nations similar to equations (116) and (117). Examine the stability of this model by performing an analysis similar to part **2**. This three-nation model can be further modified if two of the nations are close allies who are not threatened by the arms build up of each other but are threatened by the expenditures of the third.

5 Continuous Models

Project 5.3. The θ-logistic Equation

This project explores the θ-logistic equation.

1. Figure 5.22 shows plots of dN/dt versus N for a θ-logistic equation for several values of θ with $r = 0.1$ and $K = 3$. First use plotting software to reproduce this graph. Next plot the solution curves $N(t)$ for these values of θ assuming $N(0) = 1$. Mark inflection points on each. Describe in words how one would expect the population growth of a species with $\theta = 5$ to differ from a species with $\theta = 0.5$. How would both of these differ from a population with $\theta = 1$? Vertebrates typically have $\theta \geq 1$ while invertebrates typically have $\theta \leq 1$.

2. Suppose we modeled a population using a logistic equation with $r = 0.1$ and $K = 3$ when, in fact, a θ-logistic with $\theta = 5$ would have been more appropriate. Further suppose we were interested in constant harvesting. Perform phase line analysis for the logistic and θ-logistic and report on the effects that using the wrong model has on the results. You should be sure to include discussions of equilibriums and maximum sustainable yields. Comment on management consequences from using the wrong model. Repeat this problem assuming that $\theta = 0.5$ would have been more appropriate.

3. Repeat part **2** using a $H(x) = hx$ harvesting function instead of a constant harvesting function. Include Yield-Effort analysis.

4. If a θ-logistic equation were used in the spruce-budworm model instead of the logistic equation, how would the conclusions change? Note that since parameters are unspecified in this model, the question is about qualitative behavior, not quantitative behavior.

Project 5.4. Reservoir Models

Reservoirs are lakes that hold water for various urban and agricultural uses including flood control. They are naturally modeled using compartmental analysis. Here we describe scenarios for two reservoirs and ask you to track their volumes.

1. Suppose that during a storm, water is flowing into a reservoir at a rate of

$$Q = 950(1 - e^{-kt})$$

where Q is the flow in cubic feet per second, k is the runoff rate which we take to be 1.5 units per hour, and t is the time in hours from the beginning of the storm. Assume the initial reservoir volume is two billion cubic feet and that the flow of water out of the reservoir is 750 cubic feet per hour. Further assume that for purposes of this analysis, all other inflows are negligible and that the reservoir will not overflow from this rainfall. Use a differential equation to model the change in the volume of water in the reservoir as a function of time (be consistent with the time units). Solve this differential equation. Plot the volume as a function of time over the first 12 hours after the storm. Find the times when the volume is at a minimum and at a maximum over the first 12 hours (endpoints are allowable times).

2. During a drought the only water entering a reservoir is due to feeder streams and rivers. Assume that the sum of all such inputs is 200 cubic feet per second. Further assume that the sum of all outflows is initially 250 cubic feet per second, but increases continuously by one cubic foot per second each day due to evaporation and increased

water demand. Finally assume that the reservoir initially holds five billion cubic feet of water. First convert all time units to days. Next write a differential equation describing the flow of water into and out of this lake. Solve the differential equation and plot the solution. Assuming the parameters remain constant right up until the last drop of water, how many days will it take for the reservoir to empty to half of its initial level? How long will it take until 1 cubic foot of water remains?

These models ideas are from [58] *Environmental Modeling,* by Jerald L. Schnoor, Copyright ©1996 John Wiley & Sons, Inc. Reprinted by permission of John Wiley & Sons, Inc.

Project 5.5. Blood Alcohol Model II

This project advances the Blood Alcohol Model in Chapter 1 (Project 1.3) to a continuous setting, and the reader should refer there for data and question statements.

Re-do the blood alcohol project, only this time treat the model as a continuous time model and analytically find the solutions instead of using computer software to find the solutions. This is similar to the drug-processing model in the text.

Project 5.6. The Michaelis-Menton Equation II

This project derives the Michaelis-Menton Equation (109) from the chemical/cellular mechanisms involved. This project is of a very different nature from others in the book in that it deals with modeling at microscopic level and with processes unfamiliar to the nonbiologist. We have, for the most part, stayed away from models of this type because the general reader has little intuition into what is happening and is merely taking our word for every assumption. We include this model since the Michaelis-Menton Equation is so important and is used in several other projects and in the text. It also introduces the reader to molecular modeling which is very important in medical research. Since we assume no chemistry or molecular biology knowledge on the reader's part, this project consists of verifying the mathematical steps in the model.

In bacterial growth models, when the nutrient concentration is low, the bacterial growth rate is proportional to the concentration; when the nutrient level is high, the growth rate is constant. The reason is that nutrients must pass through the cell wall of a bacterium through a chemical process (called transport) involving receptor sites. Receptors are a certain type of molecule on the cellular membrane which bind with the nutrient molecules on one side of the cell wall and break down on the other side of the cell wall to let nutrient through. This process is shown in Figure 5.51. When the nutrient concentration is low, there are ample receptors so the rate of processing is proportional to the amount of nutrient. When the nutrient concentration is high, the receptors are working at maximum capacity so increasing the nutrient has no additional effect. If n is the concentration of the nutrient, then the bacteria growth rate as a function of concentration $K(n)$ will be shown below to be expressed by the equation

$$K(n) = \frac{Kn}{A+n}$$

where K and A are positive constants. The reader should recognize this as a form of equation (109).

5 Continuous Models

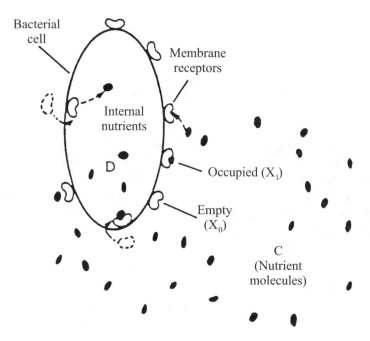

FIGURE 5.51 Nutrients Passing through a Cell Wall Using Receptors Bound to the Cell Wall. Figure reproduced from *Mathematical Models in Biology* by Edelstein-Keshet, 1988, McGraw-Hill, Inc. Used with Permission of The McGraw-Hill Companies.

In this project, we will deal with the bacteria-nutrient model as outlined, but notice that if the word "bacteria" were replaced by "liver cell" and the word "nutrient" were replaced by "drug," then this discussion would also apply to the pharmacokinetic use of the Michaelis-Menton equation.

As mentioned, the passage of nutrient molecules into a cell is a complex process. Let N denote a nutrient molecule. The nutrient molecule links with an unoccupied receptor molecule X_0 on the cell membrane to form an occupied receptor molecular complex X_1. This occupied receptor molecular complex either breaks back down into the original constituents N and X_0 (the transport fails) or produces a product P on the other side of the cell membrane and an unoccupied receptor molecule X_0 (the transport is successful).

This process is represented by the reaction equations

$$N + X_0 \underset{k_{-1}}{\overset{k_1}{\rightleftarrows}} X_1 \qquad (120)$$

$$X_1 \overset{k_2}{\to} P + X_0 \qquad (121)$$

where the k_i are constants of proportionality involved in the rates of reactions described below.

We let the symbols n, x_0, x_1, and p denote the concentrations of N, X_0, X_1, and P. Two empirical laws are used in setting up compartmental diagrams and/or differential equations. First, if a single reactant is involved (such as $X_1 \to P + X_0$), then the rate of

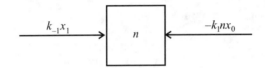

FIGURE 5.52 Nutrient Compartmental Diagram.

the reaction is proportional to the concentration of the reactant. Second, if two reactants are involved (as in $N + X_0 \rightarrow X_1$), then the rate of the reaction is proportional to the product of the concentrations. These laws are special cases of a more general law called the *law of mass action* which was mentioned in the predator-prey discussion. A compartmental diagram for the nutrient is shown in Figure 5.52, and from it the following differential equation can be formulated:

$$\frac{dn}{dt} = -k_1 n x_0 + k_{-1} x_1.$$

1. Using the reaction equations, perform compartmental analysis to derive the following set of differential equations.

$$\frac{dn}{dt} = -k_1 n x_0 + k_{-1} x_1 \qquad (122)$$

$$\frac{dx_0}{dt} = -k_1 n x_0 + k_{-1} x_1 + k_2 x_1 \qquad (123)$$

$$\frac{dx_1}{dt} = k_1 n x_0 - k_{-1} x_1 - k_2 x_1 \qquad (124)$$

$$\frac{dp}{dt} = k_2 x_1 \qquad (125)$$

2. Add equations (123) and (124) together to obtain

$$\frac{dx_0}{dt} + \frac{dx_1}{dt} = 0.$$

Explain why this implies that $x_0 + x_1$ is always a constant. We denote this constant r which gives the equation

$$x_0 + x_1 = r. \qquad (126)$$

This is a receptor or enzyme conservation law.

3. Use equation (126) to eliminate x_0 from the system of differential equations. (Equivalently we could eliminate x_1.) The result should be

$$\frac{dn}{dt} = -k_1 r n + (k_{-1} + k_1 n) x_1 \qquad (127)$$

$$\frac{dx_1}{dt} = k_1 r n - (k_{-1} + k_2 + k_1 n) x_1 \qquad (128)$$

$$\frac{dp}{dt} = k_2 x_1. \qquad (129)$$

4. Frequently the number of nutrient molecules is much larger than the number of receptor molecules, and one assumes that the receptors are working at capacity which

is a steady state as long as the conditions remain the same. This is formalized as the *quasi-steady-state* hypothesis which states that

$$\frac{dx_1}{dt} \approx 0,$$

which, in this case, results in the condition

$$k_1 rn - (k_{-1} + k_2 + k_1 n)x_1 = 0. \tag{130}$$

Show that this assumption results in the equation

$$\frac{dn}{dt} = -\frac{Kn}{A+n} \tag{131}$$

where $K = k_2 r$ and $A = \frac{k_{-1}+k_2}{k_1}$. Note that this is the desired Michaelis-Menton equation.

This project is from *Mathematical Models in Biology* by Edelstein-Keshet, 1988, McGraw-Hill, Inc. [15]. Used with permission of The McGraw-Hill Companies. For those interested in molecular modeling this is a very good textbook. Another related treatment is found in *Modeling dynamic phenomena in molecular and cellular biology* by Lee A. Segel [59].

Project 5.7. A Finance Model

In Project 1.4, we saw that P_0 invested in a bank account earning a simple annual interest rate of r compounded n times per year for t years is worth $P(t) = P_0(1+\frac{r}{n})^{nt}$. In many precalculus and calculus books, it is shown that, as $n \to \infty$, this equation becomes $P(t) = P_0 e^{rt}$, which is the solution of $\frac{dP}{dt} = rP$. We discussed in the text that a discrete exponential model (not base e) can be expressed as a continuous model (base e) and vice versa, which matches exactly at times corresponding to the whole numbers of time steps. Thus, when convenient to do so, a discrete exponential situation can be modeled with differential equations which give accurate answers where the domains make sense.

1. If a bank account has a simple annual rate of r percent compounded monthly, find the appropriate interest rate to use for a continuous model.

2. People with sizable investments often use interest income to live on either partially or wholly. Suppose that an account earns at an annual rate of r percent compounded continuously and a person is drawing income of H dollars per year withdrawn continuously (impossible, but a modeling assumption). Use phase line analysis to analyze the behavior of the account. Discuss the meaning of any equilibrium points and their stability. If $r = 10\%$ (a reasonable rate for long-term stock investments), and $H = \$10,000$, how long should an initial investment of \$50,000 be left untouched so that when withdrawals begin the capital is not depleted?

3. In general, suppose that $F(t)$ represents the net flow of money into and out of an account over time, not counting the income due to interest. Verify that the governing differential equation has the form $\frac{dP}{dt} = rP + F(t)$. Express the solution of this differential equation in terms of an integral. (Hint: This is a linear differential equation with a forcing term and can be treated as a linear differential equation with nonconstant coefficients.)

4. Create several (at least three) scenarios putting values to r and $F(t)$ in part **3** and solve to get a closed form solution. In each case determine the amount of money in the account after 5, 10, and 20 years.

Project 5.8. Carbon Cycle

One class of environmental models tracks the flow of some element or compound through an ecosystem. This project examines the flow of carbon through several types of forest conditions.

Part 1. A litter model. We begin with a very simple model, tracking carbon levels in litter on a forest floor. In this context, litter is naturally-occurring debris such as leaves, branches, and deadfalls, not beer bottles or fast-food wrappers. A boundary for the system is set up, and only litter within this region is considered. Thus carbon is measured in density units: grams of carbon per square meter (g C/m^2).

Here are the modeling assumptions: 1) Carbon continuously enters the system through litterfall at a constant rate z; 2) Carbon continuously leaves the system through two avenues—carbon dioxide produced in respiration and the conversion of litter into humus (called humification). Even though the carbon in humus has not left the physical system boundary, it is no longer in the litter, just as carbon in the trees before the leaves fell was in the physical boundary, but not yet in the litter. 3) The rate of litterfall is constant; the rate of carbon removal from both avenues is proportional to the amount of carbon present. 4) Initially there is no carbon in the litter. This approximates the situation after a ground fire (a forest fire which burns the underbrush, but does not kill the trees).

a. Set up a compartmental diagram for this model. Set up the differential equation. Solve it analytically.

b. For a temperate forest, it is reasonable to assume a rate of litterfall of 240 g C/m^2/yr and a proportionality constant for carbon removal of 0.4/yr. Using this information, graph a solution curve for 50 years either by numerically solving or by graphing the solution to part **a** using these parameters.

Part 2. Carbon in a Terrestrial Biosphere. This model extends the model from part **1** to model carbon's flowing and being stored in an entire ecosystem. This ecosystem is assumed to have seven components: plants (subdivided into leaves, branches, stems, and roots); litter lying on the ground; humus; and stable humus charcoal. The amount of carbon in each of these components is given by the variables $x_1, x_2, \ldots, x_7$, respectively. The atmosphere is, of course, another component, but due to its immense size, it is considered to have a constant carbon content, unchanged either by giving carbon to plants or by absorbing carbon from the litter, humus, or stable humus charcoal. Technically the atmosphere is outside the system, so carbon to plants is carbon entering the system, and carbon out of the litter/soil components is carbon leaving the system. The parameter z denotes the carbon entering the system, and the partition parameters p_1 through p_4 indicate the percentage of z which goes into leaves, branches, stems, and roots, respectively. The transfer coefficients k_{ij} give the rate of carbon flow from x_i to x_j (k_{i0} will denote the transfer from x_i to the atmosphere). A compartmental diagram for this system is shown in Figure 5.53.

5 Continuous Models

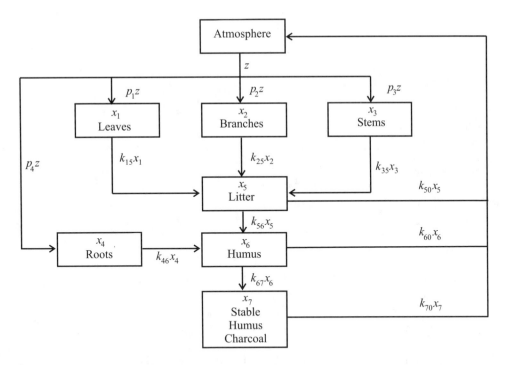

FIGURE 5.53 Terrestrial Carbon System.

Model parameters for a variety of ecosystem types are shown in Table 5.5. *Tropical forest* refers to tropical forest, forest plantation, shrub-dominated savannas, and chaparral. Temperate Forests comprise temperate forests, boreal forests, and woodlands. Note that k_{56} and k_{50} are not given, only an outflow of carbon from litter. As in part **1**, carbon is lost from the litter state by either humification or respiration. The humification factor h gives the fraction of carbon leaving the litter state that goes into humus; naturally, $1 - h$ indicates the fraction of carbon leaving the litter that does not go to humus, and hence goes to the atmosphere by respiration. Similar comments apply to k_{67} and k_{60} involving the amount of carbon leaving the humus and the coefficient of carbonization c which is the fraction going to form stable humus charcoal. The unit Gt is a Gigatonnes, or a billion tonnes, where 1 tonne (metric ton) = 1,000 kg = 1 Mg.

a. Use the compartmental diagram to set up the system of equations for this model.

b. For this part, use the parameters for the tropical forest ecosystem. First find the fixed points of this system. To do this, first write the answer to part **a** in the matrix form $X' = AX + B$. Fixed points occur when $X' = 0$, so solve the matrix equation $AX + B = 0$. Numerically solve the system to get graphs of each variable as a function of time. Run the system until all values are within 95% of their stable values. Compare the relative stabilization times for each component (how long it takes to get within 95% of the stable value).

c. Repeat **b** for at least two other ecosystems (your choice) and compare the results.

	Tropical forest	Temperate forest	Grass-land	Agri-cultural area	Human area	Tundra and semi-desert
Carbon Entering System z (Gt C/yr)	27.8	8.7	10.7	7.5	0.2	2.1
Partition coefficients						
p_1 (Leaves)	0.3	0.3	0.6	0.8	0.3	0.5
p_2 (Branches)	0.2	0.2	0.0	0	0.2	0.1
p_3 (Stems)	0.3	0.3	0.0	0	0.3	0.1
p_4 (Roots)	0.2	0.2	0.4	0.2	0.2	0.3
Flows						
Leaves to litter	1.0	0.5	1.0	1.0	1.0	1.0
Branches to litter	0.1	0.1	0.1	0.1	0.1	0.1
Stems to litter	0.033	0.0166	0.02	0.02	0.02	0.02
Roots to humus	0.1	0.1	1.0	1.0	0.1	0.5
Leaving litter	1.0	0.5	0.5	1.0	0.5	0.5
Leaving humus	0.1	0.02	0.025	0.04	0.02	0.02
Charcoal to atmosphere	0.002	0.002	0.002	0.002	0.002	0.002
Humification h	0.4	0.6	0.6	0.2	0.5	0.6
Carbonization c	0.05	0.05	0.05	0.05	0.05	0.05
Areas ($10^{12} m^2$)	36.1	17.0	18.8	17.4	2.0	29.7

TABLE 5.5 Parameters of Carbon Storage and Flow in Terrestrial Ecosystems. Data from *Climatic Change*, Volume 6, 1984, A Simulation Study for the Global Carbon Cycle, Including Man's Impact on the Biosphere by J. Goudriaan and P. Ketner, pp. 167–192. Reprinted with Kind Permission from Kluwer Academic Publishers.

This model is from Richard John Huggett's, *Modelling the Human Impact on Nature: System Analysis of Environmental Problems* [28].

Project 5.9. Predator-Prey Variations

There are many variants of the classical predator-prey model. Here we examine two. They both involve replacing exponential parts of the model with logistic parts.

Part 1. Consider a predator-prey system with all of the assumptions of the classical model except that the prey follows a logistic model instead of an exponential.

    a. Verify that this scenario is expressed by the following system of differential equations. You should be able to express a and b in terms of the logistic parameters

r and K.

$$\frac{dx}{dt} = ax - bx^2 - cxy$$
$$\frac{dy}{dt} = exy - fy. \qquad (132)$$

b. Noting that a, b, c, e, and f are all positive, perform phase plane analysis on this system. You should draw representative nullclines, mark fixed points, and draw flow arrows. There are several cases; we are interested in only the case where there is a fixed point with both coordinates positive.

c. Use Jacobian analysis to analyze the fixed point with positive coordinates.

d. Pick values for a, b, c, e, and f which yield a fixed point with positive coordinates, and use a differential equation solver to plot trajectories in the regions around the fixed point. Numerically verify the results from part **c**.

Part 2. Next consider a predator-prey system with all of the assumptions of the classical model except that 1) the prey follows a logistic model instead of an exponential and 2) in the absence of prey, the predators follow a logistic rather than dying exponentially. This would be the situation if the predators survived adequately without the prey, utilizing other forms of food. If prey are available, however, the predators grow at a rate proportional to the number of prey.

a. Verify that this scenario is expressed by the following system of differential equations:

$$\frac{dx}{dt} = ax - bx^2 - cxy$$
$$\frac{dy}{dt} = exy + fy - gy^2. \qquad (133)$$

b. Repeat **b**, **c**, and **d** from part **1** for this system of equations.

Project 5.10. Drugs in the Body

This project follows the analysis in Example 5.2.4, only using different drug delivery methods. With the exception of drug delivery, we make the same assumptions for the rest of the model (in particular $P(A) = 0.1A$).

1. Assume that, instead of giving 120 mg intervenously at a rate of 10 mg/hour, two shots of 60 mg each are given, one initially and one after 6 hours. Assume that a shot instantaneously distributes the full dosage (60 mg) to the body. Track the level of drug in the body over 24 hours. The authors intend for you to split up the domain appropriately. If you know other techniques for dealing with such things, they may be used, but are not necessary.

2. Explore how the graph for part **1** changes as you move the second shot closer and farther from the first. Solve the problem assuming that just one shot of 120 mg is given initially.

3. Next assume that the drug is given as a pill which gives off the medication as it dissolves and it is designed to dissolve slowly. As the pill dissolves, less and less of the medication is released. Assume that at any instant, the rate of release of the drug is

$$D(t) = 11.57 \exp(-\frac{t}{12})$$

milligrams/hour. How much of the drug is released by the pill over a 24-hour period? Track the amount of drug in the body over a 24-hour period.

4. Compare graphs of Example 5.2.4 and parts **1**, **2**, and **3**. What are the advantages and disadvantages of each type of delivery method? Under what conditions is one preferable to another?

Project 5.11. Two Models for the Population of the United States

This project uses the United States population data in Table 5.1.

1. Replicate the calculations in Section 5.2.2 for both the exponential model and the model with linear r.

2. Build a new model using a quadratic $r(t)$ function. Compare it to the two models discussed in Section 5.2.2. Your comparisons should include SS_{Res} and R^2 computed for the entire data set. Comment on the question of whether there is justification for using a quadratic function instead of a linear function to fit the rate data.

3. Using the methods in Section 4.8.1, fit a logistic function to the United States population data. Compare the logistic model to the three others you built in parts **1** and **2** using SS_{Res} and R^2. What is the predicted carrying capacity of the United States? What year did the logistic curve switch from concave up to concave down? What year will the United States be within 90% of its carrying capacity? What year will it be within 99% of its carrying capacity?

4. Read *An Assessment of Several of the Historically Most Influential Theoretical Models Used in Ecology and of the Data Provided in Their Support* by Charles Hall [24]. Especially pay attention to the discussion of the logistic model. Comment on how your work in part **3** should be interpreted.

6
Continuous Stochasticity

Looking back over the material covered thus far, the reader should notice that we studied discrete deterministic, discrete stochastic, and continuous deterministic modeling (with empirical modeling thrown in for good measure). A natural place to conclude our course is with continuous stochastic models.

We address two applications of continuous stochastic models: queueing theory and birth/death processes. The alert reader will notice two things. Generally these models are easy to set up (they are all variations on a theme), but after the simplest cases, they become extremely hard to solve analytically. Even the most basic queueing model requires material far beyond the scope of this text. This trend continues with other applications. To deal with equations which are unsolvable for one reason or another, we turn to simulations which are also discussed in this chapter. Thus, even though analysis is introduced where relevant, the intent is for the reader to be able to set up and simulate situations of this type.

6.1 Some Elements of Queueing Theory

Consider a service facility such as a checkout counter at a supermarket; people arrive at this facility at various times and are eventually served. Customers who arrive, but are not served, form a waiting line or *queue*. How long does a customer wait on average? How many people are in line on average? These and other questions form the basis of an area of applied probability theory known as queueing theory.

Before we develop some of the aspects of queueing theory, consider the following example. Suppose that customers constantly arrive at a checkout counter at a rate of A per minute and are serviced at a rate of S per minute. A queue does not form if the time in service ($1/S$ minutes) is less than the time until the arrival ($1/A$ minutes) of the next customer. Suppose, for example, that customers arrive at a rate of one per minute and are serviced at a rate of two customers per minute. The service time for each customer is 1/2 minute or 30 seconds. Since the server is idle when the next customer arrives, no queue is formed. Notice here that the customer's time in the queue and at the checkout counter is $1/S$, the time it takes for service to be completed.

A different situation results if the time to complete service ($1/S$) is longer than the time between customer arrivals ($1/A$). For instance, if customers arrive $A = 2$ per minute or once every $1/2$ minute and the service rate is $S = 1$ per minute, then customers arrive every 30 seconds and find the server busy. A queue will form and grow in size if no changes are made.

It is easy to determine the length of this queue at any point in time. If system was initially empty, then in T minutes, $A \times T$ customers arrived, and $S \times T$ customers completed service. Thus

$$\text{Queue Size after } T \text{ minutes} = \text{Number of Arrivals in } T \text{ minutes}$$
$$- \text{ Number of Departures in } T \text{ minutes}$$
$$= AT - ST = (A - S)T. \quad (134)$$

The linear equation (134) is only valid when $A > S$ or $1/A < 1/S$, so that a queue will form. It is evident from this expression that as T increases, the queue size grows. Using the rates $A = 2$ and $S = 1$, a queue forms and, after 10 minutes, there is a queue of size

$$(A - S)T = 10.$$

The *waiting time* or how long a customer must wait, depends upon how many are ahead of the customer in line. If there are N customers in the queue and the customer being served is finished when the next customer arrives, then this arriving customer must wait N/S minutes before being served. Similar expressions are possible for customers arriving midway through the current customer's being served.

Using the same rates, after 10 minutes, there are 10 customers in the queue and one just completing service. Thus an arriving customer waits $10/1 = 10$ minutes before being served and $1/S = 1$ minute for service to be completed.

Since constant arrival and service rates rarely occur (one example might be assembly line construction), the applicability of the above discussion is limited. It does, however, provide a basis upon which we build an understanding of queueing.

In many applications of queueing theory, it is not realistic to prescribe constant arrival and service times. For example, if we consider a grocery store checkout counter, we can not say with certainty when a customer will arrive, nor how many items that customer will purchase, and hence how long the customer will be in service. It is, therefore, reasonable to consider the arrival and service times as occurring randomly and to consider their underlying probability distributions. This approach is developed in the next section and depends upon some ideas from probability which we considered in Chapter 2.

6.2 Service at a Checkout Counter

In any study of queueing theory, it is essential to understand both the arrival patterns of customers and how they are served. We start with a model of service for one customer

6 Continuous Stochasticity

at a checkout counter. The model we build here is simple, but its development and subsequent analysis forms the basis for most of what we do in this chapter.

To model an individual at a checkout counter, we observe that at any time t, there are only two possible states that this individual is in; namely "being served" or "finished being served." Further, once an individual has finished being served, that person stays in that state. Using the terminology from Chapter 3, "finished being served" is an absorbing state.

If we let $N(t)$ denote the number of individuals at the checkout counter at time t, then $N(t)$ has only the two possible values of 0 or 1. (In the current development, we are interested in only one customer at a checkout counter. We consider arrivals to the checkout counter and the behavior of the queue in subsequent sections.) The state "being served" corresponds to $N(t) = 1$ since there is one individual at the counter being served. Similarly, $N(t) = 0$ corresponds to the state that the customer has "finished being served." Since the states of "being served" or "finished being served" depend only upon the state that the individual is currently in, this system forms a two-state absorbing Markov chain.

To obtain the time-dependent transition probabilities for this Markov chain, let

$$X_t = \begin{pmatrix} p(t) \\ q(t) \end{pmatrix}$$

be the time-dependent distribution vector for the states "being served" and "finished being served;" that is,

$$p(t) = P[N(t) = 1] \quad \text{and} \quad q(t) = P[N(t) = 0].$$

The state diagram for this two state chain is shown in Figure 6.1. Notice that $q(t) = 1 - p(t)$, so in the terminology of Chapter 2, $N(t)$ is a Bernoulli random variable for each t.

Since $p(t)$ represents the probability of a person's being served at time t, it is reasonable to expect that the service will be eventually completed (we hope!) so

$$\lim_{t \to \infty} p(t) = 0$$

which gives

$$\lim_{t \to \infty} X_t = \begin{pmatrix} 0 \\ 1 \end{pmatrix}.$$

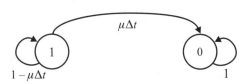

FIGURE 6.1 State Diagram for Service at a Checkout Counter.

To determine the transition matrix for this Markov chain, a small time interval Δt is chosen so that the following assumption is satisfied: if a customer is being served at time t, then the probability of service's being completed in Δt is proportional to Δt, that is

$$P[\text{service is completed in } \Delta t] = \mu \Delta t.$$

Since service's being completed is the complement of service's not being completed, we also have

$$P[\text{service not completed in } \Delta t] = 1 - \mu \Delta t.$$

The transition matrix for each Δt time step is given by

$$A = \begin{pmatrix} 1 - \mu \Delta t & 0 \\ \mu \Delta t & 1 \end{pmatrix},$$

where the 1 in row two, column two, represents the absorbing state "finished being served." Thus

$$X_{t+\Delta t} = A X_t. \tag{135}$$

Writing this in matrix form gives

$$\begin{pmatrix} p(t + \Delta t) \\ q(t + \Delta t) \end{pmatrix} = \begin{pmatrix} 1 - \mu \Delta t & 0 \\ \mu \Delta t & 1 \end{pmatrix} \begin{pmatrix} p(t) \\ q(t) \end{pmatrix}.$$

Performing matrix multiplication yields

$$p(t + \Delta t) = (1 - \mu \Delta t) p(t).$$

We rearrange this so that we have a difference quotient on the left-hand side

$$\frac{p(t + \Delta t) - p(t)}{\Delta t} = -\mu p(t).$$

By letting Δt go to zero, the difference quotient becomes a derivative, and we obtain

$$p'(t) = \lim_{\Delta t \to 0} \frac{p(t + \Delta t) - p(t)}{\Delta t}$$
$$= -\mu p(t).$$

This is the exponential differential equation with solution

$$p(t) = P[N(t) = 1] = p_0 e^{-\mu t}.$$

Since we assumed that there is an individual initially being served, $N(0) = 1$. This implies that $p(0) = 1$, which gives $p_0 = 1$. Finally, we have

$$p(t) = P[N(t) = 1] = e^{-\mu t}. \tag{136}$$

Thus the probability that an individual remains at the checkout counter decays to zero exponentially as time increases. Since $q(t) = 1 - p(t)$, the probability of completing service is

$$q(t) = 1 - e^{-\mu t}.$$

6 Continuous Stochasticity

Next let T denote the time at which service is completed. More generally, it is considered to be the *time until transition* from one state to another. Since service may be completed at any time, T is a continuous random variable with range $0 < T < \infty$. It should be evident that the probability that service is completed after time t is the same as the probability that service has not been completed at time t. In symbols

$$P[T > t] = P[N(t) = 1].$$

Taking complements gives

$$P[T \leq t] = 1 - e^{-\mu t}. \tag{137}$$

Recall that the left-hand side of expression (137) is the cumulative distribution function of the continuous random variable T. With this particular right-hand side, it is known as the *exponential* distribution. It is the only continuous distribution which satisfies the *memoryless property* which symbolically is written

$$P[T > s + t \text{ given } T > s] = P[T > t] \text{ for all } s, t \geq 0. \tag{138}$$

This property states that if you were being served at time s, then the probability of your still being served t time units later (at $s + t$) is the same as if you walked up to a counter at time 0 and were still being served t time units later (at time t). Thus the probability of service's getting completed within t time units is always the same, whether you just got there or whether you have been served for a long time. It may seem unreasonable to you if you go to a bank to get some banking done, you know how much you have to do and how long it will take. When you are halfway done, you are halfway done. There is nothing random about it. The correct point of view, however, is not to look at yourself, but at the people in front of you. You do not know how much banking they have. When someone has been at the window for five minutes, you have no idea if that person will be finished in another five minutes, or in 15 minutes. A mathematical model has a similar lack of knowledge about your business. In Exercise 11, the memoryless property, (138) is shown to characterize the exponential distribution.

Recall again from Chapter 2 that the density function of a random variable is the derivative of the cumulative distribution function. Thus the density function of T is given by

$$f(t) = \frac{d}{dt} P[T \leq t] = \mu e^{-\mu t}, \ 0 \leq t < \infty.$$

The parameter μ is given explicit meaning by computing the average or expected value of the random variable T. Recall from Chapter 2 (one last time), the expected value of a random variable T is

$$E(T) = \int_0^\infty t f(t) dt = \int_0^\infty t \mu e^{-\mu t} dt = \frac{1}{\mu}.$$

Thus the average value of T is $\frac{1}{\mu}$. We interpret the service rate μ as

$$\mu = \frac{1}{\text{Mean Time Until Transition}}.$$

6.3 Standing in Line—The Poisson Process

In the previous section, we considered how an individual is served at a checkout counter. In this section we consider arrivals. In general, the analysis presented is applicable to any physical situation in which a certain type of event is recurring in time or space, such as customer arrivals, telephone calls at a switchboard, or defects in a long piece of wire. As was done in the previous section, we consider a time dependent function, $N(t)$. More specifically, let $N(t)$ denote the number of arrivals that occur in a given interval $[0, t]$, and suppose that the following modeling assumptions are valid.

i. The probability that an arrival occurs in a short interval $[t, t + \Delta t]$ is proportional to the length of the interval Δt. In symbols,

$$P[N(t + \Delta t) - N(t) = 1] = \lambda \Delta t \qquad (139)$$

for some constant $\lambda > 0$.

ii. The probability that an arrival occurs in $[t, t + \Delta t]$ does not depend on the time of previous arrivals; that is,

$$P[N(t + \Delta t) - N(t) = n \mid N(s) = m] = P[N(t + \Delta t) - N(t) = n]$$

for all $0 \leq s \leq t$. This is a memoryless property.

iii. The occurrences of arrivals in non-overlapping intervals are independent.

iv. The probability of two or more arrivals in $[t, t + \Delta t]$ is negligible.

Assumptions **i** - **iv** are natural in light of how customers arrive at a checkout counter. More specifically, assumption **i** is similar to the assumption for how customers are serviced and is one common way of modeling the probability of an arrival's increasing as a function of time. For instance, doubling Δt doubles the probability of an arrival. Assumptions **ii** and **iii** are interpreted as "arrivals do not depend upon the specific time of occurrence." Assumption **iv** is interpreted as "two customers can not arrive at the same instant," which makes physical sense if Δt is small enough. As in the previous section, we ultimately let Δt go to zero, so that our assumptions here are for "sufficiently" small Δt.

As assumed earlier, let

$$P_n(t) = P[N(t) = n].$$

Here $N(t)$ has possible values $0, 1, 2, \ldots$ since it is possible to have any number of customers arrive.

If we consider a time step of size Δt, then $N(t)$ forms a Markov chain, with the state diagram shown in Figure 6.2. The transition matrix for this chain is

$$A = \begin{pmatrix} 1 - \lambda \Delta t & 0 & 0 & 0 & \cdots \\ \lambda \Delta t & 1 - \lambda \Delta t & 0 & 0 & \cdots \\ 0 & \lambda \Delta t & 1 - \lambda \Delta t & 0 & \cdots \\ \vdots & \vdots & \vdots & \vdots & \ddots \end{pmatrix}, \qquad (140)$$

which is an infinite dimensional matrix. There is a wealth of theory about infinite dimensional matrices. Our purpose for introducing A is a matter of notational convenience. If we extend the definition of matrix multiplication in the obvious way to include infinite

6 Continuous Stochasticity

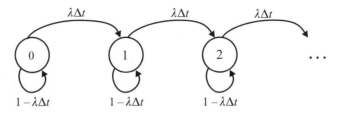

FIGURE 6.2 State Diagram for Arrivals at a Checkout Counter.

dimensional matrices, we have

$$X_{t+\Delta t} = AX_t \tag{141}$$

as the state equation for each time step of size Δt. This equation is the same as (135) of the previous section, the only difference being that X_t is now defined as

$$X_t = \begin{pmatrix} P_0(t) \\ P_1(t) \\ \vdots \\ P_n(t) \\ \vdots \end{pmatrix}.$$

Since the matrix A has mostly zero entries, the state equation (141) gives, upon matrix multiplication, the following (infinite) system of equations

$$P_0(t + \Delta t) = (1 - \lambda \Delta t) P_0(t)$$
$$P_1(t + \Delta t) = \lambda \Delta t P_0(t) + (1 - \lambda \Delta t) P_1(t)$$
$$\vdots$$
$$P_n(t + \Delta t) = \lambda \Delta t P_{n-1}(t) + (1 - \lambda \Delta t) P_n(t).$$
$$\vdots$$

Rearranging to make difference quotients and then letting Δt go to zero, we obtain a system of differential equations. For $n \geq 1$,

$$P_n'(t) = \lim_{\Delta t \to 0} \frac{P_n(t + \Delta t) - P_n(t)}{\Delta t}$$
$$= \lambda [P_{n-1}(t) - P_n(t)], \tag{142}$$

and for $n = 0$,

$$P_0'(t) = -\lambda P_0(t). \tag{143}$$

Equation (143) is the exponential differential equation with the initial condition that there are initially no arrivals, or that $P_0(0) = 1$. This gives

$$P_0(t) = e^{-\lambda t}.$$

Similarly, letting $n = 1$,
$$P_1'(t) = \lambda[P_0(t) - P_1(t)]$$
$$= \lambda[e^{-\lambda t} - P_1(t)]. \quad (144)$$

Equation (144) is a first-order linear differential equation, which we solved in Chapter 5 by multiplying by the integrating factor $e^{\lambda t}$. Equation (144) becomes
$$\frac{d}{dt}[e^{\lambda t} P_1(t)] = \lambda$$

so that
$$e^{\lambda t} P_1(t) = \lambda t + c,$$

where c is an arbitrary constant. The initial condition $P_0(0) = 1$ implies $P_n(0) = 0$ for all $n \geq 1$, thus
$$P_1(t) = \lambda t e^{-\lambda t}.$$

It can be shown by induction (see Exercise 12) that the system of differential equations given by (142) has solution
$$P_n(t) = e^{-\lambda t} \frac{(\lambda t)^n}{n!} \quad \text{for } t \geq 0, n = 0, 1, 2, \ldots.$$

This time-dependent probability distribution is known as a *homogeneous Poisson process* with rate λ. A process is called *homogeneous* if the rate λ does not depend upon the time parameter t. (A Poisson process with a time-dependent rate $\lambda(t)$, is called *nonhomogeneous*.)

The mean of the Poisson process is λt (see Exercise 14). The mean λt is interpreted in the following manner: since the Poisson process is the time-dependent distribution for the number of occurrences at time t, with λ being the average number of occurrences per unit time, then there would be λt customers on average during a time interval of length t. For instance, if customers arrive at a checkout counter at an average rate of $\lambda = 2$ per minute, then there would be on average $10 = 2 \times 5$ customers in a given five-minute interval. This discussion also implies that the average number of arrivals is the same for each time interval of length t, which is the memoryless property. Thus we may consider the process as restarting after each arrival. Earlier we mentioned that the memoryless property characterized the exponential distribution. So, in fact the inter-arrival times are exponentially distributed.

To see more directly that the inter-arrival times are exponentially distributed, let W denote the time the first arrival occurs in a Poisson process. Since W denotes the time of the first arrival, the event $W > t$ takes place if and only if no events of the Poisson process occur in the time interval $[0, t]$. Thus,
$$P[W > t] = P[N(t) = 0],$$

but,
$$P[N(t) = 0] = P_0(t) = e^{-\lambda t},$$

and this is an exponential distribution.

6.4 Example: Crossing a Busy Street

The Poisson process provides a time-dependent model for the occurrence of events. One interesting application of the Poisson process is its use in determining how busy a street must be before a crossing control is required.

Suppose that λ is the average number of cars per hour that pass a point on a road. For simplicity, assume that these cars all travel in the same direction. This is not an unreasonable assumption, if one considers a two-lane road with a median. If traffic is flowing freely, then the hypotheses for the Poisson process are approximately met; thus a Poisson process is an appropriate model for traffic flow.

To determine if the road needs a crossing control, we must determine if there is sufficient time between cars for a pedestrian to cross safely. Thus, the time "gaps" between car arrivals must be considered. Let G_k be the time between the $(k-1)$st and the kth car as seen in Figure 6.3

Thus G_k is the inter-arrival time between the $k-1$ and kth events in a Poisson process with rate λ. As noted in the previous section, the inter-arrival times for occurrences from a Poisson process follow an exponential distribution; thus

$$P[G_k > t] = e^{-\lambda t}.$$

Let T be the average time it takes a pedestrian to cross the road. The probability p_k that a pedestrian crosses the road between the $(k-1)$st and kth cars is

$$\begin{aligned} p_k &= P[G_1 \leq T, G_2 \leq T, \ldots, G_{k-1} \leq T, G_k > T] \\ &= P[G_1 \leq T] P[G_2 \leq T] \ldots P[G_{k-1} \leq T] P[G_k > T] \\ &= (1 - e^{-\lambda T})^{k-1} e^{-\lambda T}. \end{aligned} \quad (145)$$

Here we used the independence of the inter-arrival times (to form the product) and the fact that these inter-arrivals are exponentially distributed.

Using (145), the average number of gaps, $\overline{G}$, is found. More specifically,

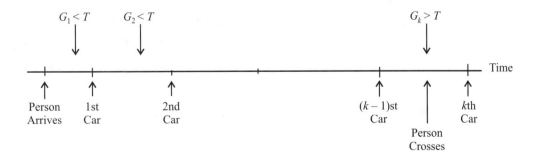

FIGURE 6.3 Crossing a Busy Street.

$$\overline{G} = \sum_{k=0}^{\infty} k p_k$$
$$= \frac{1}{\lambda} \sum_{k=0}^{\infty} k(1-e^{-\lambda T})^{k-1} \lambda e^{-\lambda T} = \frac{-1}{\lambda} \sum_{k=0}^{\infty} \frac{d}{dT}(1-e^{-\lambda T})^k$$
$$= \frac{-1}{\lambda} \frac{d}{dT} \sum_{k=0}^{\infty} (1-e^{-\lambda T})^k = \frac{-1}{\lambda} \frac{d}{dT} \left(\frac{1}{1-(1-e^{-\lambda T})} \right)$$
$$= e^{\lambda T}.$$

The second-to-the-last expression is obtained from summing the geometric series,
$$\sum_{k=0}^{\infty} (1-e^{-\lambda T})^k.$$

If we let α denote the average length of the gaps that occur while a pedestrian is waiting to cross, then a pedestrian waits an average of

Average wait = Average number of gaps × Average time of gaps = $(e^{\lambda T} - 1)\alpha$.

We now determine if a crossing control is required. We let τ be the longest average time that a pedestrian is expected to wait without taking foolish risks in crossing a street. A crossing control is required if (the average wait is larger than the average expected wait)
$$(e^{\lambda T} - 1)\alpha \geq \tau.$$

Notice that α is not larger than T; otherwise at least one time gap would have exceeded T and the pedestrian would have been able to cross the road. Thus,
$$(e^{\lambda T} - 1)T \geq \tau.$$

Solving for λ gives
$$\lambda \geq \frac{1}{T} \ln(1 + \frac{\tau}{T}). \tag{146}$$

Thus (146) provides an expression for the minimum average number of cars that are allowed to pass if pedestrians are not delayed unreasonably. That is, if more than $\frac{1}{T} \ln(1 + \frac{\tau}{T})$ cars pass per unit of time, a light is needed.

Example. According to Gerlough and Barnes [18], pedestrians walk, on average, about 3.5 feet per second. Suppose that a road is 100 feet wide, then the time it takes a pedestrian to cross the road is

$$T = 100/3.5 = 28.5714 \text{ seconds}.$$

Suppose that τ, the maximum wait reasonably expected for a pedestrian, is 60 seconds. Then, using (146), the maximum safe traffic flow in cars per second is

$$\frac{1}{T} \ln(1 + \frac{\tau}{T}) = 0.0395991,$$

which in cars per hour is about 142.5.

6 Continuous Stochasticity

The maximum safe traffic flow in cars per hour is expressed as a function of the width w of the road and the maximum reasonable wait time τ. If w is the road width, then, using 3.5 seconds as the walking rate of pedestrians, the time it takes a pedestrian to cross the road is

$$T = \frac{w}{3.5} \text{ seconds} = \frac{w}{12600} \text{ hours,}$$

so that,

$$m(w, \tau) = \frac{12600}{w} \ln(1 + \frac{12600\tau}{w})$$

is the maximum safe traffic flow.

A graph of $m(w, \tau)$, the maximum safe traffic flow in cars per hour, is shown in Figure 6.4 for $0 < w < 200$ feet and $1 < \tau < 10$ minutes. A reasonable estimate of τ is about one minute. A graph of m for $\tau = 1$ and the $0 < w < 200$ feet is shown in Figure 6.5.

Notice in each of the Figures 6.4 and 6.5 how quickly the maximum traffic flow decreases with increasing road width. For example, a traffic flow of 2,000 cars per hour is judged safe if the road is 20 feet wide; however, a traffic flow of 600 cars per hour is unsafe for a road 50 feet wide. Thus, traffic controls are required on major highways where traffic flow is heavy and the width of the road is large. This model provides some insight into the behavior of traffic and gives a criteria for traffic control. This model relies upon the arrivals of cars being regarded as a Poisson process, an assumption that

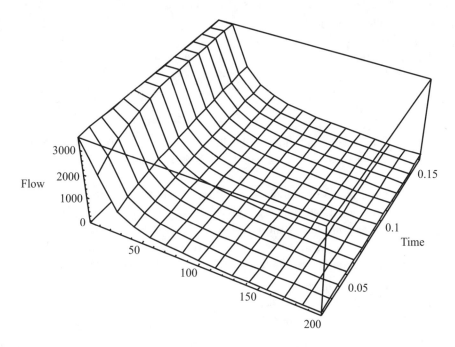

FIGURE 6.4 The Maximum Safe Traffic Flow in Cars per Hour.

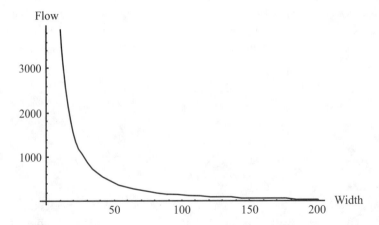

FIGURE 6.5 The Maximum Safe Traffic Flow in Cars per Hour with $\tau = 1$ Minute.

we can verify by analyzing some traffic data with a chi-square goodness-of-fit test. An example of this method is outlined in the projects.

6.5 The Single-Server Queue

Earlier in this chapter, we considered how a customer arrives and how a customer is serviced at a checkout counter. In this section, these ideas are combined to model the formation of a queue and its behavior.

Suppose that a checkout counter has only one cashier. Using the assumptions we gave for service times, we assume that the servicing time is exponentially distributed with a mean service rate of μ minutes. Further, suppose that customer arrivals occur independently and that the inter-arrival time of the customers is exponentially distributed with a mean arrival rate of λ customers per minute, so that the number of customers arriving forms a Poisson process.

To model the queue that is formed at the checkout counter, let $N(t)$ be the number of customers in line or being served at time t. The time-dependent random variable $N(t)$ assumes any of the integer values $0, 1, 2, \ldots$ at time t.

Now, as we did in the derivation of the Poisson process, we shall assume that the probability of an arrival in an interval of short duration Δt is $\lambda \Delta t$. Similarly, the probability of a departure during an interval of length Δt is $\mu \Delta t$. As done previously, we assume that Δt is so small that at most one event occurs during an interval of length Δt. In particular, there are three mutually exclusive events which might occur:

i. Exactly one arrival in $(t, t + \Delta t)$
ii. Exactly one departure in $(t, t + \Delta t)$
iii. No arrivals or departures in $(t, t + \Delta t)$.

Using these assumptions and the fact that arrivals and departures do not depend upon the number of customers in the queue, we construct the state diagram shown in Figure 6.6.

6 Continuous Stochasticity

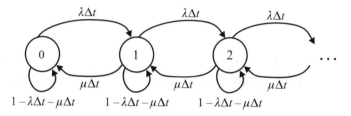

FIGURE 6.6 State Diagram for the Number of Customers in a Queue.

The time step for this Markov chain is Δt and the transition matrix is

$$A = \begin{pmatrix} 1 - \lambda \Delta t & \mu \Delta t & 0 & 0 & 0 & \cdots \\ \lambda \Delta t & 1 - \mu \Delta t - \lambda \Delta t & \mu \Delta t & 0 & 0 & \cdots \\ 0 & \lambda \Delta t & 1 - \mu \Delta t - \lambda \Delta t & \mu \Delta t & 0 & \cdots \\ \vdots & \vdots & \vdots & \vdots & \vdots & \ddots \end{pmatrix}. \quad (147)$$

Let $P_n(t) = P[N(t) = n]$ be the probability that there are n customers in the queue at time t. Thus letting

$$X_t = \begin{pmatrix} P_0(t) \\ P_1(t) \\ \vdots \\ P_n(t) \\ \vdots \end{pmatrix},$$

we obtain the state equation

$$X_{t+\Delta t} = AX_t \quad (148)$$

for the single-server queue.

The state equation (148) gives upon matrix multiplication the following (infinite) system of equations for the single-server queue:

$$P_0(t + \Delta t) = (1 - \lambda \Delta t) P_0(t) + \mu \Delta t P_1(t)$$
$$P_1(t + \Delta t) = (1 - \lambda \Delta t - \mu \Delta t) P_1(t) + \lambda \Delta t P_0(t) + \mu \Delta t P_2(t).$$
$$\vdots$$
$$P_n(t + \Delta t) = (1 - \lambda \Delta t - \mu \Delta t) P_n(t) + \lambda \Delta t P_{n-1}(t) + \mu \Delta t P_{n+1}(t).$$
$$\vdots$$

We now obtain a system of differential equations by letting Δt go to zero; more specifically, for $n \geq 1$,

$$P'_n(t) = \lim_{\Delta t \to 0} \frac{P_n(t + \Delta t) - P_n(t)}{\Delta t}$$
$$= -(\lambda + \mu) P_n(t) + \lambda P_{n-1}(t) + \mu P_{n+1}(t) \quad (149)$$

and
$$P_0'(t) = -\lambda P_0(t) + \mu P_1(t). \tag{150}$$

Many classical methods are available for solving the system (149) and (150); however, they involve ideas which are beyond the scope of this text. It is known that the solution to (149) and (150) is

$$P_n(t) = \rho^{(n-i)/2} I_{n-i}(at) + \rho^{(n-i-1)/2} I_{n+i+1}(at) + (1-\rho)\rho^n \sum_{k=n+i+2}^{\infty} \rho^{-k/2} I_k(at) \tag{151}$$

where
$$\rho = \frac{\lambda}{\mu} \tag{152}$$

$a = 2\mu\rho^{1/2}$, and $I_k(x)$ is the modified Bessel function of the first kind of order k, which is defined to be

$$I_k(x) = \sum_{m=0}^{\infty} \frac{(x/2)^{k+2m}}{(k+m)!m!} \quad k \geq -1.$$

The interested reader is encouraged to pursue the development and solution given by L. Kleinrock [32], who remarks after obtaining the solution (151) that "this expression is most disheartening. What it has to say is that an appropriate model for the *simplest interesting* queueing system leads to an ugly expression for the time-dependent behavior of its state probabilities." Due to the nature of the solution (151), we do not consider the analytic time-dependent solutions further. In the next section, we use the system of equations (149) and (150) to obtain steady-state (time-independent) behavior. These are used in Section 6.5 on the single-server queue.

One final remark before pursuing these ideas. The applicability of the single-server queue and the nature of its solution has prompted researchers to continue the analysis of the single-server queue and its methods of solution. Recently, A. Krinik [35], [36] obtained some elementary methods of solutions based upon combinatorial techniques. His elegant methods provide insight into the behavior of this system and are a direction for further investigations.

6.5.1 Stationary Distributions

As we saw in the previous section, the time-dependent solution to the single-server queue is unmanageable, and any further complication of this model yields an even more complex solution. Even though we may be able to solve analytically for the time-dependent probabilities $P_n(t)$, it is not clear how useful these functions are in aiding our understanding of the behavior of the single-server queue. Consequently, we may ask whether the system of time-dependent probabilities $P_n(t)$ settles down and displays no more "transient" behavior. This idea is analogous to the idea of finding the fixed points for deterministic systems.

In the deterministic systems we considered, a fixed point is a solution to the system that "stays fixed" or, equivalently, a point at which the rates of change of the system are zero. In the deterministic models we considered, the fixed points are ordered n-tuples.

6 Continuous Stochasticity

In the case we are currently considering, the fixed points are "stationary" probability distributions. More specifically, the system of differential equations

$$P_0'(t) = -\lambda P_0(t) + \mu P_1(t). \tag{153}$$

and

$$\begin{aligned} P_n'(t) &= \lim_{\Delta t \to 0} \frac{P_n(t + \Delta t) - P_n(t)}{\Delta t} \\ &= \lambda P_{n-1}(t) - (\lambda + \mu) P_n(t) + \mu P_{n+1}(t) \text{ for } n \geq 1 \end{aligned} \tag{154}$$

has a fixed point or stationary distribution, provided its rates of change are zero, that is, if

$$P_n'(t) = 0 \text{ for } n \geq 0. \tag{155}$$

We assume that condition (155) is valid, and we proceed. Condition (155) is equivalent to assuming that

$$\lim_{t \to \infty} P_n(t) = P_n$$

exists, where P_n does not depend upon t. Now applying condition (155) to the system of equations (153) and (154), we obtain

$$-\lambda P_0 + \mu P_1 = 0$$

$$\lambda P_{n-1} - (\lambda + \mu) P_n + \mu P_{n+1} = 0 \text{ for } n \geq 1. \tag{156}$$

The first equation in (156) yields, upon solving for P_1,

$$P_1 = \left(\frac{\lambda}{\mu}\right) P_0. \tag{157}$$

The expression (156), when $n = 1$ is

$$\lambda P_0 - (\lambda + \mu) P_1 + \mu P_2 = 0,$$

so that solving for P_2 and using (157) gives

$$P_2 = \left(\frac{\lambda}{\mu}\right)^2 P_0.$$

Continuing in this manner, an induction argument (Exercise 16) shows

$$P_n = \left(\frac{\lambda}{\mu}\right)^n P_0. \tag{158}$$

It remains to determine P_0. Since queue size must be a non-negative integer, it follows that the probability of the queue's size being a non-negative integer is one, hence,

$$P_0 + P_1 + P_2 + \ldots = 1.$$

Thus,

$$P_0 + \left(\frac{\lambda}{\mu}\right) P_0 + \left(\frac{\lambda}{\mu}\right)^2 P_0 + \ldots = 1$$

which is

$$P_0 \sum_{n=0}^{\infty} \left(\frac{\lambda}{\mu}\right)^n = 1. \tag{159}$$

The series in (159) is geometric with ratio (λ/μ). This geometric series will converge provided

$$\left(\frac{\lambda}{\mu}\right) < 1. \tag{160}$$

This is the required condition for the stationary distribution to exist. It is a natural condition, since a stationary distribution will exist provided the arrival rate (μ) is less than the departure rate (λ) so that queue lengths will not grow without bound.

Now assuming $\lambda < \mu$,

$$\sum_{n=0}^{\infty} \left(\frac{\lambda}{\mu}\right)^n = \frac{1}{1 - \left(\frac{\lambda}{\mu}\right)},$$

so that

$$P_0 = 1 - \left(\frac{\lambda}{\mu}\right). \tag{161}$$

Using this expression in (158) gives

$$P_n = \left(\frac{\lambda}{\mu}\right)^n \left(1 - \left(\frac{\lambda}{\mu}\right)\right) \tag{162}$$

as the stationary distribution for the single-server queue.

The expression (162) represents the stationary or steady-state distribution for the single-server queue. It is reasonable to ask how well this represents the behavior of a queue since the arrival of customers varies. It is not unreasonable to assume that the transition probabilities $P_n(t)$ of the single-server queue are essentially constant over short time intervals (such as a lunch hour); this assumption implies the stationary distribution is valid over these periods. This next section makes use of some of these ideas.

6.5.2 Example: Traffic Intensity—When to Open Additional Checkout Counters

In the previous section, we obtained the stationary distribution for the single-server queue. In the derivation of this expression, it was necessary to assume that the arrival rate was less than the departure rate, that is $\lambda < \mu$. Let

$$\rho = \left(\frac{\lambda}{\mu}\right). \tag{163}$$

By assumption, $\rho < 1$. The parameter ρ is called the *traffic intensity* of the queue. Using ρ in (162) gives

$$P_n = \rho^n (1 - \rho).$$

Readers familiar with probability theory will recognize this as a geometric distribution. Using (163), we obtain the average number $\overline{N}$ of customers in the queue. This important

6 Continuous Stochasticity

measure is obtained as follows

$$\overline{N} = E(N) = \sum_{n=0}^{\infty} n P_n$$

$$= (1-\rho) \sum_{n=0}^{\infty} n\rho^n = (1-\rho)\rho \sum_{n=0}^{\infty} \frac{d}{d\rho} \rho^n$$

$$= (1-\rho)\rho \frac{d}{d\rho} \sum_{n=0}^{\infty} \rho^n = (1-\rho)\rho \frac{d}{d\rho} \frac{1}{1-\rho}$$

$$= \frac{\rho}{1-\rho}. \tag{164}$$

The behavior of the expected number in the system is shown in Figure 6.7.

Notice how quickly $\overline{N}$ grows as ρ approaches 1. As ρ goes from 0.75 to 0.9, the average queue length $\overline{N}$ goes from three to 10. Average queue lengths of up to three are probably acceptable at a checkout counter; but queue lengths of 10 are not, so that additional checkout counters would be opened.

The average number of people in a queue provides one measure of a queue's behavior, but it is perhaps not the most useful. A queue of 10 people might be acceptable if each took one minute to complete service; however, a queue of two people is unacceptable if you were second in line and had to wait an hour. The average time $\overline{T}$ that a customer spends in the system is often a more reliable (and practical) indicator of a queues behavior.

To obtain $\overline{T}$, we use a fact about queues. This fact, known as Little's result, states that *the average number of customers in a queueing system is equal to the average arrival rate of customers to that system times the average time spent in that system.* In our notation, Little's result is

$$\overline{N} = \lambda \overline{T}. \tag{165}$$

Intuitively, we note that Little's result is obtained by observing that an arriving customer should find the same average number of customers $\overline{N}$ in the system as he leaves behind upon departing. This latter quantity is just the arrival rate λ times the customers average

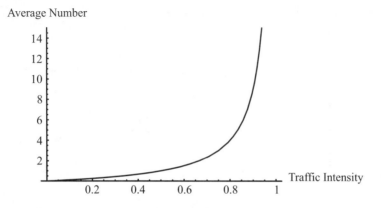

FIGURE 6.7 The Average Number $\overline{N}$ of Customers.

time $\overline{T}$ in the system. This is an informal verification of this result which, according to Kleinrock [33], existed as a "folk theorem" without proof in queueing theory for many years. The result which bears his name was first formally established by J.D.C Little in 1961 [38].

Using (165), we solve for $\overline{T}$ to obtain $\overline{T} = \overline{N}/\lambda$. Now using our expression (164), for the average number $\overline{N}$ of customers in a single-server queue, it follows that

$$\overline{T} = \frac{\overline{N}}{\lambda} = \left(\frac{\rho}{1-\rho}\right)\left(\frac{1}{\lambda}\right)$$
$$= \frac{\frac{1}{\mu}}{1-\rho}. \tag{166}$$

The dependence of the average time in the system on the traffic intensity ρ is shown in Figure 6.8. Notice that $\overline{T} = 1/\mu$ when $\rho = 0$, that is, the average time a customer spends in the system when there in no one in the queue, is the average service time, $1/\mu$. The behavior of the average number of customers and the average time shown in Figures 6.7 and 6.8 is rather dramatic. As ρ approaches one, both of these quantities grow in an unbounded manner. This behavior is not surprising if one recalls that $\rho < 1$ is required for a stationary distribution to exist. What is surprising, however, is that $\overline{N}$ and $\overline{T}$ behave so badly when we are near (but below) one. This implies that the queueing system pays an extreme penalty if it is run near (but below) capacity.

Using the stationary distribution, we also find the probability of finding at least k customers in the system. That is,

$$P[\text{at least } k \text{ customers in the system}] = \sum_{n=k}^{\infty} p_n = \sum_{n=k}^{\infty} (1-\rho)\rho^n$$
$$= (1-\rho)\rho^k(1 + \rho + \rho^2 + \ldots)$$
$$= \rho^k. \tag{167}$$

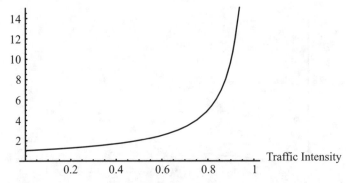

FIGURE 6.8 The Average Time $\overline{T}$ in the System.

Probability of at least *k* in the system

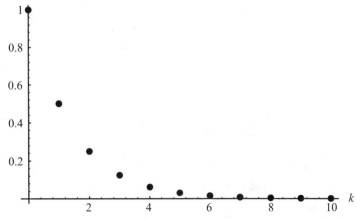

FIGURE 6.9 The Probability of at Least k Customers.

Thus the probability of exceeding some limit on the number of customers in the system is a decreasing function of that number and goes to zero rapidly as shown in Figure 6.9.

In the next section, we simulate a single-server queue and use some of the tools we developed to analyze the queue's behavior.

6.6 A Queueing Simulation

In Section 6.5, we studied the single-server queue. Equation (151) is the time-dependent behavior of this queueing system; however, as we noted before, this solution is not easily understood. In this section, we develop a simulation model for the single-server queue. Using this simulation, we verify aspects of the theory we obtained earlier.

To begin our discussion of the simulation, recall that in the development of the theoretical model, Δt was infinitesimally small (in fact, the limit as Δt went to zero was taken). In the simulation model, Δt is small, but not infinitesimal ($\Delta t = 0.01$ is our choice). We wish to simulate a queue for a period of $T = 10$ minutes. The queue's time scale is in minutes; thus our simulation with a Δt of 0.01 represents 100 simulation time steps per minute of the queueing system. The simulation will run for $T/\Delta t = 10/0.01 = 1000$ simulation time steps.

To build a single-server simulation model, we first consider the structure of a queue. A queue has an inherent order; the first customer to arrive at an empty queue is the first one served, the second customer is the second served, and so on. A customer arriving with n people in the queue must wait until these n people are served before being served. Often this is referred to as First In First Out or FIFO. This structure suggests that we simulate a queue with an array or list structure, where the first item in the list represents the customer who is served next. Each arrival is placed at the end of the list. Departures from the queue are from the front of the list and allow each item (customer) in the list to move up one position. If the reader has taken a data structures course, the idea of a queue may be familiar. You may even have constructed a queue from a structure called a linked list. If you have experience with these structures using a programming language, you are encouraged to use them to build your simulations. We use *Mathematica* to be consistent with the rest of the book, but simulations written in *Mathematica* run much more slowly than simulations written in a pure programming language like C.

To simulate arrivals and departures, recall the assumption that during an interval of length Δt only one of three events can occur, namely, either an arrival, or a departure, or no arrivals or departures. It was also assumed that the probabilities of an arrival or departure during Δt are proportional to Δt.

As we did in Chapter 2 to simulate a catastrophe for the sandhill cranes, we simulate an arrival or departure using a uniform random variable. An arrival occurs during Δt if the value of a uniformly distributed random variable over $[0, 1]$ is less than $\lambda \Delta t$. Similarly, a departure occurs if the value of a uniformly distributed random variable is less than $\mu \Delta t$. Most software packages and graphing calculators have the ability to generate a uniform random variable. Typically, a uniform random variable is generated by a computer with the command

$$\text{Uniform}(0, 1).$$

Thus, to generate an arrival during a time step of Δt, let

$$\text{Arr} = \text{Uniform}(0, 1).$$

An arrival occurs if Arr has a value less than $\lambda \Delta t$. This is determined using an "If" statement of the form

If Arr $< \lambda \Delta t$, Then arrival occurred, Else no arrival.

Similarly, letting

$$\text{Dep} = \text{Uniform}(0, 1),$$

on each time step Δt allows departures to be computed with a similar "If" statement. The "If" statement for departures is

If Dep $< \mu \Delta t$, Then departure occurred, Else no departure.

This model essentially involves flipping two weighted coins, one for arrivals and one for departures. The time interval Δt affects the weights. Cut Δt in half and the probability of tossing a head is cut in half. Our assumption is that Δt is so small that only one event can occur, either an arrival or a departure, but not both. Actually it is possible to flip heads on both coins, but the assumption is that Δt is chosen so small that the probability of this happening is essentially zero. In practice, however, we may find that if Δt is that small, the simulation may take an unreasonable amount of time to run.

Thus we must balance between choosing Δt small enough so that the probability of a simultaneous arrival and departure is essentially zero and choosing Δt large enough so that the simulation runs in a reasonable amount of time. The occurrence of simultaneous events skews the results. If practical constraints require a choice of Δt large enough so that simultaneous arrivals and departures occur more than rarely, there are ways to

6 Continuous Stochasticity

force the assumption that simultaneous events do not occur. Simply put, if two heads are thrown, we flip again.

One way of implementing this is with a While loop structure. A While loop structure repeats a set of instructions while a test statement is true. The usual form of a While statement is

$$\text{While(test)}$$
$$\text{body}$$
$$\text{Endwhile}$$

which evaluates the instructions in *body* as long as *test* is true.

Thus, to guarantee that only one event occurs during a particular Δt, we use the following structure, presented in pseudocode:

arrivalflag = 1

departflag = 1r

    While (arrivalflag + departflag $\neq 0$ AND arrivalflag + departflag $\neq 1$)

        arrivalflag = 0

        departflag = 0

        Dep = Uniform$(0, 1)$

        Arr = Uniform$(0, 1)$

        If Dep $< \mu \Delta t$, Then departflag = 1

        If Arr $< \lambda \Delta t$, Then arrivalflag = 1.

    Endwhile

The body of the While loop is shown indented under the While statement. The While loop executes these instructions as long as the test statement

$$\text{arrivalflag + departflag} \neq 0 \text{ and arrivalflag + departflag} \neq 1,$$

is true. Before entering the While loop, the variables *arrivalflag* and *departflag* are set to one, which makes the test statement initially true. During the execution of the While loop's body, the variable *arrivalflag* is set to one if an arrival occurred and zero otherwise. Similarly, the variable *departflag* is one if a departure occurs and zero otherwise. The test statement prevents an arrival and a departure from occurring simultaneously. Thus the computer keeps "flipping" coins until the result has at most one head.

Now that we have developed the components of the queueing simulation, the simulation routine is obtained by putting these components together.

The required information for a simulation of a single-server queue is the arrival and departure rates λ and μ, the length of the simulation T, and the simulation time step Δt. It is strongly recommended that Δt be taken as 0.01 or smaller.

The pseudocode for the simulation of a single-server queue follows.

For $i = 1$ to $T/\Delta t$
    arrivalflag = 1
    departflag = 1
    While arrivalflag + departflag $\neq 0$ and arrivalflag + departflag $\neq 1$,
        arrivalflag = 0
        departflag = 0
        Dep = Uniform (0, 1)
        Arr = Uniform (0, 1)
        If Dep $< \mu \Delta t$, Then departflag = 1
        If Arr $< \lambda \Delta t$, Then arrivalflag = 1
    If arrivalflag = 1, Then Join Queue
    If departflag = 1, Then Depart Queue
    Endwhile Endfor

The instructions "Join Queue" and "Depart Queue" are self explanatory for the operation of the queueing simulation. How they should be implemented depends on the data structure used for the queue.

As we discussed in the previous section, the average time that customers spend in the queue and the average queue length are of interest in understanding the behavior of the queueing system. To obtain these quantities from our simulation, we must keep track of the arrival and departure times of individuals in the simulation and the queue length at each time step Δt.

To obtain the number served during the simulation, we simply need a counter that is incremented whenever a customer departs. Similarly, to obtain the average queue length during the simulation, record the length of the queue at the end of each time step and then find the average of these values. To find the average time for the queue, each arriving customer is given a "time stamp" of the time of their arrival. In the pseudocode above, since the simulation starts at time zero, an arrival occurring on the jth iteration of the loop is given the arrival time stamp $j\Delta t$. This information is stored with each customer. The time a customer spent in the queue, if departing on the kth $(k > j)$ iteration of the loop, is given by

$$\text{time customer in queue} = (k - j)\Delta t.$$

The average time in the queue is the sum of these customer times divided by the total number of customers served. A *Mathematica* version of this pseudocode is found in the *Mathematica* appendix.

We now consider a simulation of the single-server queue and compare the simulation results to the values predicted by the model of the previous section.

Example. Suppose that customers arrive at a checkout counter at an average rate of $\lambda = 4$ per hour and are serviced at an average rate of $\mu = 7$ per hour. Simulate the single-server queue for a ten-hour period.

We are taking the simulation time step Δt to be 0.01.

6 Continuous Stochasticity

Number Served	Average Queue Time	Average Queue Size
44	0.341818	1.504

TABLE 6.1 Results of One Run of the Single-Server Queue Simulation with $\lambda = 4$ and $\mu = 7$.

Performing one simulation of the single-server routine with these values gives the results summarized in Table 6.1.

The predict values of $\overline{N}$ and $\overline{T}$ are computed using equations (164) and (166) with $\rho = \lambda/\mu = 4/7$. The results are:

$$\overline{N} = \frac{4/7}{1 - 4/7} = 4/3 = 1.333\bar{3}, \quad \text{and} \quad \overline{T} = \frac{1/7}{1 - 4/7} = 1/3 = 0.333\bar{3}.$$

These compare favorably to the simulated values of $\bar{N} = 1.504$ and $\bar{T} = 0.3418$.

As with all simulations, it is a good idea to run several trials to verify that the results obtained can be replicated and were not due to chance. Table 6.2 summarizes 10 simulations of the single-server queue model with $\lambda = 4$, $\mu = 7$, and $T = 10$.

Examining the data in Table 6.2, we note that there is a large range to the simulated average queue times and average queue sizes. The range of the simulated average queue time is $0.616765 - 0.181351 = 0.435414$, whereas the range of the average queue size is $2.566 - 0.69 = 1.876$. This large difference is due to the random factors that occur during a short simulation time of $T = 10$. Notice here, though, that the average value of the 10 simulated average queue times is 0.338499, which compares very well to the predicted average queue time of $\overline{T} = 1/3$. Further, note that the average value of the 10 simulated average queue sizes is 1.2872, which is also close to the predicted value of the average queue size of $\overline{N} = 4/3$.

Equations (164) and (166) produce the steady-state or long term average behavior of the single-server queue. It is not too surprising that the simulated values obtained above

Number Served	Average Queue Time	Average Queue Size
33	0.454545	1.508
41	0.198293	0.813
45	0.248222	1.117
34	0.271176	1.030
34	0.616765	2.566
33	0.325758	1.075
46	0.387826	1.784
37	0.181351	0.671
32	0.505625	1.618
35	0.195429	0.690

TABLE 6.2 Results of Ten Runs of the Single-Server Simulation with $\lambda = 4$ and $\mu = 7$.

Number Served	Average Queue Time	Average Queue Size
2,041	0.387242	1.58206
2,071	0.314951	1.28780
2,119	0.301245	1.32171

TABLE 6.3 Results of Three Runs of the Single-Server Queue Simulation with $\lambda = 4$, $\mu = 7$, and $T = 500$.

do not precisely match these predicted values; after all, the simulations were run for only $T = 10$. If we increase the size of T, the simulated values of $\overline{T}$ and $\overline{N}$ will approach the predicted values. Table 6.3 summarizes the values for three simulation runs of time $T = 500$.

Computing the average of the three simulated average queue times gives 0.334479, while the average of the three simulated average queue sizes is 1.397190. Each of these averages is closer to its predicted values than the results from the $T = 10$ simulation were. Notice that the range of the simulated values is also smaller.

6.7 A Pure Birth Process

In Chapter 2, we considered the effects of demographic stochasticity on a population. Recall that we observed that population parameters often vary from year to year due to reasons ranging from small environmental variations to changes in the number of fertile females. Usually, though, the rates are close to the average, and large deviations are rare. For this reason, we assumed that the birth and death rates were normally distributed. In this section, we look at a different type of demographic stochasticity. The important difference is that we use a continuous model rather than a discrete model. Thus change is happening in the population all the time, not just during breeding periods or over winters. Another difference is that with the sandhill crane model in Chapter 2, for each time step we generated random variables from normal distributions for the birth and death rates. Now we have a fixed birth rate which can be thought of as giving the probability of flipping heads with an unfair coin. Every time step we flip the coin. If heads comes up, an individual is born. Since this is a continuous model, the time step is very small and the coin is flipped all the time. Thus the demographic stochasticity in this model comes from the randomness involved with a coin toss (Bernoulli trial), whereas the randomness in the Chapter 2 model comes from the normal distribution. Finally, we consider a model involving only births. This simplifies the analysis. Deaths can be added in an obvious manner, though the analysis becomes more complicated. Thus the model in this section is called a *pure birth* process.

Throughout this text, we extensively considered the exponential model, the continuous version governed by the differential equation

$$\frac{dx}{dt} = rx(t).$$

6 Continuous Stochasticity

This deterministic model works with the aggregate population and assumes not simply that each individual may reproduce but that it actually *does* reproduce. Further, if the average number of births per individual over some time period is 0.8, then this model gives 0.8 offspring to each individual. We understand this as an average over the entire population, but this averaging process eliminates the natural variance in the population. We now consider a model that assumes there is a probability that a particular individual will reproduce in a given time interval.

We assume that the probability of reproduction for one individual in a very short time interval of length Δt is proportional to Δt, that is

$$P[\text{one individual gives birth in } \Delta t] = b\Delta t.$$

If we continue the coin flip analogy $b\Delta t$ is the probability of our coin coming up heads and Δt is how frequently we toss the coin. Notice that by throwing the coin twice as frequently, Δt is cut in half, as is the probability of tossing a head. Using complements, we find

$$P[\text{one individual does not give birth in } \Delta t] = 1 - b\Delta t.$$

These probabilities are for each individual in the population; we need to know the probability of one birth's occurring for the population as a whole.

Suppose that there were $n = 2$ individuals in the population. If we assume that births to each individual occur independently, then the probability of no births' occurring during the time interval Δt is

$$P[\text{no births occur in } \Delta t] = (1 - b\Delta t)^2. \tag{168}$$

Thus, expanding the right-hand side of (168), we obtain

$$(1 - b\Delta t)^2 = 1 - 2b\Delta t + (\Delta t)^2.$$

Since Δt is assumed small, $(\Delta t)^2$ is even smaller, in fact, essentially negligible. Thus we assume that

$$P[\text{no births occur in } \Delta t] = 1 - 2b\Delta t. \tag{169}$$

Hence, the probability of one birth occurring in Δt in a population of size $n = 2$ is

$$P[\text{one birth occurs in } \Delta t] = 2b\Delta t. \tag{170}$$

A similar argument (see Exercise 17) using the binomial theorem shows that in a population of size n,

$$P[\text{one birth occurs in } \Delta t] = nb\Delta t. \tag{171}$$

As we assumed earlier, the interval of time Δt is sufficiently small, so that at most one birth can occur with probability given by equation (171).

Suppose that the population has n_0 individuals initially. Letting $N(t)$ be the number of individuals in the population at time t. Then $N(0) = n_0$, and we seek to determine

$$P_n(t) = P[N(t) = n].$$

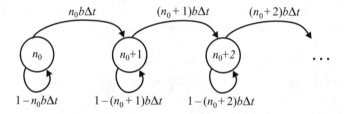

FIGURE 6.10 State Diagram for a Pure Birth Process.

Here $N(t)$ has possible values $n_0, n_0 + 1, n_0 + 2, \ldots$ since no deaths occur in this population.

The state diagram for the Markov chain that $N(t)$ forms is shown in Figure 6.10. The transition matrix for this chain is

$$A = \begin{pmatrix} 1 - n_0 b \Delta t & 0 & 0 & 0 & \cdots \\ n_0 b \Delta t & 1 - (n_0 + 1) b \Delta t & 0 & 0 & \cdots \\ 0 & (n_0 + 1) b \Delta t & 1 - (n_0 + 2) b \Delta t & 0 & \cdots \\ \vdots & \vdots & \vdots & \vdots & \ddots \end{pmatrix}, \quad (172)$$

Following our usual method, we obtain the state equations for this Markov chain by letting

$$X_t = \begin{pmatrix} P_0(t) \\ P_1(t) \\ \vdots \\ P_n(t) \\ \vdots \end{pmatrix},$$

so that we obtain the state equation

$$X_{t+\Delta t} = A X_t. \quad (173)$$

Performing the matrix multiplication gives the system of equations

$$P_{n_0}(t + \Delta t) = (1 - n_0 b \Delta t) P_{n_0}(t)$$

$$P_{n_0+1}(t + \Delta t) = n_0 b \Delta t P_0(t) + (1 - (n_0 + 1) b \Delta t) P_{n_0+1}(t)$$

$$\vdots$$

$$P_{n_0+k}(t + \Delta t) = (n_0 + k - 1) b \Delta t P_{n_0+k-1}(t) + (1 - (n_0 + k) b \Delta t) P_{n_0+k}(t)$$

$$\vdots$$

Proceeding as before, we let Δt go to zero and obtain a system of differential equations. For $n = n_0$,

$$P'_{n_0}(t) = -n_0 b P_0(t), \quad (174)$$

6 Continuous Stochasticity

and for $n \geq n_0 + 1$,
$$P_n'(t) = (n-1)bP_{n-1}(t) - nbP_n(t). \tag{175}$$

The equation (174) is the exponential differential equation with the initial condition $P_{n_0}(0) = 1$. Since there are initially n_0 individuals in the population, this gives
$$P_{n_0}(t) = e^{-n_0 bt}. \tag{176}$$

Equation (176) predicts the probability that the population is still at size n_0 at time t, that is, the probability that no births occurred in the interval $[0, t]$. Note that this probability is always positive, but decreases as time increases, asymptotically approaching 0 as t increases without bound. Using (176) in the equation (175), we obtain
$$P_{n_0+1}'(t) = e^{-n_0 bt} - (n_0+1)bP_{n_0+1}(t) \tag{177}$$

which is a first-order linear differential equation (recall from Chapter 5). If we multiply by the integrating factor $e^{b(n_0+1)t}$, we have
$$e^{b(n_0+1)t}P_{n_0+1}'(t) + (n_0+1)be^{b(n_0+1)t}P_{n_0+1}(t) = n_0 be^{bt},$$

which is equivalent to
$$\frac{d}{dt}\left[e^{b(n_0+1)t}P_{n_0+1}(t)\right] = n_0 be^{bt}.$$

Integrating this expression gives
$$e^{b(n_0+1)t}P_{n_0+1}(t) = n_0 e^{bt} + c,$$

where c is an arbitrary constant. The assumption that there are n_0 individuals in the population initially implies that $P_{n_0+1}(0) = 0$, so that $c = 0$. Hence, the solution to the equation (177) is
$$P_{n_0+1}(t) = n_0 e^{-bn_0 t}(1 - e^{-bt}). \tag{178}$$

Once $P_{n_0+1}(t)$ is known, it is used to find $P_{n_0+2}(t)$. More specifically, using equation (178) in equation (175) with $n = n_0 + 2$ gives the differential equation
$$P_{n_0+2}'(t) + (n_0+2)bP_{n_0+2}(t) = (n_0+1)bn_0 e^{-bn_0 t}(1 - e^{-bt}). \tag{179}$$

Equation (179) is again a first-order linear differential equation. If we multiply each side of the equation by $e^{b(n_0+2)t}$, integrate, and use the initial condition that $P_{n_0+2}(0) = 0$, we obtain the solution
$$P_{n_0+2}(t) = \frac{(n_0+1)n_0}{2} e^{-n_0 bt}(1 - e^{-bt})^2.$$

We continue this procedure. The general formula, which may be checked by induction, is
$$P_n(t) = \binom{n-1}{n_0-1} e^{-n_0 bt}(1 - e^{-bt})^{n-n_0} \text{ for } n \geq n_0. \tag{180}$$

The equation (180) is the probabilistic model for a pure birth process. It gives the probability distribution of the size of the population at time t. While the deterministic model gives a single number as the prediction for the population size at time t, the probabilistic model gives the relative likelihood of each different possible population size at time t.

The deterministic model was much simpler to obtain than the probabilistic model. Is there a connection between these two models? The answer to this question is found by considering the expected value (or average) of the probabilistic model.

To compute the expected value, let $m(t) = E(N(t))$ and recall the definition of the expected value of a discrete random variable from Chapter 2. Hence

$$m(t) = \sum_{n=n_0}^{\infty} nP_n(t).$$

Differentiating this expression gives the equation

$$m'(t) = \sum_{n=n_0}^{\infty} nP_n'(t). \tag{181}$$

If we substitute the equation (175) in for $P_n'(t)$ in (181), we obtain

$$m'(t) = \sum_{n=n_0}^{\infty} nP_n'(t)$$

$$= \sum_{n=n_0}^{\infty} n((n-1)bP_{n-1}(t) - nbP_n(t))$$

$$= b\sum_{n=n_0}^{\infty} (n(n-1)P_{n-1}(t) - n^2 P_n(t)). \tag{182}$$

Expanding the sum in the expression (182) and recalling that $P_n(t) = 0$ for $n < n_0$ gives

$$m'(t) = b(-n_0^2 P_{n_0}(t) + (n_0+1)n_0 P_{n_0}(t) - (n_0+1)^2 P_{n_0+1}(t)$$
$$+ ((n_0+2)(n_0+1)P_{n_0+1}(t) - (n_0+2)^2 P_{n_0+2}(t) + \ldots)$$
$$= b(n_0 P_{n_0}(t) + (n_0+1)P_{n_0+1}(t) + (n_0+2)P_{n_0+2}(t) + \ldots)$$
$$= b\sum_{n=n_0}^{\infty} nP_n(t)$$
$$= bm(t). \tag{183}$$

Equation (183) is the exponential differential equation, the initial condition following from the initial population size's being n_0. That is, since $P_{n_0}(0) = 1$, we have $m(0) = n_0$. The solution to the equation (183) with this initial condition is then

$$m(t) = n_0 e^{bt}. \tag{184}$$

Equation (184) is also the deterministic model for the population. Thus, the deterministic model coincides with the mean of the probabilistic model. Recall that we observed this numerically in the sandhill crane model with demographic stochasticity.

6 Continuous Stochasticity

The variance of the pure birth process is

$$v(t) = n_0 e^{bt}(e^{bt} - 1),$$

and is obtained similarly using the above approach (see Exercise 20). The variance provides a measure of spread for this model and is unavailable if the deterministic model is used alone.

6.7.1 A Remark on Simulating the Pure Birth Process

In the previous section we considered the pure birth process. It is not difficult to develop simulation models for both it and the more complicated birth-death processes considered in the projects section. Recall that in the pure birth process, the probability that a birth occurs during an interval of length Δt is

$$P[\text{one birth occurs in } \Delta t] = nb\Delta t,$$

where n is the current population size and b is the average birth rate. This probability is simulated in the way we simulated arrivals and departures to a queue. In particular, an event, such as a birth, is generated each time step by letting

$$\text{birth} = \text{Uniform}(0, 1).$$

A birth occurs if this random variable has a value less than $nb\Delta t$. This is determined using the following "If" statement:

$$\text{If birth} < nb\Delta t, \text{ Then a birth occurred Else no birth.}$$

The consequences of these simulations are considered in the project section.

6.8 For Further Reading

For a book introducing many issues in continuous and discrete stochastic models see:

Howard M. Taylor and Samuel Karlin, *An Introduction to Stochastic Modeling*, Revised Edition, Academic Press, San Diego, 1994.

Everything worth knowing about queues can be found in these two volumes:

Leonard Kleinrock, *Queueing Systems Volume 1: Theory*, Wiley Interscience, New York, 1975.

Leonard Kleinrock, *Queueing Systems Volume 2: Computer Applications*, Wiley Interscience, New York, 1976.

6.9 Exercises

11. a. Show that if T is an exponentially distributed random variable with mean $\frac{1}{\mu}$, then

$$P[T > t] = e^{-\mu t}.$$

b. Show that if T is an exponentially distributed random variable with mean $\frac{1}{\mu}$, then T satisfies the memoryless property; that is, show

$$P[T > s+t \text{ given } T > s] = e^{-\mu t} = P[T > t] \text{ for all } s, t \geq 0.$$

c. A useful fact about real-valued functions is that if ϕ is a function defined on $[0, \infty]$ which is nonincreasing and satisfies both $\phi(0) \neq 0$ and

$$\phi(s+t) = \phi(s)\phi(t) \text{ for } s, t \geq 0,$$

then

$$\phi(t) = e^{-\theta t}$$

for some $\theta \geq 0$. Using this fact, show that if a random variable T satisfies the memoryless property

$$P[T > s+t \text{ given } T > s] = P[T > t] \text{ for all } s, t \geq 0,$$

then T is exponentially distributed. Thus, this fact together with part **b** gives the memoryless property as a characterization of the exponential distribution.

12. Show by induction, that the system of differential equations

$$P'_n(t) = \lambda[P_{n-1}(t) - P_n(t)],$$

and for $n = 0$,

$$P'_0(t) = -\lambda P_0(t),$$

has solution given by the Poisson process

$$P_n(t) = e^{-\lambda t}\frac{(\lambda t)^n}{n!} \text{ for } t \geq 0, n = 0, 1, 2, \ldots.$$

13. Given the Poisson process,

$$P_n(t) = e^{-\lambda t}\frac{(\lambda t)^n}{n!} \text{ for } t \geq 0, n = 0, 1, 2, \ldots.$$

Show that for each t,

$$\sum_{n=0}^{\infty} P_n(t) = 1.$$

This is the key step in showing that the formula given for the Poisson process is a valid probability distribution.

14. a. Using the power series representation for $e^{-\lambda t}$, show that for a fixed value of t,

$$\sum_{x=0}^{\infty} x \frac{(\lambda t)^x e^{-\lambda t}}{x!} = \lambda t.$$

b. Show that the mean of the Poisson process with parameter λ is λt.

15. Using

$$\lim_{n \to \infty} \left(1 - \frac{\lambda}{n}\right)^n = e^{-\lambda},$$

6 Continuous Stochasticity

show that for a fixed value of x and if $\lambda = np$, then

$$\lim_{n\to\infty} \binom{n}{x} p^x(1-p)^{n-x} = \frac{(\lambda)^x e^{-\lambda}}{x!}.$$

Thus, the Poisson distribution can be considered as a large sample size approximation of the binomial distribution considered in Chapter 2.

16. Show by induction that the system of equations

$$-\lambda P_0 + \mu P_1 = 0$$

$$\lambda P_{n-1} - (\lambda + \mu)P_n + \mu P_{n+1} = 0 \text{ for } n \geq 1.$$

has solution

$$P_n = \left(\frac{\lambda}{\mu}\right)^n P_0.$$

17. Using the binomial theorem

$$(x+y)^n = \sum_{k=0}^{n} \binom{n}{k} x^k y^{n-k},$$

expand $(1 - b\Delta t)^n$ to show that if $(\Delta t)^j$ is assumed to be negligible for $1 < j \leq n$, then

$$(1 - b\Delta t)^n \approx 1 - nb\Delta t.$$

18. Show by induction that the system of differential equations

$$P'_{n_0}(t) = -n_0 b P_0(t), \text{ for } n = n_0,$$

$$P'_n(t) = (n-1)b P_{n-1}(t) - nb P_n(t), \text{ for } n \geq n_0 + 1,$$

has solution

$$P_n(t) = \binom{n-1}{n_0-1} e^{-n_0 bt}(1 - e^{bt})^{n-n_0} \text{ for } n \geq n_0.$$

19. Given the pure birth process,

$$P_n(t) = \binom{n-1}{n_0-1} e^{-n_0 bt}(1 - e^{-bt})^{n-n_0} \text{ for } n \geq n_0.$$

Show that

$$\sum_{n=n_0}^{\infty} P_n(t) = 1.$$

This is the key step in showing that $P_n(t)$ is a valid probability distribution on the set of integers greater than or equal to n_0.

20. To compute the variance of the pure birth process, we need first to find the second moment. Let $s(t) = E(N^2(t))$ so that

$$s(t) = \sum_{n=n_0}^{\infty} n^2 P_n(t).$$

Differentiate this expression to obtain

$$s'(t) = \sum_{n=n_0}^{\infty} n^2 P'_n(t).$$

a. Using the differential equation (175) of the pure birth process, replace $P'_n(t)$ in the expression for $s'(t)$ to obtain

$$s'(t) = \sum_{n=n_0}^{\infty} n^2 P'_n(t)$$

$$= b \sum_{n=n_0}^{\infty} (n^2(n-1)P_{n-1}(t) - n^3 P_n(t)).$$

b. Expanding the sum in **a** and using the differential equation (175) of the pure birth process, show that

$$s'(t) = 2bs(t) + 2bm(t),$$

where

$$m(t) = n_0 e^{bt}$$

is the mean of the pure birth process. Using the initial condition $s(0) = n_0^2$, show that

$$s(t) = n_0 e^{bt}((n_0+1)e^{bt} - 1).$$

c. Using the solution obtained in **b** and the fact that the mean of the pure birth process is $m(t) = n_0 e^{bt}$, show that the variance

$$V(N(t)) = s(t) - m^2(t)$$

of the pure birth process is

$$V(N(t)) = n_0 e^{bt}(e^{bt} - 1).$$

6.10 Projects

Project 6.1. A Basic Queuing Simulation

You are to construct a simulation for a bank so that its managers can study the time customers have to wait in line. The bank gives you the following assumptions and simulation requirements.

i. The time between customers' arriving varies from a minimum of $Arrive_{\min}$ to a maximum of $Arrive_{\max}$, and any time in between is equally likely (in other words a random number from a uniform distribution).

ii. The amount of time required to serve a customer varies from a minimum of $Service_{\min}$ to a maximum of $Service_{\max}$, again with any time in between equally likely.

iii. The bank opens its doors and stays open for T minutes. After T minutes, the doors close to prevent new arrivals, but customers currently in line continue to be served. Thus the simulation may run longer than T minutes.

6 Continuous Stochasticity

iv. There is one bank teller. In part **3**, we expand the problem for n tellers, so keep this in mind when designing the simulation.

v. A typical day begins with the bank opening. The first customer will arrive between $Arrive_{min}$ and $Arrive_{max}$ minutes later. The customer goes straight to the teller. The next customer arrives. If the first customer is finished, then the new customer goes straight to the window. If the first customer is still being served, the new customer starts a line or queue. Whenever a customer arrives, that person will go to the teller if the teller is unoccupied or to the end of the line if the teller is occupied. Whenever a teller finishes serving the current customer, the customer at the front of the queue leaves the queue and goes to the counter. If there is currently no queue formed, then when the current customer is through with service, the teller is idle until the new customer arrives.

1. Build a simulation that satisfies these requirements. While this can be constructed in a program such as *Mathematica*, we recommend using a programming language to build the simulation. The simulation parameters should be easy to change so that bank managers can study different scenarios. The simulation should keep track of and report the number of customers, the average wait per customer (the time a customer waits is the time spent in line and does not count the time being served), the percent of time that the teller was occupied, and the length of the simulation (time until the last customer leaves).

This simulation is similar to the example in the text, but differs in several important regards. The simulation in the book had a clock that advanced in small time units Δt. If $\Delta t = 0.01$, a five-minute simulation would require stepping through the code loop 500 times. This model has a clock that advances from event to event. For example, if, in the first five minutes, a customer arrives at 2.3 minutes and departs at 4.2 minutes, the simulation loop needs to be run only twice. Thus this simulation will run much faster. You can implement this in any way you see fit. One way is to compute the next arrival time whenever an arrival occurs and to compute the next departure time whenever a customer goes to the teller. The next event will be the smaller of these two times.

Another difference between this simulation and the example in the text is that we are assuming uniform inter-arrival times instead of exponential; similarly, we have different service assumptions. Thus this is a different model and will give different results. Our reason for these assumptions, however, is to start with the simplest possible simulation. Your simulation can easily be modified to generate an exponential random number for the next arrival. We leave it to the inquisitive mind to figure out how.

2. Run your model for $T = 480$ minutes with the following sets of parameters:

a. $Arrive_{min} = 1$, $Arrive_{max} = 3$, $Service_{min} = 1$, and $Service_{max} = 15$.
b. $Arrive_{min} = 1$, $Arrive_{max} = 3$, $Service_{min} = 1$, and $Service_{max} = 3$.
c. $Arrive_{min} = 1$, $Arrive_{max} = 15$, $Service_{min} = 1$, and $Service_{max} = 3$.

Comment on why the results behave as they do.

3. Revise your model to have n tellers. When more than one teller is open, a decision needs to be made about which teller the next arrival should go to. Number your tellers and always send a customer to the available teller with the smallest number. This corresponds to a bank where the queue starts at one end of a row of tellers. Think about how you would model a queue that begins at the middle of a row of tellers.

4	1	2	4	1	1	6	2	3	5	3
7	6	3	5	3	0	2	3	4	5	3
2	6	9	4	6	4	6	6	5	6	5
3	5	1	7	2	6	4				

TABLE 6.4 Number of Car Arrivals at a Traffic Light during One Light Cycle. First Hour of Observation.

3	5	5	7	6	1	6	6	13	6	1
1	2	5	2	8	4	6	6	1	1	1
6	5	4	7	2	1	0	8	7	3	2
5	6	4	4	5	2	0	3	8	7	4
7	7	12	3	4						

TABLE 6.5 Number of Car Arrivals at a Traffic Light during One Light Cycle. Second Hour of Observation.

4. Repeat part **2**, but with $n = 2$ and again with $n = 3$.

Project 6.2. Modeling Arrivals at a Traffic Light

In this project, we verify that the number of arrivals at a traffic light arise from a Poisson process. The data in Tables 6.4 and 6.5 were collected by student Mark M. Young at a traffic light in Russellville, Kentucky. He counted the number of cars that arrived during each cycle of the traffic light. The data is presented in two separate tables to reflect two periods of observation. The data in Table 6.4 were collected between 11:00 am and 12:00 noon, while the data in Table 6.5 were collected between noon and 1:00 pm.

We apply the chi-square goodness-of-fit test to determine if these observations arise from a Poisson process.

1. For each of the data sets in Tables 6.4 and 6.5, calculate the average number of cars that arrived during a light cycle. Note that the average number of arrivals during the first hour is smaller than during the second hour. Also calculate the average number of arrivals for both hours of data.

2. Testing for a Poisson Distribution. For each value of t, the Poisson process

$$P_n(t) = e^{-\lambda t} \frac{(\lambda t)^n}{n!} \quad \text{for } t \geq 0, n = 0, 1, 2, \ldots.$$

gives a probability distribution for the number of occurrences of an event. This distribution is known as the Poisson distribution. In particular,

$$P(n \text{ occurrences}) = e^{-\tilde{\lambda}} \frac{(\tilde{\lambda})^n}{n!} \quad \text{for } t \geq 0, n = 0, 1, 2, \ldots.$$

6 Continuous Stochasticity

where $\tilde{\lambda} = \lambda t$ represents the average number of occurrences, is the Poisson distribution. If we assume that the length of each light cycle is the same, then the number of arrivals at a traffic light theoretically should follow a Poisson distribution.

Create a frequency distribution for the data in Table 6.4. This gives the observed values. Use the average value calculated in part **1** for $\tilde{\lambda}$ for the number of arrivals during the first hour to construct a Poisson distribution for the number of arrivals. This gives the expected values. Apply the chi-square goodness-of-fit test to determine how well the observed data fits a Poisson distribution. (The categories are $n = 0, 1, 2, \ldots$.)

Repeat this procedure for the data in Table 6.5. Is the fit as good? Why or why not? Recall that the data were collected at different times of the day.

Combine the data in Tables 6.4 and 6.5 and create a frequency distribution. Does a Poisson distribution fit the data? Is this fit better than the individual fits considered above? Why or why not?

3. Simulation Use a computer to simulate arrivals at a traffic light by generating random data from a Poisson distribution. Use the mean values obtained in part **1**. How does your simulated data compare to the data in Tables 6.4 and 6.5?

Project 6.3. Modeling Inter-arrival Times

In Project 6.2, a Poisson process was fit to the number of arrivals at a traffic light and the goodness-of-fit was measured with the X^2 statistic. Recall from the text that if the number of arrivals follows a Poisson process, then the time between arrivals (the inter-arrival time) follows an exponential distribution (and conversely).

In this project, we verify that the inter-arrival times at the traffic light follow an exponential distribution. The data in Tables 6.6 and 6.7 are the inter-arrival times of the car arrivals reported in Tables 6.4 and 6.5. Mark M. Young also collected this data.

1. For each of the data sets in Tables 6.6 and 6.7, calculate the average inter-arrival time of the cars. Note that the average inter-arrival time during the first hour is larger than during the second hour. Calculate the average inter-arrival time for the combined data.

2. Testing for an Exponential Distribution. Group the data into two-second intervals. Create a frequency distribution for the data in Table 6.6. Use the average value calculated in part **1** for the inter-arrival time during the first hour to construct the exponential distribution for the inter-arrivals. Recall that the exponential distribution is $f(t) = \frac{1}{\theta}\exp(\frac{-t}{\theta})$ where θ is the mean value of t (inter-arrival time in this case). Apply the chi-square goodness-of-fit test to determine how well the observed frequency fits an exponential distribution. Repeat using one-second intervals? Does the fit get better or worse? Why? What about other interval sizes?

Repeat this procedure for the data in Table 6.7. Is the fit as good? Why or why not? Recall that the data were collected at different times of the day. Does changing the interval widths improve the fit?

Combine the data in Tables 6.6 and 6.7 and create a frequency distribution. Does an exponential distribution fit the data? Is this fit better than the individual fits considered above? Why or why not? Experiment with the interval widths.

3. Collect your own data set for a traffic light. Keep track of the number of arrivals during each light cycle and the times between each cars arrival. Repeat the above analysis for

0.58	10.90	14.90	14.77	8.05	5.83	28.36	2.96
6.93	12.60	0.55	9.94	10.51	7.56	3.82	30.79
9.91	16.61	7.18	10.26	2.56	2.04	2.29	17.30
5.59	5.41	6.38	2.62	2.57	0.00	4.45	4.24
8.11	8.75	3.17	2.20	12.87	12.97	9.17	2.96
15.58	12.20	10.57	5.35	1.02	2.62	25.01	4.54
2.96	18.39	3.76	7.05	5.01	3.06	8.64	9.67
9.42	6.24	12.67	4.27	3.20	3.43	5.36	5.07
7.41	0.00	13.34	1.82	2.51	0.00	4.21	4.58
5.36	0.00	19.01	3.96	28.64	3.18	6.12	3.39
1.99	4.99	14.98	5.15	2.97	4.14	4.56	19.49
0.00	5.64	7.25	7.04	0.19	3.03	3.85	0.00
13.23	11.24	0.00	4.71	7.00	5.08	8.76	5.83
3.37	0.00	25.87	6.01	0.00	0.00	11.00	3.37
3.98	0.00	6.07	0.00	4.80	0.00	5.20	13.45
15.37	0.00	5.35	4.79	18.28	6.47	2.50	3.55
22.45	5.28	2.61	1.74	8.16	2.32	3.17	24.37
5.45	3.39	7.29	2.49	21.31	4.40	0.00	21.23
4.09	3.14	3.52	6.16	3.03	5.06	16.49	4.82
5.13	7.05	0.00	3.56	2.64	4.38	10.73	7.20
2.28	9.02	7.91	10.95	0.00	12.01	0.00	3.03

TABLE 6.6 Inter-arrival Times (in seconds) of Cars at a Traffic Light. First Hour of Observation.

2.44	13.37	2.89	11.74	3.20	2.05	6.95	3.02
6.02	0.00	0.00	4.08	6.42	8.28	0.00	11.17
2.56	3.27	1.91	4.30	11.18	5.62	0.00	14.29
2.82	12.20	2.22	1.96	7.17	2.49	0.00	4.07
5.21	0.00	12.96	5.05	2.60	0.00	3.76	0.00
3.65	0.00	18.88	7.06	8.37	8.96	3.36	0.00
1.87	5.19	1.83	3.89	6.74	8.42	3.39	4.63
2.76	27.24	0.00	25.52	0.00	6.52	7.93	0.00
1.70	3.84	1.86	3.03	0.00	12.85	8.62	5.35
2.00	8.77	0.00	0.00	11.34	3.92	0.00	8.67
5.11	5.49	2.56	3.54	3.00	12.20	4.14	27.43
4.35	4.87	3.53	8.74	0.00	6.03	9.09	3.55
3.49	3.47	4.51	6.61	16.54	8.69	12.80	7.64
10.28	12.21	0.00	9.15	13.66	3.76	0.00	0.00
4.88	15.97	0.00	14.90	16.19	7.20	2.07	4.85
6.62	8.58	4.94	1.32	35.59	2.52	2.00	6.76
6.86	5.46	13.97	4.48	2.76	0.00	3.61	7.06
2.97	6.86	2.23	0.00	11.26	25.59	2.32	0.00
16.39	2.93	8.37	0.00	0.00	4.03	0.00	0.00
7.62	0.00	9.18	0.00	0.00	4.04	8.26	0.00
5.45	14.49	4.21	23.23	4.11	5.53	10.07	3.85

TABLE 6.7 Inter-arrival Times (in seconds) of Cars at a Traffic Light. Second Hour of Observation.

your data. Does the number of arrivals follow a Poisson distribution? Are the inter-arrival times exponentially distributed? Collect data for different times of the day. Are some fits better than others? Why?

Project 6.4. A Simple Death Process

In the text we considered the pure birth process. In this project, we consider a *pure death process*. In this rather macabre process, individuals persist only until they die and there are no replacements. The assumptions are similar to those in the pure birth process, but now each individual, if still alive at time t, is removed in $(t, t + \Delta t)$ with probability $\mu \Delta t$. If we let $N(t)$ be the number of individuals alive at time t, then we obtain a system of differential equations similar to those obtained in the pure birth process. Suppose initially, there are n_0 individuals, that is

$$N(0) = n_0.$$

1. Draw the state diagram for the pure death process and find its transition matrix.

2. Using the transition matrix, show that the differential equations for the time-dependent probability distribution of the pure death process are:

$$P'_{n_0}(t) = -n_0 P_{n_0}(t)$$
$$P'_{n_0-1}(t) = n_0 P_{n_0}(t) - (n_0 - 1) P_{n_0-1}(t)$$
$$\vdots \tag{185}$$
$$P'_0(t) = P_1(t)$$

where

$$P_k(t) = P[N(t) = k \mid N(0) = n_0].$$

3. Show that the system of differential equations (185) obtained in part **1** has solution

$$P_k(t) = \binom{n_0}{k} e^{-\mu k t}(1 - e^{-\mu t})^{n_0 - k} \text{ for } k = n_0, n_0 - 1, \ldots, 1, 0, \tag{186}$$

by first finding a solution for $P_{n_0}(t)$ and then using this solution to find the subsequent $P_k(t)$'s.

4. **An alternate method of solution.** Assume that $n_0 = 1$. Now $P_1(t)$ is the probability that this single individual is still alive at time t; show that

$$P_1(t) = e^{-\mu t}.$$

(This is the same model obtained in Section 6.2 for service at a checkout counter.) If there are $n_0 > 1$ individuals alive at $t = 0$, argue that the number alive at t is a binomial random variable with parameters n_0 and $P_1(t)$ so that (186) follows immediately.

5. **The mean and variance of the pure death process.** Using the mean and variance of a binomial random variable with $P_1(t) = e^{-\mu t}$ as the probability of a success, show that the mean of the pure death process is

$$E(N(t)) = n_0 e^{-\mu t}, \tag{187}$$

and that the variance is

$$V(N(t)) = n_0 e^{-\mu t}(1 - e^{-\mu t}). \tag{188}$$

Compare the expression for the mean of the pure death process with that of an exponentially decreasing deterministic population.

6. Extinction. In the pure death process the population either remains constant or it decreases. It may eventually reach zero in which case we say that the population has gone *extinct*. Show that the probability the population is extinct at time t is

$$P[N(t) = 0 \mid N(0) = n_0] = (1 - e^{-\mu t})^{n_0}. \tag{189}$$

Use this expression to show that extinction is inevitable.

7. Simulation. The pure death process can be simulated in the same manner as we simulated the arrivals and departures to a queue. In particular, the event "a death occurs" can be generated each time step by letting

$$\text{death} = \text{Uniform}(0, 1)$$

so that a death occurs if this random variable has a value less than $k\mu\Delta t$, where $0 < k \leq n_0$ is the current population size. This event is computed using the following "If" statement:

If death $< k\mu\Delta t$, Then a death occurred, Else no death.

Using a small time step size—$\Delta t \leq 0.01$ is recommended, run a simulation of the pure death process. Use an initial population size of $n_0 = 25$ and $\mu = 0.5$ and run your simulation for a simulation time period of $T = 10$ years. Report the population size at each half year of the simulation run. Plot a graph of the population size. How does this graph compare to the deterministic exponential model (187)? Run the simulation 25 times and compute the average population size of the 25 simulations at each half year. Plot a graph of the average population size. How does this graph compare to the deterministic model?

Experiment with the simulation. Try changing the initial population size n_0 and/or the death rate μ.

8. Estimating Extinction Using the simulation, compute the probability of extinction. The probability of extinction can be estimated from the simulation data. To estimate this probability, count the number of simulation runs in which the population went extinct and form the ratio with the total number of simulation runs. How well does this quantity compare with the theoretical probability of extinction (189)? Increase the number of simulations; does this estimated probability of extinction approach the probability given by (189)? Experiment with extinction. Try changing the initial population size n_0 and/or the death rate μ. Which affects the probability of extinction more, n_0 or μ?

Project 6.5. A Birth-Death Process

Next we consider a *birth-death process*. In this process, individuals are born and die. The assumptions are similar to those in the pure birth process, but now each individual will die in $(t, t + \Delta t)$ with probability $\mu \Delta t$ and will give birth in $(t, t + \Delta t)$ with

6 Continuous Stochasticity

probability $\lambda \Delta t$. If we let $N(t)$ be the number of individuals alive at time t, then we obtain a system of differential equations similar to those obtained in the pure birth process. Suppose initially that there are n_0 individuals, that is

$$N(0) = n_0.$$

1. Draw the state diagram for the birth-death process and find its transition matrix.

2. Using the transition matrix, show that the differential equations for the time-dependent probability distribution of the birth-death process are:

$$P'_n(t) = (n-1)\lambda P_{n-1}(t) - n(\lambda + \mu)P_n(t)$$
$$+ (n+1)\mu P_{n+1}(t) \quad \text{for } n \geq 1 \tag{190}$$
$$P'_0(t) = \mu P_1(t) \quad \text{for } n = 0$$

where

$$P_k(t) = P[N(t) = k \mid N(0) = n_0].$$

3. Verify, by substitution that the system of differential equations (190) obtained in **1** has as solution for $\lambda \neq \mu$

$$P_0(t) = \frac{\mu(1 - e^{-(\mu-\lambda)t})}{\mu - \lambda e^{-(\mu-\lambda)t}} \quad \text{and}$$

$$P_n(t) = \lambda^n \mu \left[\frac{(1 - e^{-(\mu-\lambda)t})}{\mu - \lambda e^{-(\mu-\lambda)t}} \right]^{n+1}$$

$$- \lambda^{n-1} \left[\frac{(\lambda - \mu e^{-(\mu-\lambda)t})}{\mu - \lambda e^{-(\mu-\lambda)t}} \right] \left[\frac{(1 - e^{-(\mu-\lambda)t})}{\mu - \lambda e^{-(\mu-\lambda)t}} \right]^{n-1} \quad \text{for } n \geq 1.$$

The solution for the case $\lambda = \mu$ is given by

$$P_0(t) = \frac{\lambda t}{1 + \lambda t} \quad \text{and}$$

$$P_n(t) = \left[\frac{\lambda t}{(1 + \lambda t)^2} \right]^{n-1} \quad n \geq 1.$$

4. The mean and variance of the birth-death process. To obtain the mean number of individuals in the population at time t, let

$$m(t) = E[N(t) \mid N(0) = n_0] = \sum_{n=0}^{\infty} n P_n(t),$$

so that after differentiating with respect to t, we obtain

$$m'(t) = \sum_{n=1}^{\infty} n P'_n(t).$$

Show that by substituting (190) for $P'_n(t)$ and rearranging terms, we obtain the differential equation

$$m'(t) = (\lambda - \mu)m(t). \tag{191}$$

Solve equation (191) with the initial condition $N(0) = n_0$ so that $m(0) = n_0$. Compare this solution to the deterministic model for a population with birth rate λ and death rate μ.

The variance of the birth-death process is obtained in a similar manner using the second moment of $N(t)$. Show that the second moment of $N(t)$,

$$S(t) = \sum_{n=0}^{\infty} n^2 P_n(t),$$

satisfies the differential equation

$$S'(t) = 2(\lambda - \mu)S(t) + (\lambda + \mu)S(t). \tag{192}$$

Using the solution to (192) with initial condition $S(0) = n_0^2$ show that the variance

$$V(N(t)) = S(t) - m^2(t)$$

is

$$V(N(t)) = n_0 \frac{(\lambda + \mu)}{(\lambda - \mu)} e^{(\lambda - \mu)t} (e^{(\lambda - \mu)t} - 1) \text{ for } \lambda \neq \mu. \tag{193}$$

Using L'Hopital's rule, show that the variance in the special case $\lambda = \mu$ is

$$V(N(t)) = 2\lambda n_0 t.$$

5. Extinction. In the birth-death process the population either remains constant, increases, or decreases. It may eventually reach zero, in which case we say that the population has gone *extinct*. Show that the probability that the population is extinct at time t is

$$P[N(t) = 0 \mid N(0) = n_0] = \begin{cases} 1 & \text{for } \lambda \leq \mu \\ \left(\frac{\mu}{\lambda}\right)^{n_0} & \text{for } \lambda > \mu. \end{cases} \tag{194}$$

Using this expression, show that extinction is inevitable if $\lambda \leq \mu$. It may seem surprising that extinction is certain when $\lambda = \mu$. To understand this, we note that $N(t)$ is always a finite distance from the absorbing state 0. The expected time to extinction is infinite. To see this, define the random variable τ which is known as the *extinction time*. Show that for $\lambda < \mu$,

$$P[\tau \leq t] = \left(\frac{\mu(1 - e^{-(\mu - \lambda)t})}{\lambda - \mu e^{-(\mu - \lambda)t}} \right)^{n_0}, \tag{195}$$

and for $\lambda = \mu$,

$$P[\tau \leq t] = \left(\frac{\lambda t}{\lambda t + 1} \right)^{n_0}. \tag{196}$$

Show, using (195), that the expected time to extinction is finite, but using (196) is infinite.

The extinction time τ can also be considered when $\lambda > \mu$; show that in this case

$$P[\tau < \infty] = \left(\frac{\mu}{\lambda}\right)^{n_0},$$

so that

$$P[\tau = \infty] = 1 - \left(\frac{\mu}{\lambda}\right)^{n_0}.$$

6 Continuous Stochasticity

Clearly, in this case, τ has no finite moments and, hence is not a "proper" random variable.

6. Simulation. The birth-death process can be simulated in the same manner as we simulated the arrivals and departures to a queue. In particular, the event "a birth occurs" is generated each time step by letting

$$\text{birth} = \text{Uniform}(0, 1),$$

and the event "a death occurs" is generated each time step by letting

$$\text{death} = \text{Uniform}(0, 1),$$

so that a birth or death occurs if its random variable has a value less than $k\lambda\Delta t$ or $k\mu\Delta t$ respectively, where $0 < k \leq n_0$ is the current population size. These events are computed using the following "If" statements:

If birth $< k\mu\Delta t$, Then a birth occurred, Else no birth,

and

If death $< k\mu\Delta t$, Then a death occurred, Else no death.

Care must be taken to allow only one of the events—"a birth occurred", "a death occurred," or "neither a birth nor a death occurred"—on each of the simulation time steps Δt.

Using a small time step size—$\Delta t \leq 0.01$ is recommended, run a simulation of the birth-death process. Use an initial population size of $n_0 = 25$ and the three cases $\lambda = 0.3$, $\mu = 0.5$, $\lambda = \mu = 0.5$ and $\lambda = 0.7$, $\mu = 0.5$. Run your simulation for a simulation time period of $T = 10$ years. Report the population size at each half year of the simulation run. Plot a graph of the population size. How does this graph compare to the deterministic exponential model (191)? Run the simulation 25 times and compute the average population size of the 25 simulations at each half year. Plot a graph of the average population size. How does this graph compare to the deterministic model?

Experiment with the simulation. Try changing the initial population size n_0 and/or the birth and death rates λ and μ.

7. Estimating Extinction Using the simulation, compute the probability of extinction. The probability of extinction can be estimated from the simulation data. To estimate this probability, count the number of simulation runs that have a population that went extinct and form the ratio with the total number of simulation runs. For the case $\lambda < \mu$, how well does this estimated probability compare with the theoretical probability of extinction (195)? Increase the number of simulations; does this estimated probability of extinction approach the probability given by (195)? Experiment with extinction. Try changing the initial population size n_0 and/or the birth rate λ (try having λ approach μ). What affects the probability of extinction the most, n_0 or λ close to μ?

Project 6.6. A Birth-Death-Immigration Process

In the birth-death process, Project 6.5, we considered a process in which individuals are born with a birth rate λ and die with rate μ. Now we add an immigration rate ν. As before, we think of λ and μ as probabilities to be applied to the individuals present; ν, on the other hand, is a fixed number and is not related to the population attained.

The assumptions are similar to those given in Project 6.5; each individual gives birth during $(t, t + \Delta t)$ with probability $\lambda \Delta t$ and dies with probability $\mu \Delta t$. Suppose initially, there are n_0 individuals; that is,

$$N(0) = n_0.$$

1. Draw the state diagram for the birth-death-immigration process and find its transition matrix.

2. Using the transition matrix, show that the differential equations for the time-dependent probability distribution of the birth-death-immigration process are:

$$P_0'(t) = -\nu P_0(t) + \mu P_1(t)$$
$$P_n'(t) = -(n\lambda + n\mu + \nu)P_n(t) + ((n-1)\lambda + \nu)P_{n-1}(t)$$
$$+ (n+1)\mu P_{n+1}(t) \text{ for } n \geq 1 \tag{197}$$

where

$$P_k(t) = P[N(t) = k \mid N(0) = n_0].$$

3. The steady-state distribution. The system of differential equations (197) is rather difficult to solve analytically. The interested reader can consult Keyfitz [31], who obtains the first two terms $P_0(t)$ and $P_1(t)$. It is possible, however, to obtain the steady-state distribution for the birth-death-immigration process. Recall that the steady-state distribution exists for each $k = 0, 1, \ldots$ when

$$\lim_{t \to \infty} P_k(t) = P_k,$$

which is equivalent for each $k = 0, 1, \ldots$ to

$$P_k'(t) = 0 \text{ for all } t.$$

Applying this condition to (197) gives the system of equations

$$-\nu P_0 + \mu P_1 = 0$$
$$-(n\lambda + n\mu + \nu)P_n + ((n-1)\lambda + \nu)P_{n-1} + (n+1)\mu P_{n+1} = 0 \quad \text{for } n \geq 1. \tag{198}$$

Solve the system of equations (198) for the $P_k's$ to obtain the steady state distribution of the birth-death-immigration process. This system of equations can be solved in the same manner as was accomplished for the steady-state distribution of the single server queue. In particular, using the identity,

$$\sum_{k=1}^{\infty} \frac{\prod_{j=0}^{k-1}(j\lambda + \nu)}{k!\mu^k} = (1 - \frac{\lambda}{\nu})^{-\nu/\lambda} - 1, \tag{199}$$

show that the steady-state distribution is

$$P_k = \frac{\prod_{j=0}^{k-1}(j\lambda + \nu)}{k!\mu^k}(1 - \frac{\lambda}{\nu})^{\nu/\lambda}. \tag{200}$$

6 Continuous Stochasticity

4. The mean and variance of the steady-state distribution of the birth-death-immigration process. The mean and variance of the steady-state distribution of the birth-death-immigration process can be obtained using the identity (199). In particular, show that the mean

$$m = E[N] = \sum_{k=0}^{\infty} k P_k$$

of the steady-state distribution of the birth-death-immigration process is

$$m = \frac{\nu}{\mu - \lambda}. \tag{201}$$

Using the second moment

$$s = E[N] = \sum_{k=0}^{\infty} k^2 P_k,$$

of the steady-state distribution of the birth-death-immigration process, show that the variance is

$$V[N] = s - m^2 = \frac{\mu\nu}{(\lambda - \mu)^2}. \tag{202}$$

5. The mean and variance of the birth-death-immigration process. To obtain the time-dependent mean number of individuals in the population at time t, let

$$m(t) = E[N(t) \mid N(0) = n_0] = \sum_{n=0}^{\infty} n P_n(t),$$

so that after differentiating with respect to t we obtain

$$m'(t) = \sum_{n=1}^{\infty} n P_n'(t).$$

Show that by substituting (197) in for $P_n'(t)$ and rearranging terms, we obtain the differential equation

$$m'(t) = (\lambda - \mu) m(t) + \nu. \tag{203}$$

Solve equation (203) with the initial condition $N(0) = n_0$ so that $m(0) = n_0$. Compare this solution with the affine model (Chapter 1) of a deterministic population with birth rate λ, death rate μ, and recruitment rate ν.

Compute $\lim_{t \to \infty} m(t)$ and compare this value with equation (201). Are these the same? Explain.

The variance of the birth-death-immigration process is obtained in a similar manner using the second moment of $N(t)$. Show that the second moment of $N(t)$,

$$S(t) = \sum_{n=0}^{\infty} n^2 P_n(t),$$

satisfies the differential equation

$$S'(t) = 2(\lambda - \mu) S(t) + (\lambda + \mu) S(t) + \nu(2m(t) + 1), \tag{204}$$

where $m(t)$ is the solution to (203). Using the solution to (204) with initial condition $S(0) = n_0^2$, find the variance

$$V(N(t)) = S(t) - m^2(t)$$

of the birth-death-immigration process. Using L'Hopital's rule, obtain the variance in the special case $\lambda = \mu$.

Compute $\lim_{t \to \infty} V(N(t))$ and compare this value with equation (202). Are these the same? Explain.

6. Extinction—A degenerate case. In the birth-death-immigration process the population either remains constant, increases, or decreases. It may eventually reach zero; however, since there is always a positive immigration rate ν, the population will never become extinct. This may also be observed by noting the existence of the steady-state distribution (200). Show that the population will become extinct as ν goes to zero, by computing $\lim_{\nu \to 0} P_k$, where P_k is given by (200).

7. Simulation. The birth-death-immigration process can be simulated in the same manner as we simulated the arrivals at and departures from a queue. In particular, the event "a birth occurs" is generated each time step by letting

$$\text{birth} = \text{Uniform}(0, 1);$$

the event "a death occurs" is generated each time step by letting

$$\text{death} = \text{Uniform}(0, 1);$$

and the event "an immigration occurs" is generated each time step by letting

$$\text{immigration} = \text{Uniform}(0, 1).$$

A birth, death, or immigration occurs if its random variable has a value less than $k\lambda\Delta t$, $k\mu\Delta t$, or $\nu\Delta t$ respectively, where $0 < k \leq n_0$ is the current population size. (Note that immigration does not depend on the current population size k.) These events are computed using the following "If" statements:

If birth $< k\mu\Delta t$, Then a birth occurred, Else no birth,

If death $< k\mu\Delta t$, Then a death occurred, Else no death,

and

If immigration $< \nu\Delta t$ Then immigration occurred Else no immigration.

Care must be taken to allow only one of the events—"a birth occurred", "a death occurred", "immigration occurred," or "neither a birth, death, nor immigration occurred"—on each of the simulation time steps Δt.

Using a small time step size ($\Delta t \leq 0.01$ is recommended) run a simulation of the birth-death-immigration process. Use an initial population size of $n_0 = 25$, $\nu = 0.5$, and the three cases $\lambda = 0.3$, $\mu = 0.5$; $\lambda = \mu = 0.5$; and $\lambda = 0.7$, $\mu = 0.5$. Run your simulation for a simulation time period of $T = 10$ years. Report the population size at each half year of the simulation run. Plot a graph of the population size. How does this graph compare to the deterministic affine model (203)? Run the simulation 25 times and compute the average population size of the 25 simulations at each half year. Plot a graph of the average population size. How does this graph compare to the deterministic model?

Experiment with the simulation. Try changing the initial population size n_0 and/or the birth-death-immigration rates λ, μ, and ν.

If you have completed Project 6.5, run a simulation of the birth-death-immigration process for decreasing values of ν. Does the behavior of the simulation resemble the behavior of the birth-death process when ν is small?

A

Chi-Square Table

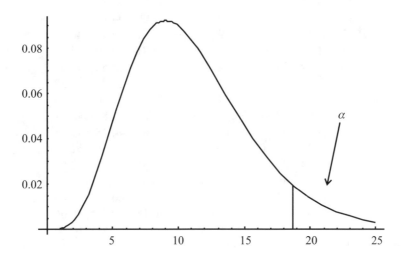

FIGURE A.1 A Chi-Square Distribution.

	α						
r	0.250	0.100	0.050	0.025	0.010	0.005	0.001
1	1.32	2.71	3.84	5.02	6.63	7.88	10.83
2	2.77	4.61	5.99	7.38	9.21	10.60	13.82
3	4.11	6.25	7.81	9.35	11.34	12.83	16.27
4	5.39	7.78	9.49	11.14	13.28	14.86	18.47
5	6.63	9.24	11.07	12.83	15.09	16.75	20.52
6	7.84	10.64	12.59	14.45	16.81	18.55	22.46
7	9.84	12.02	14.07	16.01	18.48	20.28	24.32
8	10.22	13.36	15.51	17.53	20.09	21.96	26.12
9	11.39	14.68	16.92	19.02	21.67	23.59	27.87
10	12.55	15.99	18.31	20.48	23.21	25.19	29.59
11	13.70	17.28	19.68	21.92	24.72	26.76	31.26
12	14.85	18.55	21.03	23.34	26.22	28.30	32.91
13	15.98	19.81	22.36	24.74	27.69	29.82	34.53
14	17.12	21.06	23.68	26.12	29.14	31.32	36.12
15	18.25	22.31	25.00	27.49	30.58	32.80	37.70
16	19.37	23.54	26.30	28.85	32.00	34.27	39.25
17	20.49	24.77	27.59	30.19	33.41	35.73	40.79
18	21.60	25.99	28.87	31.53	34.81	37.16	42.31
19	22.72	27.20	30.14	32.85	36.19	35.58	43.82
20	23.83	28.41	31.41	34.16	37.57	40.01	45.32
21	24.93	29.62	32.67	35.48	38.93	41.40	46.80
22	26.04	30.81	33.92	36.78	40.29	42.80	48.27
23	27.14	32.01	35.17	38.08	41.64	44.18	49.73
24	28.24	33.20	36.42	39.36	42.98	45.56	51.18
25	29.34	34.38	37.65	40.65	44.31	46.93	52.62
30	34.80	40.26	43.77	46.98	50.98	53.67	59.70
40	45.62	51.80	55.76	59.34	63.69	66.77	73.40
50	56.33	63.17	67.50	71.42	76.15	79.49	86.66

TABLE A.1 A Table of Cutoff Values χ_0^2 for a Chi-Square Distribution. $P(X^2 \geq \chi_0^2) = \alpha$, with r Degrees of Freedom.

B

F-Table

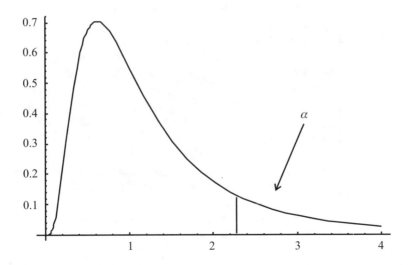

FIGURE B.1 F distribution.

	r_1						
r_2	1	2	3	4	5	6	7
1	161	199	216	225	230	234	237
2	18.5	19.0	19.2	19.2	19.3	19.3	19.4
3	10.1	9.55	9.28	9.12	9.01	8.94	8.89
4	7.71	6.94	6.59	6.39	6.26	6.16	6.09
5	6.61	5.79	5.41	5.19	5.05	4.95	4.88
6	5.99	5.14	4.76	4.53	4.39	4.28	4.21
7	5.59	4.74	4.35	4.12	3.97	3.87	3.79
8	5.32	4.46	4.07	3.84	3.69	3.58	3.50
9	5.12	4.26	3.86	3.63	3.48	3.37	3.29
10	4.96	4.10	3.71	3.48	3.33	3.22	3.14
11	4.84	3.98	3.59	3.36	3.20	3.09	3.01
12	4.75	3.89	3.49	3.26	3.11	3.00	2.91
13	4.67	3.81	3.41	3.18	3.03	2.92	2.83
14	4.60	3.74	3.34	3.11	2.96	2.85	2.76
15	4.54	3.68	3.29	3.06	2.90	2.79	2.71
16	4.49	3.63	3.24	3.01	2.85	2.74	2.66
17	4.45	3.59	3.20	2.96	2.81	2.70	2.61
18	4.41	3.55	3.16	2.93	2.77	2.66	2.58
19	4.38	3.52	3.13	2.90	2.74	2.63	2.54
20	4.35	3.49	3.10	2.87	2.71	2.60	2.51
21	4.32	3.47	3.07	2.84	2.68	2.57	2.49
22	4.30	3.44	3.05	2.82	2.66	2.55	2.46
23	4.28	3.42	3.03	2.80	2.64	2.53	2.44
24	4.26	3.40	3.01	2.78	2.62	2.51	2.42
25	4.24	3.39	2.99	2.76	2.60	2.49	2.40
30	4.17	3.32	2.92	2.69	2.53	2.42	2.33
40	4.08	3.23	2.84	2.61	2.45	2.34	2.25
50	4.04	3.18	2.79	2.56	2.40	2.29	2.20
∞	3.84	3.00	2.60	2.37	2.21	2.10	2.01

TABLE B.1 Table of Cutoff Values for the F-Distribution for $\alpha = 0.05$ with r_1 Residual (Numerator) Degrees of Freedom and r_2 Regression (Denominator) Degrees of Freedom. $P(F \geq F_0) = \alpha$.

B F-Table

r_2	\multicolumn{8}{c}{r_1}							
	8	9	10	11	12	13	14	15
1	239	241	242	243	244	245	245	246
2	19.4	19.4	19.4	19.4	19.4	19.4	19.4	19.4
3	8.85	8.81	8.79	8.76	8.74	8.73	8.71	8.70
4	6.04	6.00	5.96	5.94	5.91	5.89	5.87	5.86
5	4.82	4.77	4.74	4.70	4.68	4.66	4.64	4.62
6	4.15	4.10	4.06	4.03	4.00	3.98	3.96	3.94
7	3.73	3.68	3.64	3.60	3.57	3.55	3.53	3.51
8	3.44	3.39	3.35	3.31	3.28	3.26	3.24	3.22
9	3.23	3.18	3.14	3.10	3.07	3.05	3.03	3.01
10	3.07	3.02	2.98	2.94	2.91	2.89	2.86	2.85
11	2.95	2.90	2.85	2.82	2.79	2.76	2.74	2.72
12	2.85	2.80	2.75	2.72	2.69	2.66	2.64	2.62
13	2.77	2.71	2.67	2.63	2.60	2.58	2.55	2.53
14	2.70	2.65	2.60	2.57	2.53	2.51	2.48	2.46
15	2.64	2.59	2.54	2.51	2.48	2.45	2.42	2.40
16	2.59	2.54	2.49	2.46	2.42	2.40	2.37	2.35
17	2.55	2.49	2.45	2.41	2.38	2.35	2.33	2.31
18	2.51	2.46	2.41	2.37	2.34	2.31	2.29	2.27
19	2.48	2.42	2.38	2.34	2.31	2.28	2.26	2.23
20	2.45	2.39	2.35	2.31	2.28	2.25	2.22	2.20
21	2.42	2.37	2.32	2.28	2.25	2.22	2.20	2.18
22	2.40	2.34	2.30	2.26	2.23	2.20	2.17	2.15
23	2.37	2.32	2.27	2.24	2.20	2.18	2.15	2.13
24	2.36	2.30	2.25	2.22	2.18	2.15	2.13	2.11
25	2.34	2.28	2.24	2.20	2.16	2.14	2.11	2.09
30	2.27	2.21	2.16	2.13	2.09	2.06	2.04	2.01
40	2.18	2.12	2.08	2.04	2.00	1.97	1.95	1.92
50	2.13	2.07	2.03	1.99	1.95	1.92	1.89	1.87
∞	1.94	1.88	1.83	1.79	1.75	1.72	1.69	1.67

TABLE B.2 Table of Cutoff Values for the F-Distribution for $\alpha = 0.05$ with r_1 Residual (Numerator) Degrees of Freedom and r_2 Regression (Denominator) Degrees of Freedom. $P(F \geq F_0) = \alpha$.

	r_1							
r_2	20	25	30	35	40	45	50	∞
1	248	249	250	251	251	251	252	254
2	19.4	19.5	19.5	19.5	19.5	19.5	19.5	19.5
3	8.66	8.63	8.62	8.60	8.59	8.59	8.58	8.53
4	5.80	5.77	5.75	5.73	5.72	5.71	5.70	5.63
5	4.56	4.52	4.50	4.48	4.46	4.45	4.44	4.37
6	3.87	3.83	3.81	3.79	3.77	3.76	3.75	3.67
7	3.44	3.40	3.38	3.36	3.34	3.33	3.32	3.23
8	3.15	3.11	3.08	3.06	3.04	3.03	3.02	2.93
9	2.94	2.89	2.86	2.84	2.83	2.81	2.80	2.71
10	2.77	2.73	2.70	2.68	2.66	2.65	2.64	2.54
11	2.65	2.60	2.57	2.55	2.53	2.52	2.51	2.40
12	2.54	2.50	2.47	2.44	2.43	2.41	2.40	2.30
13	2.46	2.41	2.38	2.36	2.34	2.33	2.31	2.21
14	2.39	2.34	2.31	2.28	2.27	2.25	2.24	2.13
15	2.33	2.28	2.25	2.22	2.20	2.19	2.18	2.07
16	2.28	2.23	2.19	2.17	2.15	2.14	2.12	2.01
17	2.23	2.18	2.15	2.12	2.10	2.09	2.08	1.96
18	2.19	2.14	2.11	2.08	2.06	2.05	2.04	1.92
19	2.16	2.11	2.07	2.05	2.03	2.01	2.00	1.88
20	2.12	2.07	2.04	2.01	1.99	1.98	1.97	1.84
21	2.10	2.05	2.01	1.98	1.96	1.95	1.94	1.81
22	2.07	2.02	1.98	1.96	1.94	1.92	1.91	1.78
23	2.05	2.00	1.96	1.93	1.91	1.90	1.88	1.76
24	2.03	1.97	1.94	1.91	1.89	1.88	1.86	1.73
25	2.01	1.96	1.92	1.89	1.87	1.86	1.84	1.71
30	1.93	1.88	1.84	1.81	1.79	1.77	1.76	1.62
40	1.84	1.78	1.74	1.72	1.69	1.67	1.66	1.51
50	1.78	1.73	1.69	1.66	1.63	1.61	1.60	1.44
∞	1.57	1.51	1.46	1.43	1.40	1.37	1.35	1.00

TABLE B.3 Table of Cutoff Values for the F-Distribution for $\alpha = 0.05$ with r_1 Residual (Numerator) Degrees of Freedom and r_2 Regression (Denominator) Degrees of Freedom. $P(F \geq F_0) = \alpha$.

C
t-Table

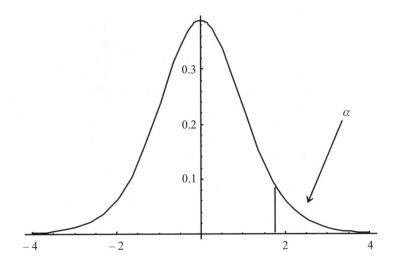

FIGURE C.1 t distribution.

	α				
r	0.10	0.05	0.025	0.010	0.005
1	3.078	6.314	12.71	31.82	63.66
2	1.886	2.920	4.303	6.965	9.925
3	1.638	2.353	3.182	4.541	5.841
4	1.533	2.132	2.776	3.747	4.604
5	1.476	2.015	2.571	3.365	4.032
6	1.440	1.943	2.447	3.143	3.707
7	1.415	1.895	2.365	2.998	3.499
8	1.397	1.860	2.306	2.896	3.355
9	1.383	1.833	2.262	2.821	3.250
10	1.372	1.812	2.228	2.764	3.169
11	1.363	1.796	2.201	2.718	3.106
12	1.356	1.782	2.179	2.681	3.055
13	1.350	1.771	2.160	2.650	3.012
14	1.345	1.761	2.145	2.624	2.977
15	1.341	1.753	2.131	2.602	2.947
16	1.337	1.746	2.120	2.583	2.921
17	1.333	1.740	2.110	2.567	2.898
18	1.330	1.734	2.101	2.552	2.878
19	1.328	1.729	2.093	2.539	2.861
20	1.325	1.725	2.086	2.528	2.845
21	1.323	1.721	2.080	2.518	2.831
22	1.321	1.717	2.074	2.508	2.819
23	1.319	1.714	2.069	2.500	2.807
24	1.318	1.711	2.064	2.492	2.797
25	1.316	1.708	2.060	2.485	2.787
26	1.315	1.706	2.056	2.479	2.779
27	1.314	1.703	2.052	2.473	2.771
28	1.313	1.701	2.048	2.467	2.763
29	1.311	1.699	2.045	2.462	2.756
30	1.310	1.697	2.042	2.457	2.750
∞	1.282	1.645	1.960	2.326	2.576

TABLE C.1 Student's t-Distribution Cutoff Points, $P(T \geq t_0) = \alpha$, with r Degrees of Freedom. For Two-Sided Testing Use Column Corresponding to Half Desired α Value.

D

Mathematica Appendix

When the authors teach this course, we use a variety of user-friendly specialty software such as spreadsheets and statistical packages. This approach enables the students to spend the bulk of their time on modeling as opposed to programming and debugging. This course could be taught entirely using a programming language such as C/C++, BASIC or TRUEBASIC, Pascal, or FORTRAN. In the process the students would become expert programmers of mathematical algorithms, but would have devoted much less effort to the modeling and interpretation process. Indeed, writing a package to compute ANOVA tables alone would be a large project, our intent is to introduce students to the ideas, not the details of the computations. An option between many specialized programs or one programming language is a mathematical programming language such as *Mathematica* or Maple. Every model in this book could be done using *Mathematica*, though most any model is more easily built using something else. On the other hand, the programs which are easier to use are limited in the different types of models they can handle. A spreadsheet program would be a good second choice for the number of types of models it can handle. This appendix, then, is intended to illustrate how to build the models in the text using *Mathematica*.[1] While comments will be made throughout the examples, this appendix is not intended to be a *Mathematica* manual or tutorial. Rather, it assumes some familiarity on the user's part with *Mathematica*, and merely demonstrates how to put the pieces together, how to use special features, and how to avoid common problems. To get familiar with *Mathematica*, the reader is directed to the book *The Beginner's Guide to Mathematica* by Gray and Glynn, the Tour of *Mathematica* in Wolfram's *Mathematica* book (which can be found in the electronic help in version 3.0), or the *Introduction to Mathematica* notebook included with the text *Calculus&Mathematica* by Davis, Porta, and Uhl. In the examples in this appendix, we have tried to use *Mathematica* as a mathematical tool rather than a programming language whenever possible. Thus

---

[1] The text's modeling website has this appendix as a collection of downloadable *Mathematica* files. There is a link to the modeling website from the MAA Bookstore on MAA Online (www.maa.org).

we have tried to perform calculations in a manner and order reminiscent of standard mathematics. Readers skilled in programming will undoubtedly see ways to optimize code or use advanced programming capabilities of *Mathematica*. It is our opinion that while fancy code may be more efficient and elegant, for beginning users it can obscure the mathematical and modeling ideas.

D.1 Chapter 1
Discrete Dynamical Systems

***Mathematica* preliminaries** When getting *Mathematica* to solve recursion relations, the concept of *caching* is very important. For example, consider the Fibonacci recurrence relation. Naively this can be set up as:

```
Clear[x,n];
x[0]=1;
x[1]=1;
x[n_]:=x[n-1]+x[n-2];
```

Now suppose we want the Fibonacci number when **n=15**. We add the **//Timing** command to keep track of the amount of time required to find **x[15]**.

```
x[15]//Timing
```
{5.7 Second, 987}

Notice that if we run it again, it takes about the same amount of time.

```
x[15]//Timing
```
{5.68333 Second, 987}

This time we again define a recursive Fibonacci function, but now we add a piece of code that seems redundant. In particular, note the **x[n]:=x[n]=...** construction. This is called caching, and every time an **x[n]** value is computed it is store in an internal lookup table. Notice that when **x[15]** is requested, **x[14]** and **x[13]** are needed. With the original code, these are computed regardless of whether they have been computed before. With caching, if **x[14]** and **x[13]** are already computed, they are merely looked up resulting in the great time savings observed below.

```
Clear[x,n];
x[0]=1;
x[1]=1;
x[n_]:=x[n]=x[n-1]+x[n-2];
```

```
x[15]//Timing
```
{0.266667 Second, 987}

If we request **x[15]** again, it needs only to be looked up so it takes even less time.

```
x[15]//Timing
```
{0.0166667 Second, 987}

D Mathematica Appendix

Warning! Since the values of **x** are stored, it is essential to use the **Clear** command every time the recursive function is changed. While clearing variables is always good practice, often it can be skipped without any problems. This is not the case when caching is being used. Any time unexpected results are being obtained, clearing the variables should be considered as a first debugging device.

Sandhill Crane

In this section we replicate the sandhill crane spreadsheet analysis.

The simple model This code should be reminiscent of the Fibonacci sequence discussed above.

```
Clear[pop,n];
pop[0]=100;
pop[n_]:=pop[n]=1.0194pop[n-1];
```

The **Table** command can be used to create a list of data. Here the data consists of ordered pairs. We give the list the name **cranes** so that it can be used later.

```
cranes=Table[{n,pop[n]},{n,0,14}]
```

{{0,100},{1,101.94},{2,103.918},{3,105.934},
 {4,107.989},{5,110.084},{6,112.219},{7,114.396},
 {8,116.616},{9,118.878},{10,121.184},{11,123.535},
 {12,125.932},{13,128.375},{14,130.865}}

The suffix **//TableForm** can be used to put the data in a nice table format.

```
cranes//TableForm
0    100
1    101.94
2    103.918
3    105.934
4    107.989
5    110.084
6    112.219
7    114.396
8    116.616
9    118.878
10   121.184
11   123.535
12   125.932
13   128.375
14   130.865
```

`ListPlot[cranes,PlotJoined->True,AxesLabel->{"Years", "Population"}];`

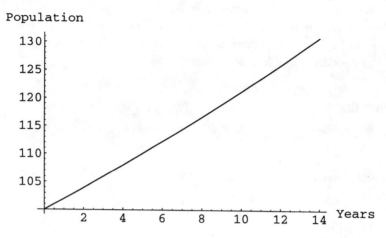

A model with three growth rates We replicate our analysis for best, medium, and worst conditions, and show all three on a single graph. One can see each part of the simple analysis replicated three times with different names. To show all graphs on the same coordinate axes, give each one a name and then use the **Show** command. Here we have made the graph of each model a different thickness so they can be identified on the combined graph.

```
Clear[popBest, popMedium, popWorst,n];
rBest=0.0194;
rMedium=-0.0324;
rWorst=-0.0382;

popBest[0]=100;
popMedium[0]=100;
popWorst[0]=100;

popBest[n_]:=popBest[n]=(1+rBest) popBest[n-1];
popMedium[n_]:=popMedium[n]=(1+rMedium)
popMedium[n-1]; popWorst[n_]:=popWorst[n]=(1+rWorst)
popWorst[n-1];

cranesBest=Table[{n,popBest[n]},{n,0,14}];
cranesMedium=Table[{n,popMedium[n]},{n,0,14}];
cranesWorst=Table[{n,popWorst[n]},{n,0,14}];

pBest=ListPlot[cranesBest,PlotJoined->True,AxesLabel
 ->{"Years","Population"}, PlotStyle->Thickness[0.005]];
pMedium=ListPlot[cranesMedium,PlotJoined->True,AxesLabel
 ->{"Years","Population"},PlotStyle->Thickness[0.01]];
pWorst=ListPlot[cranesWorst,PlotJoined->True,AxesLabel
 ->{"Years","Population"},PlotStyle->Thickness[0.015]];
Show[pBest,pMedium,pWorst];
```

D Mathematica Appendix

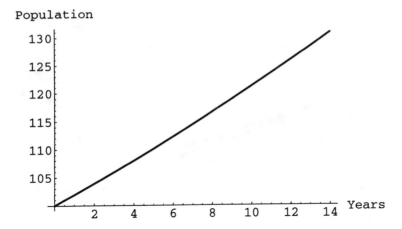

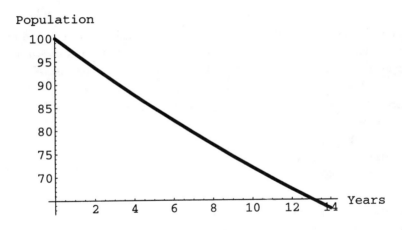

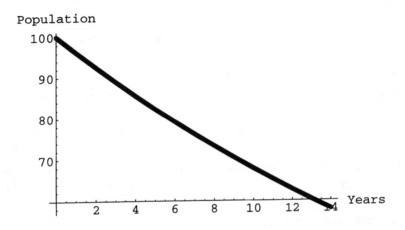

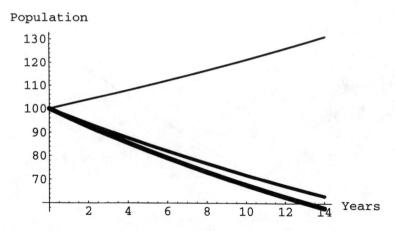

Hacking Chicks To modify a model, we can either make the change and rerun it, or, if we are interested in keeping the results, cut and paste the old code and modify it appropriately. Here we hack five chicks to the medium and worst cases.

```
Clear[popBest, popMedium, popWorst,n];
chicks=5;
yearsToRun=120;
rMedium=-0.0324;
rWorst=-0.0382;

popMedium[0]=100;
popWorst[0]=100;

popMedium[n_]:=popMedium[n]=(1+rMedium)popMedium[n-1]
 +chicks;
popWorst[n_]:=popWorst[n]=(1+rWorst)popWorst[n-1]+chicks;

cranesMedium=Table[{n,popMedium[n]},{n,0,yearsToRun}];
cranesWorst=Table[{n,popWorst[n]},{n,0,yearsToRun}];

pMedium=ListPlot[cranesMedium,PlotJoined->True,AxesLabel
 ->{"Years","Population"},PlotStyle->Thickness[0.01]];
pWorst=ListPlot[cranesWorst,PlotJoined->True,AxesLabel
 ->{"Years","Population"},PlotStyle->Thickness[0.015]];

Show[pMedium,pWorst];
```

D Mathematica Appendix 383

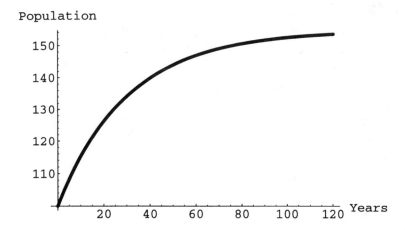

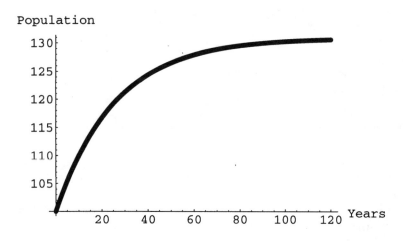

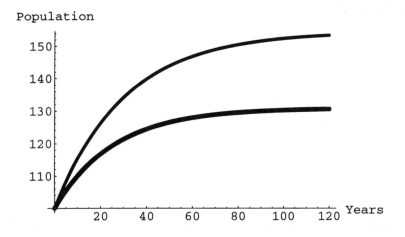

Annual Plants

In this section, we use *Mathematica* to numerically analyze the annual plant model. We first build a recursive model, then use the symbolic algebraic capabilities to find a closed-form solution. Define parameters.

```
Clear[p,alpha1, beta1, gamma1, sigma1, n];
alpha1=.5;
beta1=.25;
gamma1=2;
sigma1=.8;
```

Define the recursion relation.

```
p[n_]:=p[n]=alpha1*sigma1*gamma1*p[n-1]+
  beta1*sigma1*(1-alpha1)*sigma1*gamma1*p[n-2];
p[-1]=100;
p[0]=95;
TableForm[Table[{n,p[n]},{n,0,10}]]
  0    95
  1    92.
  2    88.8
  3    85.76
  4    82.816
  5    79.9744
  6    77.2301
  7    74.58
  8    72.0208
  9    69.5494
 10    67.1629
ListPlot[Table[{n,p[n]},{n,0,10}], PlotJoined ->True,
  AxesLabel ->{"Years", "Number of Plants"}];
```

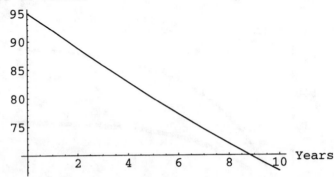

To find the closed-form solution, we need to solve the characteristic equation.

```
Solve[l² - alpha1 sigma1 gamma1 l - beta1 sigma1
  (1 - alpha1) sigma1 gamma1 == 0, l]
```

D Mathematica Appendix

```
{{1 ->-0.165685},{1 ->0.965685}}
{{1 ->-0.165685},{1 ->0.965685}}
```

Use the initial conditions to find the constants.

```
Solve [{95 == C1 + C2, 100 == - C1/0.165685 + C2/0.965685 },{C1, C2}]
{{C1 ->-0.229708, C2 ->95.2297 }}
```

To get these numbers in usable form, we will use the **Copy Output From Above** menu item and then cut and paste the values into the desired equation form.

```
{{C1 ->-0.229708, C2 ->95.2297}}
pp[n_] = -0.22970773009195789837`*(-0.16568542494923801952`)^n
   + 95.2297077300919578984`*(0.96568542494923801950`)^n;
```

The following table compares the output from the recursive equation and the closed-form equation.

```
TableForm[Table[{n,p[n], pp[n]}, {n,0,10}]]
  0    95        95.
  1    92.       92.
  2    88.8      88.8
  3    85.76     85.76
  4    82.816    82.816
  5    79.9744   79.9744
  6    77.2301   77.2301
  7    74.58     74.58
  8    72.0208   72.0208
  9    69.5494   69.5494
 10    67.1629   67.1629
```

D.2 Chapter 2

Descriptive Statistics

The X-Files *Mathematica* has functions to perform many statistical operations, but these functions are not generally loaded when *Mathematica* starts. Rather, related functions are bundled into packages which can be loaded as a group before they are needed. Measures of center and spread are in a package called **'DescriptiveStatistics'**, and the package can be loaded with the following command.

```
Needs["Statistics`DescriptiveStatistics`"];
```

Once loaded the descriptive statistic functions are available. Consider the data:

```
season1 = {7.4,6.,6.8,5.9,6.2,5.6,5.6,6.2,6.1,5.1,6.4,6.4,
   6.2,6.8,7.2,6.8,5.8,7.1,7.2,7.5,8.1,7.7,7.4,8.3}
{7.4,6.,6.8,5.9,6.2,5.6,5.6,6.2,6.1,5.1,6.4,6.4,
   6.2,6.8,7.2,6.8,5.8,7.1,7.2,7.5,8.1,7.7,7.4,8.3}
Mean[season1]
6.65833
```

```
Median[season1]
```
6.6
```
Mode[season1]
```
{6.2,6.8}
```
Variance[season1]
```
0.69558
```
StandardDeviation[season1]
```
0.834014
```
SampleRange[season1]
```
3.2
```
InterquartileRange[season1]
```
1.25
```
Quantile[season1, 0.5]
```
6.4
```
Quantile[season1, 0.25]
```
6.
```
Quantile[season1, 0.75]
```
7.2

To make bar charts of the data, the packages **Statistics`DataManipulation`** and **Graphics`Graphics`** are needed.

```
Needs["Statistics`DataManipulation`"]
```

The **Frequencies** command counts the number of occurrences of each piece of data.

```
freq = Frequencies[season1]
```
{{1,5.1},{2,5.6},{1,5.8},{1,5.9},{1,6.},{1,6.1},
 {3,6.2},{2,6.4},{3,6.8},{1,7.1},{2,7.2},{2,7.4},
 {1,7.5},{1,7.7},{1,8.1},{1,8.3}}

We complete the histogram by making a bar chart of the frequencies. This is a slightly different method of histogram construction than mentioned in the text. There the bars covered the intervals of interest; here the bars are centered on the midpoints of the intervals and the bar widths have no meaning. Other packages may use still other methods.

```
Needs["Graphics`Graphics`"]

BarChart[freq];
```

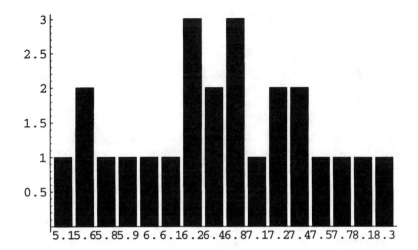

The BinCounts function can be used to bin the data. Here we group the data into 7 bins.

```
freq = BinCounts[season1, {5.09,8.3,(8.31-5.09)/7}]
{1,5,6,3,3,4,2}
```

We also need the midpoint of each interval.

```
midpoints =
Table[5.-((8.3-5.1)/(2*7))+n(8.3-5.1)/(7),{n,1,7}]
{5.22857,5.68571,6.14286,6.6,7.05714,7.51429,7.97143}
BarChart[Transpose[{freq,midpoints}]];
```

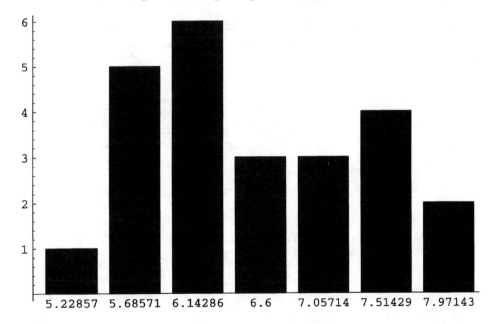

Next we look at constructing box plots of data. *Mathematica* does not come with a box plot function so we create one using graphics elements. Note that we use the

Statistics`DescriptiveStatistics` package. If it has already been loaded, it does not need to be reloaded.

```
season1 = {7.4,6.,6.8,5.9,6.2,5.6,5.6,6.2,6.1,5.1,6.4,
  6.4,6.2,6.8,7.2,6.8,5.8,7.1,7.2,7.5,8.1,7.7,7.4,8.3}
```

```
{7.4,6.,6.8,5.9,6.2,5.6,5.6,6.2,6.1,5.1,6.4,6.4,
  6.2,6.8,7.2,6.8,5.8,7.1,7.2,7.5,8.1,7.7,7.4,8.3}
```

```
Needs["Statistics`DescriptiveStatistics`"];
SingleBoxplot[data_, center_, width_]:=(
  Clear[a,b,c,d,e,f,g];
  a = Min[data]; b = Quartiles[data] [[1]];
  c = Quartiles[data] [[2]]; d = Quartiles[data] [[3]];
  e = Max[data]; f = center - width; g = center + width;
  Graphics[{
    Line[{{f,e},{g,e}}],
    Line[{{f,d},{g,d}}],
    Line[{{f,c},{g,c}}],
    Line[{{f,b},{g,b}}],
    Line[{{f,a},{g,a}}],
    Line[{{f,b},{f,d}}],
    Line[{{g,b},{g,d}}],
    Line[{{center,a},{center,e}}]}])
```

```
s1 = SingleBoxplot[season1,1,0.25];
Show[s1,Axes->True,
  Ticks ->{{1}, Automatic}, AxesOrigin ->{0.5, Automatic}];
```

```
season2 = {9.8,9.3,8.7,8.2,8.5,9.2,9,9.1,8.6,9.9,8.5,9.7,8.8,
  10.2,10.8,9.8,10.7,9.6,10.2,9.8,7.9,8.5,8.1,9,9.6}
```

```
{9.8,9.3,8.7,8.2,8.5,9.2,9,9.1,8.6,9.9,8.5,9.7,8.8,
  10.2,10.8,9.8,10.7,9.6,10.2,9.8,7.9,8.5,8.1,9,9.6}
```

```
s2 = SingleBoxplot[season2,1,0.25];
Show[s2,Axes->True,
  Ticks ->{{1},Automatic},AxesOrigin ->{0.5,Automatic}];
```

D Mathematica Appendix

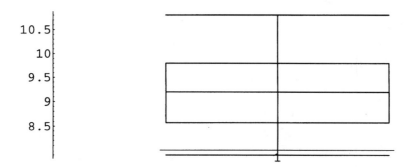

`SingleBoxPlot` can be used to make side-by-side box plots.

```
s1 = SingleBoxplot[season1, 1, 0.25];
s2 = SingleBoxplot[season2, 2, 0.25];
Show[s1, s2, Axes ->True,
  Ticks ->{{1,2}, Automatic}, AxesOrigin ->{0.5, Automatic}];
```

Stochastic Simulations

The Sandhill Cranes   Random numbers can be generated from any standard distribution and used in simulations. Before these are used, the package `Statistics'Continuous Distributions'` must be loaded.

```
Needs["Statistics'ContinuousDistributions'"];
```

We simulate the crane population with demographic stochasticity. Here the birth rate is normal with mean 0.5 and standard deviation 0.03, and the death rate is normal with mean 0.1 and standard deviation 0.08.

Here is one simulation.

```
Clear[n,t];
n[0] = 100;
n[t_] := n[t] = (Random[NormalDistribution[0.5,0.03]] -
  Random[NormalDistribution[0.1,0.08]]+1) n[t-1];
ListPlot[Table[{t, n[t]},{t, 0, 10}], PlotJoined ->True];
```

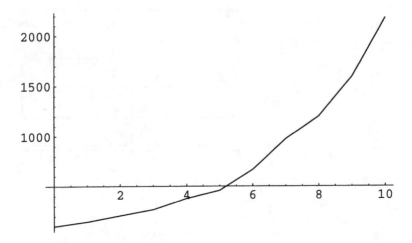

Next we create a table of simulations and show them on one plot.

```
Clear[s];
s[i_] := (Clear[n, t]; n[0] = 100;
   [t_] := n[t] = (Random[NormalDistribution[0.5, 0.03]] -
   Random[NormalDistribution[0.1,0.08]]+1)n[t-1];
   ListPlot[Table[{t,n[t]},{t,0,10}],
    PlotJoined ->True, DisplayFunction ->Identity]);
Show[
Table[s[i],{i,1,100}], DisplayFunction ->$DisplayFunction,
AxesLabel ->{"Years", "Population"}];
```

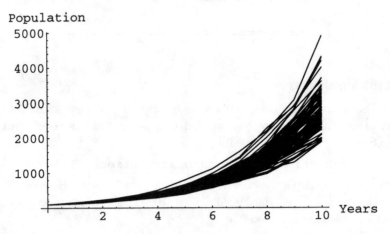

We will not replicate the environmental stochasticity model, but indicate that uniform random numbers from $[a, b]$ can be generated with **Random[Real, {a,b}]**. Thus, to generate random numbers from the unit interval, we use:

Random[Real, {0, 1}]

0.105798

D Mathematica Appendix

The model can be completed with an `If` statement. The syntax is `If[test, trueresult, falseresult]`. The following will return 0.3 40% of the time and 0.5 the remaining 60%.

    If[Random[Real, {0, 1}] <0.4, 0.3, 0.5]

    0.5

Instead of Tables and p-values

Finding the Area Under a Probability Density Curve. Whenever we are doing a statistical test, we obtain a value for a statistic and we want to know the probability of getting that value or larger by chance. As an example, suppose we are doing a chi-square goodness-of-fit test and get $\chi^2 = 47.3653$ and we have 11 degrees of freedom.

The *Mathematica* package `Statistics'ContinuousDistributions'` contains functions for most commonly used distributions.

    Needs["Statistics'ContinuousDistributions'"]

To use the chi-square distribution with eleven degrees of freedom we enter:

    fdist = ChiSquareDistribution[11]

    ChiSquareDistribution[11]

The probability of a random value from a chi-square with df = 11 being less than 47.3653 is the value of the cumulative density function (cdf) at 47.3653. Thus we use *Mathematica*'s CDF function.

    CDF[fdist, 47.3653]

    0.999998

Finally, the probability of being larger than 47.3653 by chance is one minus the value we just obtained. This is sometimes called the p-value.

    pval = 1 - CDF[fdist, 47.3653]

    1.852×10^{-6}

We can convert this distribution into a probability density function (PDF) and use it like any other function. Here is a plot of the chi-square distribution function with 11 degrees of freedom.

    f[x_] = PDF[fdist, x];
    Plot[f[x], {x, 0, 25}];

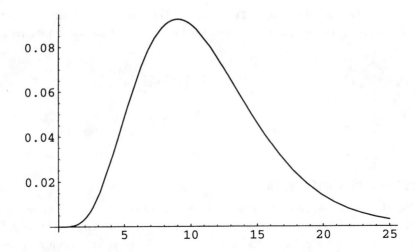

D.3 Chapter 3
Matrix Analysis

Analysis of the Teasel Model In this section, we demonstrate matrix analysis using *Mathematica*. We begin by entering the Teasel transition matrix. A matrix is a list of lists. Each row of the matrix is a list.

```
A={{0,0,0,0,0,322.38},
   {0.966,0,0,0,0,0}, {0.013,0.01,0.125,0,0,3.448},
   {0.007,0,0.125,0.238,0,30.17},
   {0.008,0,0,0.245,0.167,0.862}, {0,0,0,0.023,0.75,0}};
```

The `Eigenvalue[]` function computes the eigenvalues of A.

```
Eigenvalues[A]
```

{2.321880145972127009,
 -0.9573855619833462649 + 1.488565742766620676 i,
 -0.9573855619833462649 - 1.488565742766620676 i,
 0.1423853248821802164 + 0.1980150112833666705 i,
 0.1423853248821802164 - 0.1980150112833666705 i,
 -0.1618796717697949118}

To compute the dominant eigenvalue, we need to find the modulus of each eigenvalue. The dominant eigenvalue is the positive real one, namely 2.3188. This tells us that teasel, in all its states (including seeds), is growing at a rate of 132% (i.e., (2.32188-1)100%). The `Abs[]` command finds the absolute value of real numbers and the modulus of complex numbers.

```
Abs[Eigenvalues[A]]
```

{2.321880145972127009, 1.769863013013297837,
 1.769863013013297837, 0.2438924464499789269,
 0.2438924464499789269, 0.1618796717697949118}

D Mathematica Appendix

The eigenvectors of A are found next.

```
m = Eigenvectors[A]
```

{{0.9183589074355381124,0.3820760111677958274,
    0.01755468821081351718,0.09989845124953175926,
    0.01741327558678372625,0.006614303970628544916},
  {0.8752760865810005216+0. i,
    -0.2584221626846726942-0.40180089800723126 i,
    0.004411770738221728170-0.003094968796608710897 i,
    0.07317065093675799936-0.01056273086321099061 i,
    -0.006947229168502451950-0.009994196823921364553 i,
    -0.002599344525224688961+0.004041522420582257060 i},
  {0.8752760865810005216+0. i,
    -0.2584221626846726942+0.40180089800723126 i,
    0.004411770738221728170+0.003094968796608710897 i,
    0.07317065093675799936+0.01056273086321099061 i,
    -0.006947229168502451950+0.009994196823921364553 i,
    -0.002599344525224688961-0.004041522420582257060 i},
  {0.1426564398494417913+0.1983920503732153821 i,
    0.9678393542915422898+0. i,
    0.02093851966635722114-0.05537956743118808652 i,
    -0.004648716315209251388-0.007049397749168426909 i,
    0.00008511911951789301171+0.0002339138569357704921 i,
    -0.00005885110912179577870+0.0001752473264249155344 i},
  {0.1426564398494417913-0.1983920503732153821 i,
    0.9678393542915422898+0. i,
    0.02093851966635722114+0.05537956743118808652 i,
    -0.004648716315209251388+0.007049397749168426909 i,
    0.00008511911951789301171-0.0002339138569357704921 i,
    -0.00005885110912179577870-0.0001752473264249155344 i},
  {0.1652061694418828398,-0.9858505267283135501,
    0.02787530768608426829,-0.005346755079541114169,
    0.0001818724531146686891,-0.00008295651244989489562}}

The eigenvector associated with the dominant eigenvalue is the first of this list.

```
m[[1]]
```

{0.9183589074355381124,0.3820760111677958274,
  0.01755468821081351718,0.09989845124953175926,
  0.01741327558678372625,0.006614303970628544916}

We divide by the sum of the entries of the eigenvector.

$$\frac{\texttt{m[[1]]}}{\sum_{i=1}^{6}\texttt{m[[1]] [[i]]}}$$

{0.6369019681003457497,0.2649780619607915444,
 0.01217455983747820449,0.0692817586848192622,
 0.01207648709290134629,0.004587164323663892829}

This tells us that eventually 64% of the Teasel will be in first-year dormant seeds, 26% will be in second-year dormant seeds, 1.2 % will be small rosettes, 6.9% will be medium rosettes, 1.2 % will be in large rosettes, and 0.4% will be in a flowering state.

We will verify this by computing the distribution of Teasel after a long period of time, say 100 years. According to theory of eigenvectors, it should make no difference what the initial stage distribution is, so we will start with the Teasel's being distributed equally among the stages. Caution: Matrix multiplication is symbolized with a period. A space or an asterisk between matrices is an operation, but not matrix multiplication. The **MatrixPower[]** command is an efficient way to compute a matrix product.

Matrix multiplication between a vector and a matrix can be done carelessly in *Mathematica*. Technically we should be transposing the row vectors to make them column vectors, but *Mathematica* interprets them the only way that makes sense. Further, a vector as a list (not a list containing one list) is acceptable.

```
Clear[X];
X[0] = {10,10,10,10,10,10}
X[t_] := MatrixPower[A,t].X[0]
```

{10,10,10,10,10,10}

After 100 years we have the following distribution. (Presumably the model is not valid for this length of time, but we are verifying the mathematics here).

X[100]

{2.579014488260051442 10^{39}, 1.072978723719316606 10^{39},
 4.929858562486495658 10^{37}, 2.805434248432140413 10^{38},
 4.890145852861199779 10^{37}, 1.857485742439716276 10^{37}}

Finally we compute the fraction in each class. Compare this to the eigenvector associated with the dominant eigenvector (scaled to have sum equal to one).

$$\frac{X[100]}{\sum_{i=1}^{6} X[100] \; [\![i]\!]}$$

{0.6369019681006821303,0.2649780619604192322,
 0.01217455983748257073,0.0692817586848438064l,
 0.01207648709289699843,0.004587164323675261839}

Next we compute the stage distribution vector over a 10-year period. Each row of the table is a year; the columns correspond to the stages.

TableForm[Table[X[t], {t,0,10}]]

10	10	10	10
3223.8	9.66	35.96	305.4
2491.9974	3114.1908	73.15404	332.9609
5364.145296	2407.2694884	130.0540908	607.837595
28924.352919666	5181.764355936	419.4223898956000001	2905.362768151
36936.8856869508	27940.924920397356	875.318739255648	4403.12081177675
92040.33378841695325	35681.0315735944728	1853.416630240283864	10029.52976153467476
296354.5451607410183	88910.96243961077684	4954.657715880404733	30997.38984205049288
510184.8027948137243	286278.4906252758237	0817.7082869625334	57816.94472169599577
1.266366771422884654 10^6	492838.5194997900576	24391.76662033202993	137197.140240615955
3.341067587457285434 10^6	1.223310301194506576 10^6	60174.35397086046763	357240.9694032244682

D.4 Chapter 4

Curve Fitting

In this section, we demonstrate the ideas of fitting curves to data using *The X-Files,* Cost of Advertising, Black Bear, and Corn Storage data.

***Mathematica* Background** This Chapter relies on two commands, the `Fit` command and the `Regress` command. The `Fit` command fits any intrinsically linear function to data. Its syntax is `Fit[data, {basis functions}, independent variables]`. If an ANOVA or regression table is desired, use the `Regress` command instead. The `Regress` command is in the `Statistics'LinearRegression` package.

Reopening *The X-Files* The following data set is the number of viewers (in millions) for the first-time airing of the television show *The X-Files* for each week of its first season, second season, and both seasons combined. The data is week number versus the number of viewers (in millions).

```
season1 = {{1,7.4}, {2,6.9}, {3,6.8}, {4,5.9}, {5,6.2},
   {7,5.6}, {8,5.6}, {9,6.2}, {10,6.1}, {11,5.1},
   {14,6.4}, {15,6.4}, {18,6.2}, {20,6.8}, {22,7.2},
   {23,6.8}, {23,5.8}, {24,7.1}, {28,7.2}, {30,7.5},
   {32,8.1}, {33,7.7}, {35,7.4}, {36,8.3}};
season2 = {{53,9.8}, {54,9.3}, {55,8.7}, {56,8.2}, {57,8.5},
   {58,9.2}, {60,9}, {61,9.1}, {62,8.6}, {65,9.9},
   {66,8.5}, {69,9.7}, {70,8.8}, {72,10.2}, {73,10.8},
   {74,9.8}, {75,10.7}, {76,9.6}, {78,10.2}, {80,9.8},
   {82,7.9}, {84,8.5}, {85,8.1}, {86,9}, {87,9.6}};
seasons = {{1,7.4}, {2,6.9}, {3,6.8}, {4,5.9}, {5,6.2},
   {7,5.6}, {8,5.6}, {9,6.2}, {10,6.1}, {11,5.1}, {14,6.4},
   {15,6.4}, {18,6.2}, {20,6.8}, {22,7.2}, {23,6.8},
   {23,5.8}, {24,7.1}, {28,7.2}, {30,7.5}, {32,8.1},
   {33,7.7}, {35,7.4}, {36,8.3}, {53,9.8}, {54,9.3},
   {55,8.7}, {56,8.2}, {57,8.5}, {58,9.2}, {60,9},
   {61,9.1}, {62,8.6}, {65,9.9}, {66,8.5}, {69,9.7},
```

{70,8.8}, {72,10.2}, {73,10.8}, {74,9.8}, {75,10.7},
{76,9.6}, {78,10.2}, {80,9.8}, {82,7.9}, {84,8.5},
{85,8.1}, {86,9}, {87,9.6}};

We begin by creating a scatterplot of the data from the first season.

```
season1Plot = ListPlot[season1,
AxesLabel ->{"Week Number", "Viewers in Millions"},
PlotStyle ->{PointSize[.02]}];
```

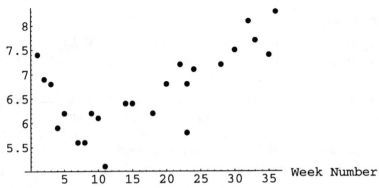

Next we fit a line to the first-season data and overlay the line and the data.

```
f[x_] = Fit[season1,{1,x},x]
```

5.89374 + 0.046611 x

```
season1ext = Plot[f[x], {x,0,36},
AxesLabel ->{"Week Number", "Viewers in Millions"}];
Show[season1Plot,season1ext];
```

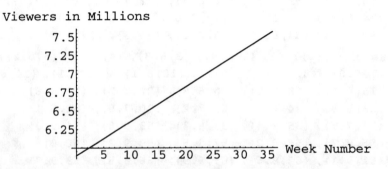

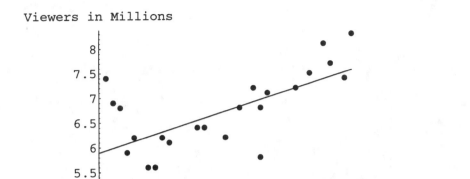

Finally we compute R^2 directly from the definition.

    data = season1;

$$\text{SSres} = \sum_{n=1}^{\text{Length[data]}} (\text{data [[n]] [[2]]} - \text{f[data[[n]] [[1]]]})^2$$

9.19358

$$\text{mean} = \frac{\sum_{n=1}^{\text{Length[data]}} \text{data [[n]] [[2]]}}{\text{Length[data]}}$$

6.69583

$$\text{SSreg} = \sum_{n=1}^{\text{Length[data]}} (\text{f[data[[n]] [[1]]]} - \text{mean})^2$$

    SStot = SSres + SSreg

15.5896

$$\frac{\text{SSreg}}{\text{SStot}}$$

0.410274

Thus R^2 is 0.41, so this linear model accounts for 41% of the variance in the data. Next we will see how the first-season model does at explaining the variation seen in both seasons.

    seasonsPlot = ListPlot[seasons,
    AxesLabel->{"Week Number", "Viewers in Millions"},
    PlotStyle->{PointSize[.02]}];

Viewers in Millions

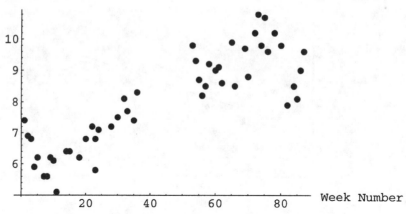

```
seasonsext = Plot[f[x],{x,0,90},
AxesLabel->{"Week Number", "Viewers in Millions"}];
Show[seasonsPlot,seasonsext];
```

Viewers in Millions

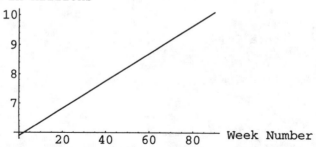

Viewers in Millions

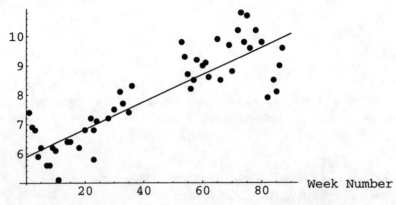

D Mathematica Appendix

```
data = seasons;
```

$$\text{SSres} = \sum_{n=1}^{\text{Length[data]}} (\text{data}[\![n]\!][\![2]\!] - f[\text{data}[\![n]\!][\![1]\!]])^2$$

29.1019

$$\text{mean} = \frac{\sum_{n=1}^{\text{Length[data]}} \text{data}[\![n]\!][\![2]\!]}{\text{Length[data]}}$$

8.00408

$$\text{SSreg} = \sum_{n=1}^{\text{Length[data]}} (f[\text{data}[\![n]\!][\![1]\!]] - \text{mean})^2$$

```
SStot = SSres + SSreg
```

85.7508

$$\frac{\text{SSreg}}{\text{SStot}}$$

0.746615

The first season model explains 75% of the variation in the data for both seasons. Interestingly, this is better than the amount of variation in the first season's data explained by the first season's model.

Cost of Advertising This section shows the complete analysis used in the Cost of Advertising example. The annual advertising expenditures (in millions) in the US from 1970 to 1989:

```
CostOfAdv = {{70,19550}, {71,20700}, {72,23210}, {73,24980},
{74,26620}, {75,27900}, {76,33300}, {77,37440},
{78,43330}, {79,48780}, {80,53550}, {81,60430},
{82,66580}, {83,75850}, {84,87820}, {85,94750},
{86,102140}, {87,109650}, {88,118050}, {89,123930}}

  {{70,19550}, {71,20700}, {72,23210}, {73,24980}, {74,26620},
{75,27900}, {76,33300}, {77,37440}, {78,43330}, {79,48780},
{80,53550}, {81,60430}, {82,66580}, {83,75850},
{84,87820}, {85,94750}, {86,102140}, {87,109650},
{88,118050}, {89,123930}}

  p1 = ListPlot[CostOfAdv, PlotStyle ->{PointSize[0.02]},
AxesLabel ->{"Year", "Expenditures in Millions"}];
```

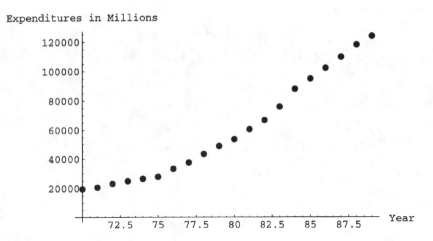

Linear Fit

```
c[x_] = Fit[CostOfAdv, {1,x},x];
p0 = Plot[c[x], {x, 70, 89}];
```

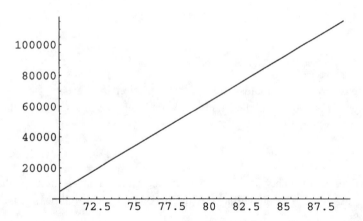

```
Show[p0, p1];
```

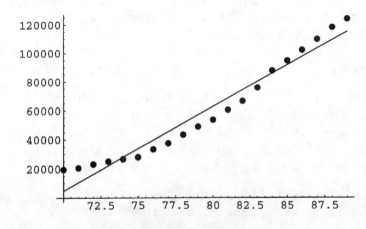

D Mathematica Appendix

```
SSres = ∑_{n=1}^{20} (CostOfAdv[[n]][[2]] -c[CostOfAdv[[n]][[1]]])^2
```

1.10502×10^9

```
meancost = (1/20) ∑_{n=1}^{20} CostOfAdv[[n]][[2]]
```

59928

```
SSreg = ∑_{n=1}^{20} (c[CostOfAdv[[n]][[1]]] -meancost)^2
```

2.24413×10^{10}

```
SStot = SSres + SSreg
```

2.35463×10^{10}

```
SSreg
─────
SStot
```

0.95307

0.95307

Exponential Fit Separate the cost data from the ordered pairs.

```
Cost = Table[CostOfAdv[[n]][[2]], {n,1,20}]
```

{19550, 20700, 23210, 24980, 26620, 27900, 33300, 37440,
43330, 48780, 53550, 60430, 66580, 75850, 87820, 94750,
102140, 109650, 118050, 123930}

We log the Cost data.

```
LoggedCost = N[Log[Cost]]
```

{9.88073, 9.93789, 10.0523, 10.1258, 10.1894, 10.2364, 10.4133,
10.5305, 10.6766, 10.7951, 10.8884, 11.0092, 11.1062,
11.2365, 11.383, 11.459, 11.5341, 11.605, 11.6789, 11.7275}

Recombine the ordered pairs.

```
LogCost = Table[{CostOfAdv[[n]][[1]], LoggedCost[[n]]}, {n,1,20}]
```

{{70, 9.88073}, {71, 9.93789}, {72, 10.0523}, {73, 10.1258},
{74, 10.1894}, {75, 10.2364}, {76, 10.4133}, {77, 10.5305},
{78, 10.6766}, {79, 10.7951}, {80, 10.8884}, {81, 11.0092},
{82, 11.1062}, {83, 11.2365}, {84, 11.383}, {85, 11.459},
{86, 11.5341}, {87, 11.605}, {88, 11.6789}, {89, 11.7275}}

If this were true exponential growth, then this graph should be linear.

```
p2 = ListPlot[LogCost, PlotStyle ->{PointSize[0.02]},
    AxesLabel ->{"Year", "Log of Expenditures"}];
```

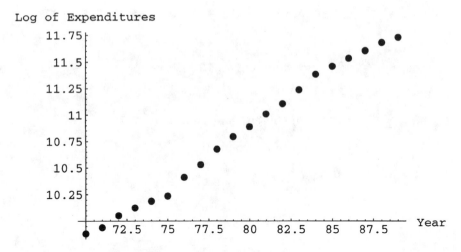

```
f[x_] = Fit[LogCost, {1,x}, x]
```
2.43694 + 0.105489x
```
p3 = Plot[f[x], {x, 70, 89}];
```

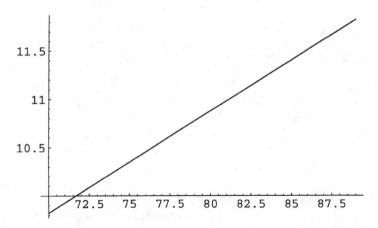

```
Show[p2, p3];
```

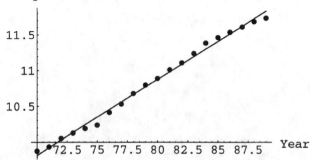

Transform back to original coordinates

D Mathematica Appendix

```
g[x_] = e^f[x]
```
$e^{2.43694+0.105489x}$

```
p4 = Plot[g[x], {x, 70, 89}];
```

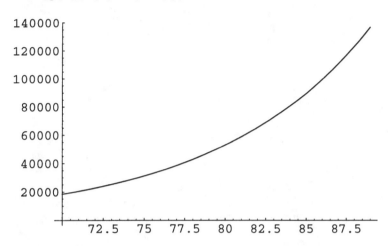

Show the model with the original data.

```
Show[p1, p4];
```

Expenditures in Millions

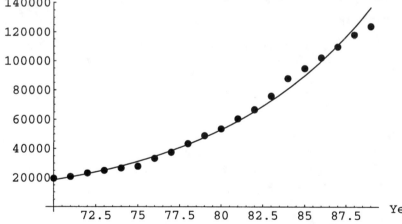

```
SSres = ∑_{n=1}^{20} (CostOfAdv[[n]][[2]] - g[CostOfAdv[[n]][[1]]])^2
```
3.07135×10^8

```
meancost = (1/20) ∑_{n=1}^{20} CostOfAdv[[n]][[2]]
```
59928

```
SSreg = ∑_{n=1}^{20} (g[CostOfAdv[[n]][[1]]] - meancost)^2
```
2.48132×10^{10}

```
SStot = SSres + SSreg
```

2.51204×10^{10}

$\dfrac{\text{SSreg}}{\text{SStot}}$

0.987773

Power Fit

```
loggeddata = N[Log[CostOfAdv]]
```

{{4.2485, 9.88073}, {4.26268, 9.93789}, {4.27667, 10.0523},
  {4.29046, 10.1258}, {4.30407, 10.1894}, {4.31749, 10.2364},
  {4.33073, 10.4133}, {4.34381, 10.5305}, {4.35671, 10.6766},
  {4.36945, 10.7951}, {4.38203, 10.8884}, {4.39445, 11.0092},
  {4.40672, 11.1062}, {4.41884, 11.2365}, {4.43082, 11.383},
  {4.44265, 11.459}, {4.45435, 11.5341}, {4.46591, 11.605},
  {4.47734, 11.6789}, {4.48864, 11.7275}}

```
p6 = ListPlot[loggeddata, PlotStyle ->{PointSize[0.02]},
   AxesLabel ->{"Log of Year", "Log of Expenditure"}];
```

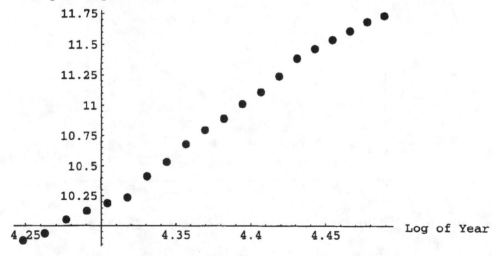

```
h[x_] = Fit[loggeddata, {1, x}, x]
```

$-25.7094 + 8.35392x$

```
p7 = Plot[h[x], {x, 70, 89}];
```

D Mathematica Appendix 405

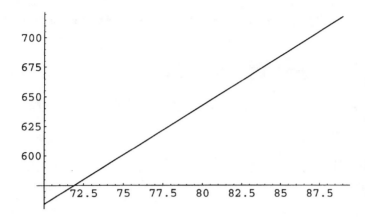

k[x_] = e^h[Log[x]]

$e^{-25.7094+8.35392\text{Log}[x]}$

p8 = Plot[k[x], {x, 70, 89}];

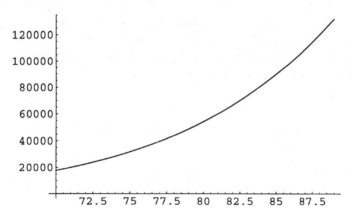

Show[p1, p8];

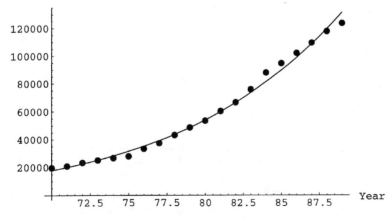

SSres = N [$\sum_{n=1}^{20}$ (CostOfAdv[[n]][[2]] - k[CostOfAdv[[n]][[1]]])^2]

1.7673×10^8

meancost = $\frac{1}{20} \sum_{n=1}^{20}$ CostOfAdv[[n]][[2]]

59928

SSreg = N [$\sum_{n=1}^{20}$ (k[CostOfAdv[[n]][[1]]] - meancost)^2]

2.35029×10^{10}

SStot = SSres + SSreg

2.36797×10^{10}

$\dfrac{\text{SSreg}}{\text{SStot}}$

0.992537

Fitting a Logistic Curve to Data: Black Bear Data This example fits a logistic curve to female black bear age/weight data. The data is read from a graph in Smith and Clark, Black Bears in Arkansas: Characteristics of a Successful Translocation, *Journal of Mammalogy,* Vol 75, No. 2, page 316. The *Mathematica* analysis directly follows Fabio Cavallini's, Fitting a Logistic Curve to Data, *College Mathematics Journal,* Vol. 24, No. 3, May 1993. This is the data, where the first coordinate is the age class in years and the second coordinate is the weight in kilograms:

data = {{1,35}, {2,55}, {3,68}, {4,70},
{5,71}, {6,75}, {7,79}, {8,82}, {9,81},
{10,80}, {11,78}, {12,99}, {13,99}, {15,82}};

A scatter plot of the data:

ListPlot[data,PlotStyle ->PointSize[.02],AxesLabel ->
 {"Age Class(Years)","Weight(Kg)"}, PlotRange ->{20,100}];

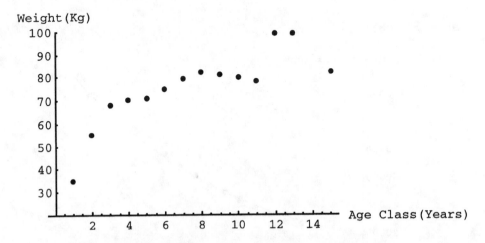

Cavallini's article performs the analysis in a very elegant manner using vectors to deal with the data. We follow the same method and notation. See the Chapter 4 of this book

D Mathematica Appendix

or Cavallini's article for details. Splitting the data into two separate lists; one for age (`T` for time), one for mass (`Y`):

```
T=Table[data[[n]][[1]],{n,1,14}]
```

{1,2,3,4,5,6,7,8,9,10,11,12,13,15}

```
Y=Table[data[[n]][[2]],{n,1,14}]
```

{35,55,68,70,71,75,79,82,81,80,78,99,99,82}

Create a vector of theoretical values in terms of parameters `r` and `t0`. The result, `H`, will be a vector

```
H=1.0/(1.0+Exp[-r(T-t0)]);
```

The error vector is

```
error=(Y.Y)-(H.Y)^2/(H.H);
```

The key step in this analysis is choosing a wise starting point for the search algorithm. Next we display a 3D plot of the error. Since `t0` must be within the data, choosing its range to be 0 to 16 is easy. A range for `r` may require some guesswork and repeated trials.

```
Plot3D[error, {r, 0.01, 1}, {t0, 0, 16},
   AxesLabel ->{"r", "t0", "e "}];
```

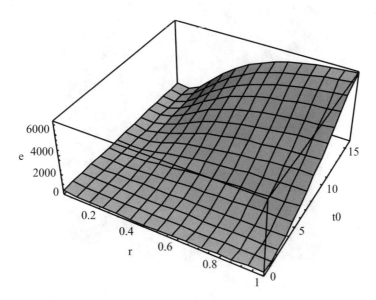

The next graph restricts the scale on the range to help identify the minimum.

```
Plot3D[error, {r, 0.01, 1}, {t0, 0, 16},
   AxesLabel ->{"r", "t0", "e"}, PlotRange ->{0, 1000}];
```

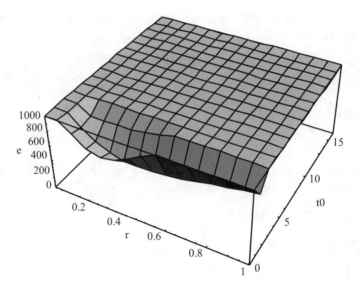

We get a sense of the shape from this plot. Next we will switch to a contour plot to flush out more details. Recall that the contour plot is of error values for different `r` and `t0` values. The 3D Plot helps us determine the range of `r` and `t0` to plot. Darker regions indicate less error.

```
ContourPlot[error,{r,0.01,1}, {t0,0,5}];
```

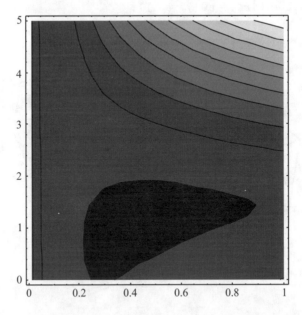

We will pick our initial guess from somewhere near the center of the black region. The following gives the error for these parameter choices.

```
error/.{r ->0.4, t0 ->1}
```
```
590.224
```

D Mathematica Appendix

Use this guess to start the **FindMinimum** algorithm. You can experiment with other r, t0 combinations to try to get a smaller error, but all we need to be is close enough for the search routine to work correctly.

```
FindMinimum[error,{r,0.4}, {t0,1}]
```

```
{585.75, {r ->0.440661, t0 ->1.15245}}
```

The minimum error is 585.75 at parameter values **r**=0.441 and **t0**=1.152. Find K for these parameter values, using Cavallini's method.

```
K=((H.M)/(H.H))/.{r ->0.440661014003951072466',
  t0 ->1.15244778447574305923'}
```

```
87.238
```

Another copy of the scatterplot of the data:

```
FigData=ListPlot[Transpose[{T,M}],PlotStyle->
  PointSize[0.02],PlotRange->{20,100}];
```

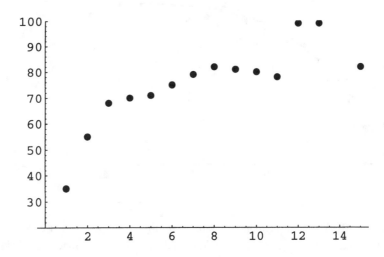

A plot of the fitted (found parameters inserted) logistic curve:

```
FigModel=Plot[K/(1+Exp[-0.440661014003951072466'
(t-1.15244778447574305923')]),
{t,Min[T],Max[T]},PlotRange->{20,100}];
```

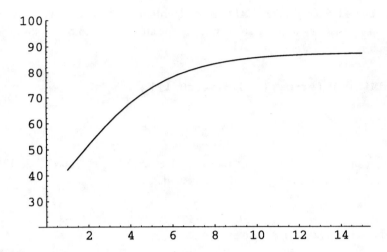

Finally the theoretical curve displayed with the data:

```
Show[FigData,FigModel,AxesLabel ->{"Age Class(Years)",
    "Weight (kg)"}];
```

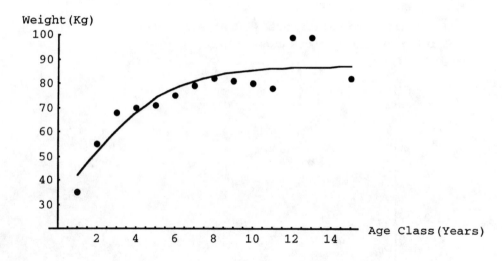

Curvilinear Regression: Corn Storage In this example, we use *Mathematica*'s ability to create Regression or ANOVA tables. This is done by replacing the `Fit[]` command with the `Regress[]` command. `Regress` is found in the `Statistics'LinearRegression` package.

```
Needs["Statistics'LinearRegression'"]

Clear[data, x];
data={81463, 43676, 24014, 10477, 83947, 42539, 20182,
    373, 104212,87433, 36952, 17953, 80581, 46134, 24807,
    9028, 97936, 52217, 26415}
```

D Mathematica Appendix

```
{81463, 43676, 24014, 10477, 83947, 42539, 20182, 7373,
   104212, 87433, 36952, 17953, 80581, 46134, 24807,
   9028, 97936, 52217, 26415}

p0=ListPlot[data];
p1=ListPlot[data, PlotJoined ->True,
   AxesLabel ->{"Quarter", "Bushels"}];
```

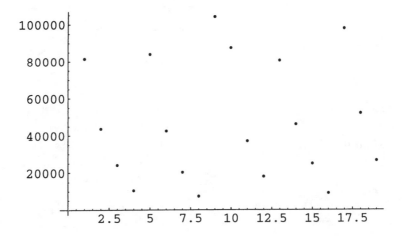

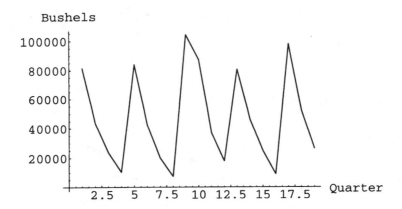

```
g[x_] = Fit[data, {1,x}, x]
```

48361.1 - 113.275x

```
p3 = Plot[g[x], {x, 0, 20}];
```

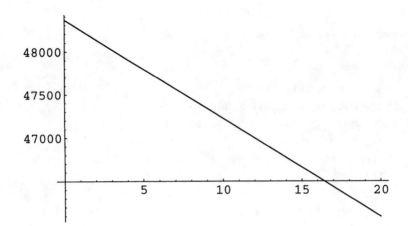

Show[p0, p3];

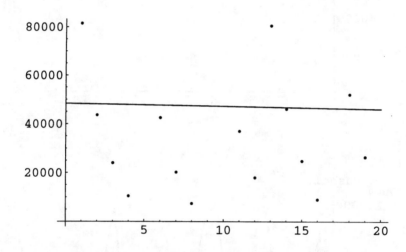

Regress[data, {1, x}, x]

```
                        Estimate    SE        TStat       PValue
{ParameterTable -> 1    48361.1     15834.1   3.05424     0.00717394
                   x    -113.275    1388.74   -0.0815669  0.935944
RSquared ->0.000391209,AdjustedRSquared ->-0.0584093,
EstimatedVariance ->1.09931 x 10^9, ANOVATable ->
         DF  SumOfSq              MeanSq              FRatio       PValue
  Model  1   7.31386 x 10^6       7.31386 x 10^6      0.00665315   0.935944
  Error  17  1.86882 x 10^10      1.09931 x 10^9
  Total  18  1.86955 x 10^10
    }
```

f1[x_]=Fit[data, {1, Cos [$\frac{\pi x}{2}$], Sin [$\frac{\pi x}{2}$] }, x]

46169.9 - 20110.9 Cos [$\frac{\pi x}{2}$] + 31576.9 Sin [$\frac{\pi x}{2}$]

p10=Plot[f1[x], {x, 0, 20}];

D Mathematica Appendix

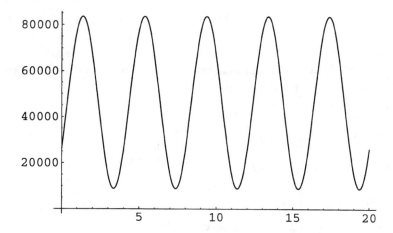

`Show[p0, p10];`

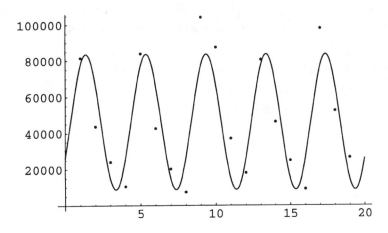

D.5 Chapter 5
Differential Equations

***Mathematica* background** There are two commands that are fundamental to this section: `DSolve` and `NDSolve`. The function `DSolve` can symbolically solve differential equations, whereas `NDSolve` finds numerical solutions. The two commands are used in nearly the same manner, except that `NDSolve` requires that all parameters be given numerical values and that initial condition(s) be specified, and the desired domain of the solution is required. The output of `NDSolve` is an interpolating polynomial, as opposed to the standard functions given by `DSolve`.

The syntax for `DSolve` is `DSolve[`*diffeqn*`, x[t], t ]` where *diffeqn* is the differential equation expressed in terms of `x'[t]`, `x[t]`, and `t`. Explicit use of the independent variable in the dependent variable name is required and is a common source of mistakes. Further the equality in *diffeqn* must be expressed as `==`. Using `DSolve` to solve a logistic differential equation is shown below.

Caution: Differential equation solving with *Mathematica* is slow, memory intensive, and occasionally causes system crashes. Save your work often and reboot if strange results occur. Differential equations with errors frequently cause *Mathematica* to grind away forever, and the **Abort Evaluation** procedure frequently does not work. If this happens, either reboot or kill the kernel.

DSolve[x'[t]==r x[t](1-x[t]/K), x[t], t]

$\left\{\left\{x[t] \to \dfrac{e^{rt} K}{e^{rt} - e^{C[1]}}\right\}\right\}$

Clear[x, t, Derivative];

Here is an example using **DSolve** to find the constant of integration. Notice the two equations (the differential equation and the initial condition) grouped in braces.

DSolve[{x'[t] == 1.3 x[t], x[0]==20}, x[t],t]

{{x[t] ->20. 2.71828182845904524$^{1.3\,t}$}}

Mathematica reports its solution as a list. To extract the solution, name the list something (like solution) and use the procedure below. The result is a function which can be used in all the ways any function is used, such as plotting.

solution = DSolve[{x'[t]==1.3 x[t], x[0]==20}, x[t], t]

{{x[t] ->20. 2.71828182845904524$^{1.3\,t}$ }}

x[t_] = x[t]/.solution[[1]]

20. 2.71828182845904524$^{1.3\,t}$

Plot[x[t], {t,0,5}];

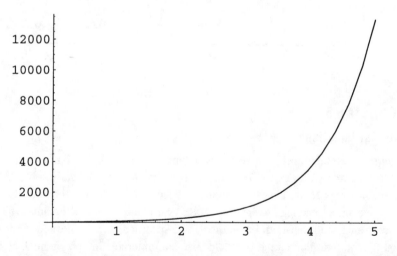

Next we do an example with **NDSolve**. Notice the only difference is that instead of t, we need the range {t, tmin, tmax}. Of course there can be no symbolic constants and the initial condition is necessary. If previous solutions involving the variables x and t have been performed this session, it is necessary to **Clear** them. The word **Derivative** causes both **x** and **x'** to be cleared. Notice we need to perform the same trick to get a solution which is usable as a function.

D Mathematica Appendix

```
Clear[x, t, Derivative];
solution=
  NDSolve[{x'[t]==1.3 x[t](1-x[t]/100)^5, x[0]==20},
    x[t], {t, 0, 20}]
{{x[t] ->InterpolatingFunction[{{0., 20.}}, <>][t]
x[t_] = x[t]/.solution[[1]]
InterpolatingFunction[{{0., 20.}}, <>][t]
Plot[x[t], {t, 0, 20}];
```

The **Options** command tells us the options a function has. Here are the options for **NDSolve**. Notice that if we desire some control over the numerical solving process, we can set the options manually.

```
Options[NDSolve]
{AccuracyGoal ->Automatic,Compiled ->True,
 DifferenceOrder ->Automatic,
 InterpolationPrecision ->Automatic,MaxRelativeStepSize ->1,
 MaxSteps ->Automatic,MaxStepSize ->∞,
 Method ->Automatic,PrecisionGoal ->Automatic,
 SolveDelayed ->False,StartingStepSize ->Automatic,
 StoppingTest ->None,WorkingPrecision ->19}
```

Here we specify that the algorithm used is Runge-Kutta. For more details, see Wolfram's *Mathematica* Book.

```
Clear[x, t, Derivative];
solution=
  NDSolve[{x'[t]==1.3 x[t](1-x[t]/100)^5,x[0]==20},
    x[t],{t,0,20}, Method ->RungeKutta]
{{x[t] ->InterpolatingFunction[{{0, 20.}}, <>][t]}}
x[t_] = x[t]/.solution[[1]]
```

```
InterpolatingFunction[{{0, 20.}}, <>][t]
Plot[x[t], {t, 0, 20}];
```

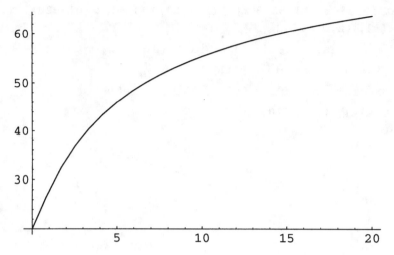

Solving systems of first-order homogeneous linear equations with constant coefficients

Definition of the initial value problem We are solving a problem of the form $X'=MX$ with the initial value $X(0)$ known

```
M={{-0.24, 0.1, 0.04}, {0.2,-0.2,0.15}, {0.05,0.1,-0.2}}
```
{{-0.24,0.1,0.04}, {0.2,-0.2,0.15}, {0.05,0.1,-0.2}}
```
X0={12000, 10000, 8000}
```
{12000, 10000, 8000}

Eigenvalues and Eigenvectors

```
Eigenvalues[M]
```
{-0.3794287560387887969, -0.2600645349462926947,
 -0.0005067090149185081806}

To find numeric values of the eigenvalues use the `N[]` function.

```
N[%]
```
0.3794287560387887969,-0.2600645349462926947,
 -0.0005067090149185081806}

```
eigenlist = Eigensystem[M]
```
{{-0.3794287560387887969, -0.2600645349462926947,
 -0.0005067090149185081806}, {{-0.4916706199968276904,
  0.8116720472887348256, -0.3153545450473940453},
 {0.5436977187447060921,0.2153794800308560728,
  -0.8111747470264594844}, {0.4044020558788450414,
  0.7729834974707147577, 0.4888307374121486281}}}

D Mathematica Appendix

Define the eigenvalues and eigenvector as the appropriate components of eigenlist. This can be done by entering them directly, but this "automated" approach makes it easy to reuse the code for other systems.

```
l1 = eigenlist[[1]][[1]]
l2 = eigenlist[[1]][[2]]
l3 = eigenlist[[1]][[3]]
```
-0.3794287560387887969
-0.2600645349462926947
-0.0005067090149185081806

```
v1 = eigenlist[[2]][[1]]
v2 = eigenlist[[2]][[2]]
v3 = eigenlist[[2]][[3]]
```
{-0.4916706199968276904, 0.8116720472887348256,
  -0.3153545450473940453}
{0.5436977187447060921, 0.2153794800308560728,
  -0.8111747470264594844}
{0.4044020558788450414, 0.7729834974707147577,
  0.4888307374121486281}

Define a vector of arbitrary constants.

```
c = {c1, c2, c3};
```

Notice that multiplying a list by a function distributes the function across the list.

e^{-2t} v1

$$\left\{ -\frac{0.4916706199968276904}{e^{2t}}, \frac{0.8116720472887348256}{e^{2t}}, -\frac{0.3153545450473940453}{e^{2t}} \right\}$$

Define the matrix Φ.

```
Phi[t_]=Transpose[{e^{l1 t} v1, e^{l2 t} v2, e^{l3 t} v3}]
```

$$\left\{ \left\{ -\frac{0.4916706199968276904}{e^{0.3794287560387887969\,t}}, \frac{0.5436977187447060921}{e^{0.2600645349462926947\,t}}, \frac{0.4044020558788450414}{e^{0.0005067090149185081806\,t}} \right\}, \left\{ \frac{0.8116720472887348256}{e^{0.3794287560387887969\,t}}, \frac{0.2153794800308560728}{e^{0.2600645349462926947\,t}}, \frac{0.7729834974707147577}{e^{0.0005067090149185081806\,t}} \right\}, \right.$$

$$\left. \left\{ -\frac{0.3153545450473940453}{e^{0.3794287560387887969\,t}}, -\frac{0.8111747470264594844}{e^{0.2600645349462926947\,t}}, \frac{0.4888307374121486281}{e^{0.0005067090149185081806\,t}} \right\} \right\}$$

The general solution

```
Phi[t].c
```

$$\left\{ -\frac{0.4916706199968276904\ c1}{e^{0.3794287560387887969\ t}} + \frac{0.5436977187447060921\ c2}{e^{0.2600645349462926947\ t}} + \right.$$

$$\frac{0.4044020558788450414\ c3}{e^{0.0005067090149185081806\ t}},\ \frac{0.8116720472887348256\ c1}{e^{0.3794287560387887969\ t}} +$$

$$\frac{0.2153794800308560728\ c2}{e^{0.2600645349462926947\ t}} + \frac{0.7729834974707147577\ c3}{e^{0.0005067090149185081806\ t}},$$

$$-\frac{0.3153545450473940453\ c1}{e^{0.3794287560387887969\ t}} - \frac{0.8111747470264594844\ c2}{e^{0.2600645349462926947\ t}} +$$

$$\left. \frac{0.4888307374121486281\ c3}{e^{0.0005067090149185081806\ t}} \right\}$$

Solution of the initial value problem

`cspec = Inverse[Phi[0]].X0`

{-5822.889478019990458, 3324.005777461279479,
18125.03634765131886}

`Phi[t].Simplify[cspec]`

$$\left\{ \frac{2862.943679831093072}{e^{0.3794287560387887969\ t}} + \frac{1807.254358299920839}{e^{0.2600645349462926947\ t}} + \right.$$

$$\frac{7329.80196186898609}{e^{0.0005067090149185081806\ t}},\ -\frac{4726.276623760518139}{e^{0.3794287560387887969\ t}} +$$

$$\frac{715.9226359691718584}{e^{0.2600645349462926947\ t}} + \frac{14010.35398779134628}{e^{0.0005067090149185081806\ t}},$$

$$\frac{1836.274662202251879}{e^{0.3794287560387887969\ t}} - \frac{2696.349545646643162}{e^{0.2600645349462926947\ t}} +$$

$$\left. \frac{8860.074883444391282}{e^{0.0005067090149185081806\ t}} \right\}$$

`soln = Simplify[%]`

$$\left\{ \frac{2862.943679831093072}{e^{0.3794287560387887969\ t}} + \frac{1807.254358299920839}{e^{0.2600645349462926947\ t}} + \right.$$

$$\frac{7329.80196186898609}{e^{0.0005067090149185081806\ t}},\ -\frac{4726.276623760518139}{e^{0.3794287560387887969\ t}} +$$

$$\frac{715.9226359691718584}{e^{0.2600645349462926947\ t}} + \frac{14010.35398779134628}{e^{0.0005067090149185081806\ t}},$$

$$\frac{1836.274662202251879}{e^{0.3794287560387887969\ t}} - \frac{2696.349545646643162}{e^{0.2600645349462926947\ t}} +$$

$$\left. \frac{8860.074883444391282}{e^{0.0005067090149185081806\ t}} \right\}$$

D Mathematica Appendix

The functions x, y, and z are defined to be the appropriate components of `soln`.

```
x[t_]=soln[[1]]
y[t_]=soln[[2]]
z[t_]=soln[[3]]
```

$$\frac{2862.943679831093072}{e^{0.3794287560387887969\,t}} + \frac{1807.254358299920839}{e^{0.2600645349462926947\,t}} +$$

$$\frac{7329.80196186898609}{e^{0.0005067090149185081806\,t}}$$

$$-\frac{4726.276623760518139}{e^{0.3794287560387887969\,t}} + \frac{715.9226359691718584}{e^{0.2600645349462926947\,t}} +$$

$$\frac{14010.35398779134628}{e^{0.0005067090149185081806\,t}}$$

$$\frac{1836.274662202251879}{e^{0.3794287560387887969\,t}} - \frac{2696.349545646643162}{e^{0.2600645349462926947\,t}} +$$

$$\frac{8860.074883444391282}{e^{0.0005067090149185081806\,t}}$$

Plot the solution curves

```
Plot[{x[t], y[t], z[t]}, {t, 0, 100}];
```

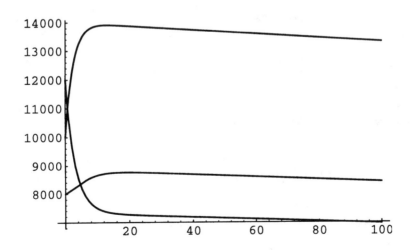

```
ParametricPlot3D[{x[t], y[t], z[t]}, {t, 0, 20},
    ViewPoint ->{2.228, -1.932, 1.659}];
```

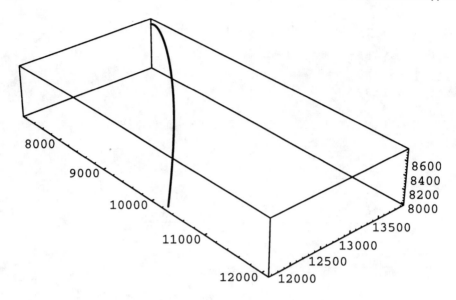

D.6 Chapter 6

Queueing Simulations

Single-Server Queue Simulation. This section gives *Mathematica* code for a simulation of a single-server queue. It makes use of list properties and Do and While statements.

```
numsim=1;
tbar=0;
cumavequelength=0;
(*Initialize parameters*)
Do[Clear[queuelength,a];
   a={};
   T=10;
   deltat=0.01;
   n=T/deltat;
(*Arrival and Departure Rates*)
   lambda=4;
   mu=7;
(*Initialize Queue Statistics*)
   sum=0;
   cumServiceTime=0;
   averageServiceTime=0;
   numServed=0;
   queuelength = Table[{ideltat, 0}, {i,0,n}];
(*Determine if an arrival or departure occurred*)
   Do[arrival = 1;
   depart = 1;
```

D Mathematica Appendix

```
While[arrival + depart ≠ 0 && arrival + depart ≠ 1,
  arrival = 0;
  depart = 0;
  If[Random[Real, {0,1}] <mu deltat && a ≠ {}, depart=1]
  If[Random[Real,{0,1}] <lambda deltat, arrival = 1];];
```

(*An Arrival occurs if the arrival variable is one *)

```
If[arrival==1, a=Join[a,{ideltat}]];
```

(*An Departure occurs if the departure variable is one *)

```
If[depart == 1,
  cumServiceTime = cumServiceTime + (i deltat - Take[a,1])[[1]];
  numServed = numServed + 1;
  a = Drop[a,1]];
queuelength = ReplacePart[queuelength, {ideltat, Length[a]}, i+1];
sum = sum + Length[a];, {i, 0, n}];

Print[Number served is numServed];
Print[Average time is cumServiceTime/numServed];
Print[Average queue size is sum/n];
ListPlot[queuelength, PlotJoined ->True];
tbar = tbar + cumServiceTime/numServed;
cumavequelength = cumavequelength + sum/n;, {j, 1, numsim}];
```

$$\text{Print[Predicted Average Time is } \frac{1}{(1 - \frac{lambda}{mu})\,mu}];$$

$$\text{Print[Simulated Average Time is } \frac{tbar}{numsim}];$$

$$\text{Print[Predicted Average Queue Size is } \frac{1}{(1 - \frac{lambda}{mu})\,mu}];$$

$$\text{Print[Simulated Average Queue Size is } \frac{cumavequelength}{numsim}];$$

References

[1] Martha L. Abell and James P. Braselton. *Modern Differential Equations: Theory, Applications, Technology*. Saunders College Publishing, Fort Worth, 1996.

[2] A.R. Behnke and J.H. Wilmore. *Evaluation and Regulation of Body Build and Composition*. Prentice-Hall, Englewood Cliffs, N.J., 1974.

[3] Edward Beltrami. *Mathematical Models in the Social and Biological Sciences*. Jones and Bartlett Publishers, Boston, 1993.

[4] C.A. Bosch. Redwoods: A population model. *Science*, 172:345–349, 1971.

[5] Mark S. Boyce. *The Jackson Elk Herd*. Cambridge University Press, Cambridge, 1989.

[6] W.E. Boyce and R.C. DiPrima. *Elementary Differential Equations*. John Wiley & Sons, Inc., New York, fourth edition, 1986.

[7] Robert L. Brown. *Introduction to the Mathematics of Demography*. ACTEX Publications, Inc., Winsted, Connecticut, second edition, 1993.

[8] Financial California State Department of Finance and Population Research Section. *California Migration 1955-1960*. Sacramento, 1964.

[9] Hal Caswell. *Matrix Population Models*. Sinauer Associates, Inc., 1989.

[10] Fabio Cavallini. Fitting a logistic curve to data. *The College Mathematics Journal*, 24(3):247–253, 1993.

[11] I. Chase. Vacancy chains. *Annual Review of Sociology*, 17:133–154, 1991.

[12] Colin Clark. *Mathematical Bioeconomics: The Optimal Management of Renewable Resources*. John Wiley & Sons, Inc., New York, second edition, 1990.

[13] James Cox, Randy Kautz, Maureen MacLaughlin, and Terry Gilbert. *Closing the Gaps in Florida's Wildlife Conservation System*. Office of Environmental Services, Florida Game and Fresh Water Fish Commission, Tallahassee, Florida, 1994.

[14] Bill Davis, Horacio Porta, and Jerry Uhl. *Calculus&Mathematica*. Addison-Wesley, Reading, Massachusetts, 1994.

[15] Leah Edelstein-Keshet. *Mathematical Models in Biology*. Birkhäuser Mathematics Series. McGraw Hill, Inc., New York, 1988.

[16] Ruth Dudley Edwards. *An Atlas of Irish History*. Methuen & Co., New York, second edition, 1981.

[17] Leroy A. Franklin. Using simulation to study linear regression. *The College Mathematics Journal*, 23(4):290–295, 1992.

[18] Daniel L. Gerlough and Frank C. Barnes. *Poisson and Other Distributions in Traffic*. Eno Foundation for Transportation, Saugatuck, Conn., 1971.

[19] Michael E. Gilpin. Do hares eat lynx? *Amer. Nat.*, 107:727–730, 1973.

[20] Michael E. Gilpin and Francisco J. Ayala. Global models of growth and competition. *Proc. Nat. Acad. Sci. USA*, 70(12, Part I):3590–3593, 1973.

[21] J.R. Ginsberg and E.J. Milner-Gulland. Sex-biased harvesting and population dynamics in ungulates: Implications for conservation and sustainable use. *Conservation Biology*, 8(1):157–166, 1994.

[22] Frank R. Giordano and Maurice D. Wier. *A First Course in Mathematical Modeling.* Brooks/Cole Pub. Co., Monterey, CA, 1985.

[23] J. Grier. Ban of DDT and subsequent recovery of reproduction in bald eagles. *Science*, 218:1232–1234, 1982.

[24] Charles A.S. Hall. An assessment of several of the historically most influential theoretical models used in ecology and of the data provided in their support. *Ecological Modelling*, 43:5–31, 1988.

[25] John Harte. *Consider a Spherical Cow: A Course in Environmental Problem Solving.* University Science Books, Mill Valley, California, 1988.

[26] Robert V. Hogg and Elliot A. Tanis. *Probability and Statistical Inference.* Prentice Hall, Englewood Cliffs, NJ, fourth edition, 1993.

[27] J.H. Hubbard and B.H. West. *Differential Equations: A Dynamical Systems Approach, Part I, One Dimensional Equations.* Springer-Verlag, New York, 1990.

[28] Richard John Huggett. *Modelling the Human Impact on Nature: System Analysis of Environmental Problems.* Oxford University Press, Oxford, 1993.

[29] Sam Kosh Kachigan. *Multivariate Statistical Analysis: A Conceptual Introduction.* Radius Press, New York, second edition, 1991.

[30] L.B. Keith. *Wildlife's Ten Year Cycle.* University of Wisconsin Press, Madison, WI, 1963.

[31] Nathan Keyfitz. *Introduction to the Mathematics of Population with Revisions.* Addison-Wesley, Reading, Mass., 1977.

[32] Leonard Kleinrock. *Queueing Systems Volume 1: Theory.* Wiley Interscience, New York, 1975.

[33] Leonard Kleinrock. *Queueing Systems Volume 2: Computer Applications.* Wiley Interscience, New York, 1976.

[34] Charles J. Krebs. *Ecology: The Experimental Analysis of Distribution and Abundance.* Harper and Row Publishers, New York, second edition, 1978.

[35] Alan Krinik. Taylor series solution of the M/M/1 queueing system. *J. of Comp. Appl. Math.*, 44:371–380, 1992.

[36] Alan Krinik, Daniel Marcus, Dan Kalman, and Terry Chang. Transient solution of the M/M/1 queueing system via randomization. In J. Goldstein, N. Gretsky, and J.J. Uhl, editors, *Stochastic Processes and Functional Analysis: In Celebration of M.M. Rao's 65th Birthday*, volume 186 of *Lecture Notes in Pure and Applied Mathematics*, pages 137–145. Marcel Dekker, 1997.

[37] P.H. Leslie. On the use of matrices in certain population mathematics. *Biometrika*, 33:183–212, 1945.

[38] J.D.C. Little. A proof of the queueing formula $L = \lambda W$. *Operations Research*, 9:383–387, 1961.

[39] G. Loery and J. Nichols. Dynamics of a black-capped chickadee population 1958–1983. *Ecology*, 66(4):1195–1203, 1985.

[40] Brian Lowry. *The Truth is Out There: The Official Guide to The X-Files.* Harper Prism, New York, 1995.

[41] D. Ludwig, D.D. Jones, and C.S. Holling. Qualitative analysis of insect outbreak systems: The spruce budworm and forest. *Journal of Animal Ecology*, 47:315–332, 1978.

[42] William Mendenhall, Richard L. Scheaffer, and Dennis D. Wackerly. *Mathematical Statistics with Applications*. Duxbury Press, Boston, second edition, 1981.

[43] B.R. Mitchell. *European Historical Statistics 1750-1970*. Columbia University Press, New York, 1975.

[44] J.D. Murray. *Mathematical Biology*. Springer-Verlag, Berlin, 1989.

[45] Raymond H. Myers. *Classical and Modern Regression with Applications*. Duxbury Press, Belmont, CA, second edition, 1990.

[46] The North Carolina School of Science and Mathematics. *Contemporary Precalculus through Applications*. Janson Publications, Inc, Dedham, Massachusetts, 1992.

[47] U.S. Bureau of the Census. *Statistical Abstract of the United States: 1988*. U.S. Government Printing Office, Washington, D.C., 108th edition, 1988.

[48] Michael Olinick. *An Introduction to Mathematical Models in the Social and Life Sciences*. Addison-Wesley Publishing Company, Reading, Massachusetts, 1978.

[49] Arnold Ostebee and Paul Zorn. *Calculus from Graphical, Numerical, and Symbolic Points of View: Volume 1*. Saunders College Publisher, Fort Worth, advance edition, 1997.

[50] Elazar J. Pedhazur. *Multiple Regression in Behavior Research*. Holt, Rinehart and Winston, Inc., Fort Worth, second edition, 1982.

[51] K.W. Penrose, A.G. Nelson, and A.G. Fisher. Generalized body composition prediction equation for men using simple measurement techniques. *Medicine and Science in Sports and Exercise*, 17(2):189, 1985.

[52] Robert Ramighetti, editor. *The World Almanac and Book of Facts 1996*. Funk and Wagnallis Corporation, Mahwah, NJ, 1996.

[53] J.C. Reynolds. Details of the geographic replacement of the red squirrel (*sciurus vulgaris*) by the grey squirrel (*sciurus carolinensis*) in eastern England. *Journal of Animal Ecology*, 54:149–162, 1985.

[54] Lewis F. Richardson. Generalized foreign policy. *British Journal of Psychology Monographs Supplements*, 23:130 – 148, 1939.

[55] Adrei Rogers. *Matrix Analysis of Interregional Population Growth and Distribution*. University of California Press, Berkeley, 1968.

[56] Clay C. Ross. *Differential Equations: An Introduction with Mathematica*. Springer-Verlag, New York, 1995.

[57] James T. Sandefur. *Discrete Dynamical Systems: Theory and Applications*. Clarendon Press, Oxford, 1990.

[58] Jerald L. Schnoor. *Environmental Modeling*. John Wiley & Sons, Inc., New York, 1996.

[59] Lee A. Segel. *Modeling Dynamic Phenomena in Molecular and Cellular Biology*. Cambridge University Press, Cambridge, 1984.

[60] Kentucky Agricultural Statistics Service. *Kentucky Agricultural Statistics: 1993-1994*. Kentucky Department of Agriculture, Louisville, KY, 1994.

[61] W.E. Siri. Gross composition of the body. In J.H. Lawrence and C.A. Tobias, editors, *Advances in Biological and Medical Physics*, volume IV. Academic Press, Inc., New York, 1956.

[62] Kimberly G. Smith and Joseph D. Clark. Black bears in Arkansas: Characteristics of a successful translocation. *Journal of Mammalogy*, 75(2):309–320, 1994.

[63] G.W. Snedecor and W.G. Cochran. *Statistical Methods*. Iowa State University Press, Ames, Iowa, sixth edition, 1967.

[64] A.M. Starfield and A.L. Bleloch. *Building Models for Conservation and Wildlife Management*. Burgess International Group, Inc., Edina, MN, second edition, 1991.

[65] Anthony M. Starfield, Karl A. Smith, and Andrew L. Bleloch. *How to Model It: Problem Solving for the Computer Age*. McGraw-Hill Publishing Company, New York, 1990.

[66] Howard M. Taylor and Samuel Karlin. *An Introduction to Stochastic Modeling*. Academic Press, San Diego, 1994.

[67] John W. Tukey. *Exploratory Data Analysis*. Addison-Wesley Publishing Company, Reading, Mass., 1977.

[68] M. Usher, T. Crawford, and J. Bunwell. An American invasion of Great Britain: The case of native and alien squirrel species. *Conservation Biology*, 6(1):108–115, 1992.

[69] M.B. Usher. A matrix approach to management of renewable resources with special reference to selection forests. *Journal of Applied Ecology*, 3:355–367, 1966.

[70] S. Utida. Cyclic fluctuations of population density intrinsic to the host-parasite system. *Ecology*, 38:442–449, 1957.

[71] Carl J. Walters, Ray Hilborn, and Randall Peterman. Computer simulation of barren-ground caribou dynamics. *Ecological Modelling*, pages 303–315, 1975.

[72] M. Weissburg, L. Roseman, and I. Chase. Chains of opportunity: A Markov model for the acquisition of reusable resources. *Evolutionary Biology*, 5:105–117, 1991.

[73] M. Zangrandi. *Growth Rate of Algae of Genera Fosliella and Pheophyllum in the Gulf of Trieste*. PhD thesis, University of Trieste, 1991.

INDEX

R^2, 82, 157–161, 163–164, 170, 172, 174, 199–200, 205, 207, 209, 212, 219–224, 228, 230, 232, 236, 322
SS_{Res}, 322
Erf, 297, 298
α, 202–206, 209, 210, 221
β, 259
γ, 259
λ, 330–332, 334, 338, 339, 343
μ, 327, 334, 338, 343
ρ, 338–340
θ-logistic, 270–272, 276, 298, 313

absolute address, 13
absorbing state, 325, 326
affine, 19, 21, 22, 26, 38
Allee effect, 246, 269, 275
almost logistic fit, 169
amplitude, 165
ANOVA table, 199
arrival rate, 334, 338, 339
asymptotic fit, 169
auxiliary equation, 255
Ayala, 270, 271

basis, 151, 167, 176, 177, 180, 198, 220–224
Bernoulli distribution, 68
Bernoulli random variable, 68–70, 80, 93, 325, 346
Bernoulli trial, 68, 92
Bernoulli trials, 97
beta-gamma plane, 261
binomial distribution, 66, 68–70, 80, 81, 88, 89, 92, 93, 97
binomially distributed, 69
birth-death process, 351, 360–363
birth-death-immigration process, 363–367
box plot, 58–60, 76, 90, 91

capture-recapture, 97–99
Carr, 219
carrying capacity, 31–37, 244, 269, 274, 278–282, 296
catalog of functions, 164
chaos, 33
characteristic equation, 26–28, 137, 255
characteristic root, 26, 27, 137, 255
chi-square, 82–89, 94, 322, 356, 357
chi-square distribution, 86
closed-form solution, 10, 21, 28, 29, 38, 42, 45, 48, 71, 78, 79, 255, 295, 304, 305
cobweb method, 23–25, 38
coefficient of determination, 159
compartmental analysis, 16–19

compartmental diagram, 17–19
compound interest, 43
computational complexity, 234–235
computer algebra system, 295
conditions for stability, 23
confidence, 82
confidence interval, 205–207
continuous model, 9, 32, 38
continuous models, 239–322
continuous stochasticity, 323–367
conversion between difference and recurrence equations, 17
correlation, 151–155, 161
correlation coefficient, 154
covariance, 151–155
critical, 202, 206, 210
critical value, 83
cubic spline, 190, 192, 230
cumulative density function, 62
curvilinear models, 164
curvilinear regression, 152, 198, 219–224, 227
cutoff, 202, 206, 210
cutoff value, 83

degrees of freedom, 84, 86, 89, 202–207, 209
demographic stochasticity, 70, 72, 75–78, 94, 97
density function, 61–67, 91
departure rate, 338
determinant, 109, 136–137, 259
deterministic, 6, 47–49, 51, 71, 73, 75, 76, 82
difference equations, 16–19, 37, 45–46, 239–240, 243, 300
    exponential, 20
    first-order, 10, 16, 21, 23
    logistic, 31, 36
difference quotient, 326
difference versus differential, 239–240, 243, 245, 300
differential equations, 239–322
    Allee effect, 269
    autonomous, 267, 283
    auxiliary equation, 255
    characteristic equation, 255
    characteristic root, 255
    dominant eigenvalue, 256
    eigenvalue, 255, 263, 264
    eigenvector, 263, 264
    exact, 252
    exponential, 243, 268, 326, 329, 346, 349, 350
    first-order systems, 258–266
    forcing term, 252
    formally solving, 255, 263
    general solution, 243, 252, 255, 263, 265

427

homogeneous, 254
integrating factor, 252
initial value problem, 243
linear, 251–349
linear constant coefficients, 253–258
linearization, 285–286, 292
logistic, 244, 268–269, 272, 278, 295, 298, 299, 304, 308
numerical solution, 295, 298–305
separable, 242–245, 249, 292
solver, 284, 294–305
systems, 283–293, 307, 308, 320
differential equations
first-order, 239, 252, 258–266, 330, 349
linear, 349
differential-equation solver, 284, 291, 294–305, 308
dimensionless analysis, 278
direction field, 298
discrete model, 9–99, 101–150
discrete stochasticity, 47–99
dominant eigenvalue, 27, 29, 30, 256, 260

Edelstein-Keshet, 287
eigenvalue, 26, 27, 29, 30, 112–120, 137–138, 255, 263, 264
eigenvector, 112–120, 137–138, 263, 264
Einstein, 83
Elvis, 31
empirical modeling, 151–237
envelope, 166, 177
environmental stochasticity, 70, 72–74, 78, 82, 87–89, 91, 94
equilibrium point, 260, 261, 280
error, 156, 183
Euler's method, 299–301
expected value, 64, 65, 83, 91, 92
exponential, 20–23, 26, 27, 30–32, 151, 159, 164, 166–168, 170, 172, 174, 177, 178, 180, 181, 232, 255, 268, 320, 346
exponential distribution, 327, 330, 331, 334

F-ratio, 201–205, 209, 210, 212, 217, 221–223
F-test, 201–204
factorial function, 10
Fibonacci recurrence relation, 10, 26–28
Fibonacci sequence, 10, 27, 228
FIFO, 341
fitting
almost logistic, 169
asymptotes, 169
exponential, 168
line, 155–157
logarithmic, 168
logistic, 232, 322
polynomial, 233–235
power, 168

fitting a line, 155–157
fitting to data, 151–237
five point summary, 58–60, 91
fixed point, 21–25, 32, 33, 35, 37, 38, 260, 261, 269, 274, 280, 283, 308, 321, 337
attracting, 23, 25
repelling, 23, 25
stable, 23–25, 33
unstable, 23–25
fixed points, 336
forcing term, 252
frequency distribution, 53, 66, 67, 82, 87, 88, 90, 91
fundamental matrix, 127, 132, 138
Fundamental Theorem of Calculus, 63

Galois, 26
Gaussian elimination, 27
Gilpin, 270, 271
goodness-of-fit, 356, 357
goodness-of-fit test, 82–89, 94
growth rate, 279, 296

harvesting, 272–277
histogram, 53, 55–57, 60–62, 67, 73, 74, 90, 91, 232
homogeneous, 25, 26, 28
hypothesis, 202–204, 209
hypothesis testing, 83

identity matrix, 109
independence, 167, 331
individual-based model, 97
inner product, 110
integrating factor, 252, 330, 349
inter-quartile, 49
inter-quartile range, 58, 59
interpolation, 151, 166, 187–197, 226–227, 230
intrinsically linear, 151, 167, 220
intrinsically nonlinear linearizable, 151, 167–180
intrinsically nonlinear nonlinearizable, 151, 176, 180–186
isocline, 283

Jacobian, 290, 308
Jacobian analysis, 321
Jernigan, 169

L'Hopital's rule, 362, 366
ladder of powers, 164
law of mass action, 287, 316
least-squares criterion, 151, 155–157, 167, 181, 183, 187, 198, 200
Leslie matrix, 111, 114, 134–136, 139, 141
level curve, 283
life table, 133–135, 138–139
light pole joke, 240

Index

linear, 164, 167–169, 219, 251, 285–286, 307, 308
linear combination, 25, 26, 151, 167, 252
linear interpolation, 187–189, 191
linear regression, 151, 152, 159, 197, 198, 205, 224
linear spline, 190
linearization, 285–286, 292, 308
Little's result, 339
logarithmic, 164, 168
logistic, 18, 31–33, 36, 164, 166, 169, 180–186, 230, 241, 244, 245, 268–269, 272, 274, 278, 295, 298, 299, 304, 308, 313, 320, 322
    fitting to data, 232, 322
Lotka, 271, 272, 287
lynx, 175–180

Markov, 262
Markov chain, 122–132, 138, 141–142, 146, 325, 326, 328, 335, 348
    absorbing, 125–127, 130–132
matrix, 107, 259, 262, 286
matrix addition, 107
matrix additive identity, 107
matrix algebra, 106–110
matrix inverse, 109
matrix model, 101–150
matrix models, 142–143
matrix multiplication, 108–109
matrix power, 109
matrix size, 107
matrix subtraction, 107
matrix transpose, 109
maximum sustainable yield, 274, 313
mean, 49, 51, 52, 58, 63–68, 70, 75, 86, 90, 92, 93, 231
mean sum of squares
    regression, 202
    residual, 202
mean time until transition, 327
median, 49, 58, 59, 76, 90
metered model, 9
Michaelis-Menton equation, 252, 309–310, 314–317
migration, 45–46, 142–143
minimum finding algorithm, 181, 183
mode, 81, 93
model
    θ-logistic, 270
    θ-logistic declining carrying capacity, 298, 305
    A.S.T., 53–56
    age structured bobcats, 139
    alligators, 230
    annual plants, 28–30
    arms race, 310–312
    bacteria-nutrient, 315

bald eagles, 229
bank queue, 354–356
barren ground caribou, 148–150
belly button, 228–229
birth-death, 360–363
birth-death-immigration, 363–367
black bear age vs. weight, 183–186
black-capped chickadees, 95–97
blood alcohol, 40–42, 309, 314
bobcats, 39–40
body fat, 236–237
California, 45–46, 142–143
capture-recapture, 97–99
carbon cycle, 318–320
changing human population growth rate, 140–141
chickadees, 231–232
compound interest, 42
computational complexity, 233–234
computer speeds, 233
corn storage, 220–224
cost of advertising, 168–175
credit card, 38–39
crossing a busy street, 331–334
discrete predator-prey, 35–37
drugs in the body, 321, 252–322
ecological succession, 140
Ehrenfest chain, 146
elk, 201, 210–212
England and Wales, 228
finance, 317–318
Great Lakes pollution, 45
hacking chicks, 14–16
harvesting, 272–276
    constant, 272
    linear effort, 274–275
hermit crabs, 141–142
host-parasite, 35–37
human female population, 101–103, 106, 110–112, 114
impala, 147–148
income class, 241, 264–266
inflation, 43
Ireland, 191–195
Jackson elk herd, 276
limousine service, 5–6
logistic, 31–33
    declining carrying capacity, 296
lynx, 175–180
M&M's, 49–53
    chi-square test, 84–85
M&M's in jar, 2–4
Markovian Squirrels, 127–130
more M&M's, 97
plant-herbivore, 43–44
Poisson forest, 94–95
predator-prey, 287–293

predator-prey variations, 320–321
projectile motion, 5
pure birth, 346–351
pure death, 359–360
redwoods, 136
reservoir, 313–314
sandhill crane, 10–15, 17, 114–120
Scot Pines, 130–132
Scottish Markovian squirrels, 141
service at checkout counter, 324–327
single-server queue, 334–346
social mobility, 241, 264–266
spruce budworm, 277–283, 313
standing in line, 328–330
stochastic bobcats, 93–94
stochastic sandhill cranes, 70–78
stochastic squirrels, 48–53
    chi-square test, 84–85
stock market, 232
teasel, 120–121
temperature distribution, 143–146
testing, 215–219
The X-files, 56–60, 90, 161–164, 230
traffic light, 356–359
US interpolation, 230–231
US population growth
    curves fitted to data, 245–250, 322
    logistic model, 322
model definition, 1
model validation, 82
modeling, 2
MSY, 274
multiple regression, 152, 198, 202, 204, 214–220, 224, 236–237
multiplying a matrix by a scalar, 108

new curves from old, 165
node, 307, 308
nonlinear fits, 180–186
normal distribution, 66–68, 70, 75, 92, 97, 207, 231, 232
normal probability plot, 97, 207, 212, 232
null hypothesis, 83, 202, 203
nullcline, 283, 284, 290, 308, 321
numerical solution, 295, 298–305

Occam's razor, 4–6, 78, 277
oscillation, 255

p-value, 84
percentile, 58–60
period, 165
periodic, 165, 166, 175
Perron-Frobenius theorem, 114
pharmacokinetic, 309, 315
phase line analysis, 266–283, 313, 317
phase plane analysis, 266, 283–286, 308, 311–312

phase shift, 165
Playboy, 31
Poisson forest, 94
Poisson process, 328–334, 356
population, 154
population distribution, 49
population-based model, 97
power, 151, 164, 168–170, 172
predator-prey, 270, 287–293, 316, 320
predictor, 214–216
probability density function, 61
probability function, 64
pure birth process, 346–351
pure death process, 359–360

quadratic spline, 190, 226–227
quartile, 58, 59
queue, 323, 334–346
queueing simulation, 341–346
queueing theory, 323–324, 334–346

random variable, 53, 55, 62–70, 72, 75, 79–81, 83, 85, 86, 91–93
range, 49, 54, 56–59, 63, 75
rational function, 166
recurrence relations, 10–11, 21
    affine, 19, 21, 22, 26, 38
    exponential, 19–21, 23, 26
    linear constant coefficients, 25–30
        general solution, 26
    systems, 35–37
regression, 152, 158, 160, 161, 163, 165, 187, 197–212, 231, 236–237
regression analysis, 162, 187, 197–212, 215, 218, 221, 224
regression assumptions, 200
regression equation, 156, 157, 162, 198, 200, 204, 210, 212, 222, 226
regression line, 156, 160, 163, 164, 168, 169, 171, 173, 175, 178, 200–202, 204–207, 209, 212, 226, 249, 250
regression table, 199, 203, 207–210, 212, 221, 223, 224
regular matrix, 124
relative address, 13
relative frequency, 54, 55, 87
residual, 156, 158, 207, 232
Richardson arms race model, 310–312
robust, 22
Runge-Kutta, 301

saddle, 260, 286, 307, 308
sample variance, 52
scalar, 107
scalar product, 110
sensitivity analysis, 22
side-by-side box plot, 60, 76, 77, 90

significance, 82
simple interpolation, 187–189
simple regression, 197–212, 214, 215
single-server queue, 334–346
singular matrix, 137
sink, 260, 307, 308
slope field, 298
source, 260, 286, 307, 308
special function, 296
spiral, 260, 286, 307, 308
spline interpolation, 190–191, 226–227
spread, 49, 51, 52, 58, 59, 76
stability, 21–25, 32, 33, 35, 38
standard deviation, 49, 51, 52, 65, 68, 70, 75, 92, 231
standard error, 204–207
Starfield, 294
state diagram, 104–106
state vector, 104
stationary distribution, 336–341
steady state, 21, 22
Stella II, 19
stochastic, 6
stochastic model, 47–99, 140
stochasticity
    continuous, 323–367
    discrete, 346
structured model, 101–150
sum of squares, 183, 199–200, 202, 209
    regression, 158, 164, 199–200, 202, 209
    residual, 158, 174–175, 199–200, 202, 209
    total, 158, 164, 199–200, 209
summarize by grouping, 54

t-ratio, 204–205, 209, 210
t-test, 204–205
tangent line, 254, 285, 286, 299
tangent plane, 286
Taylor's theorem, 23
temperature distribution, 143–146
Thomas algorithm, 146, 234–235
trace, 136–137, 259
traffic intensity, 338–341
transformation, 151, 167–169
tridiagonal matrix, 234–235
Tukey, 89, 224

uniform distribution, 66–67, 70, 72, 91, 92

variance, 51, 52, 65–68, 70, 82, 92, 93, 152, 158–164, 170, 172, 199, 200, 207, 209, 214–218, 221
Volterra, 271, 272, 287

waiting time, 324
William of Occam, 4

yield-effort, 274, 275, 313
Young, 356, 357

zero matrix, 107